D0843231

Ions in Aqueous Systems

Ions in Aqueous Systems

An Introduction to Chemical Equilibrium and Solution Chemistry

Therald Moeller
Professor of Chemistry
Arizona State University

Rod O'Connor
Professor of Chemistry
University of Arizona

McGraw-Hill Book Company
New York St. Louis San Francisco Düsseldorf Johannesburg
Kuala Lumpur London Mexico Montreal New Delhi
Panama Rio de Janeiro Singapore Sydney Toronto

Ions in Aqueous Systems

Library of Congress Catalog Card Number

70–147163

07–042647–3

1 2 3 4 5 6 7 8 9 0 M A M M 7 9 8 7 6 5 4 3 21

This book was set in Times Roman, and was printed and bound by The Maple Press Company. The drawings were done by John Cordes, J. & R. Technical Services, Inc. The editors were William P. Orr and Anne Marie Horowitz. Matt Martino supervised production.

Contents

APPENDIXES

Preface

In agreement with many other teachers, we believe that an essential component of the combined discussion-laboratory presentation of elementary chemistry must be a study of the reactions of representative anions and cations in aqueous solution. This component provides an almost unique opportunity to integrate necessary descriptive fact with the fundamental principles that govern chemical behavior and the equilibria that are important in ionic systems. This type of integration is a prerequisite to proper comprehension of the more involved topics that are discussed in succeeding courses in chemistry. It is our firm conviction that no laboratory discipline yet introduced into elementary chemistry provides as effective a vehicle as qualitative analysis for combining fact with theory in this way.

Qualitative analysis is uniquely illustrative of solution chemistry and the principles of chemical equilibrium. It provides, through the breadth of its approach, an opportunity for the student to test experimentally the validity and applicability of the concepts and theoretical methods discussed outside the laboratory. It offers the undeniable challenge of so interpreting observation as to establish a previously unknown composition. It gives to student and instructor alike a logical and understandable relationship between laboratory and discussion. In short, qualitative analysis makes meaningful the definition of chemistry as a science concerned with matter, the changes that matter undergoes, and the laws that account for these changes.

We believe that the success of this approach will be determined by the presentation that is offered. The qualitative laboratory is no longer an end unto itself; rather, it is a means of achieving a much broader result. Historically, qualitative analysis was an instructional development that reflected the early importance of and emphasis upon the analysis of minerals. The classic approach was based upon a series of procedures that, when followed routinely, enabled the analyst to achieve his sole objective, namely, determination of the composition of the sample in question. Although lacking in instructional value by its overemphasis on routine without questioning the *why* of a given operation, this procedure did uncover many useful reactions of separation and identification. More recently, interest in ions has so centered in the theoretical interpretation of reaction kinetics and bonding that the laboratory evaluation of their reactions has faded in importance. Although lacking in a practical aspect, this procedure has provided explanations of many of the specific and general reactions discovered in the early days of qualitative analysis. It is only proper to conclude that a balanced and integrated combination that allows the one to complement and supplement the other can have maximum instructional value. The present volume has been written to emphasize this principle.

A reasonable combination of fact with theory requires that particular attention be paid to the following:

1. Both theory and descriptive fact must be treated logically, systematically, and comprehensively.
2. Laboratory operations must be selected so that they both illustrate the principles developed and follow logically the sequence in which those principles are discussed.
3. Laboratory operations must be approached as a series of experiments that lead to reasonable answers, rather than as a series of routine exercises.
4. The importance of laboratory technique must be given constant emphasis.
5. The quantitative aspects of both the theoretical and practical approaches must be emphasized throughout the presentation.
6. Student interest must be developed and maintained through use of the concept of continuing challenge.

For convenience, this book is divided to treat in order the fundamentals of solution theory and chemical equilibrium, and the laboratory investigation of behaviors of the common anionic and cationic species that illustrate best the conceptual and theoretical topics. The book is so written that assignments can be made simultaneously from both parts to maximize overlap of principle with practice.

In the part on solution chemistry, topics are introduced in a logical order of generally increasing difficulty. By judicious cross-referencing and review, each topic is related to those already considered and to the practical operations of the laboratory. Continuing emphasis is placed upon the solving of numerical problems to implement the thesis that the student can adequately understand the fundamentally quantitative nature of chemistry only if he can solve problems illustrative of the concepts under consideration. We believe that his confidence in the validity of these concepts increases in direct proportion to his familiarity with the simple mathematical operations involved. Many illustrative numerical examples are included in addition to the usual exercises at the ends of the chapters. The logical dimensional approach to problem solving is emphasized throughout.

The laboratory studies are introduced through a series of exercises on techniques and useful operations. These exercises are designed to acquaint the student with the approaches he will find necessary in subsequent operations and, thereby, emphasize the importance of correct technique and improve his confidence in ability to solve laboratory problems. Procedures for evaluating the chemistry of and for identifying common anions and cations follow in that order. The anions precede the cations for the following reasons:

1. The useful chemistry of the cations is often dependent upon the anions present. Prior knowledge of the anions is thus important.
2. The laboratory study of the anions is less amenable to reduction to routine than that of the cations. The student is thus encouraged to develop a reasoning and logical experimental approach to analysis and to learn to interpret his observations.
3. The chemistry of the anionic species illustrates less directly than that of the cationic species the general principles of solution chemistry. The student has the opportunity to develop his background in principles before it is necessary to apply them for maximum understanding of the laboratory work.

Within the framework of anion or cation chemistry, whatever order of group presentation that seems most logical to the instructor can be followed. For the cations, the order of groups VI, V, I, II, III, and IV has the advantage of proceeding from relatively simple to much more complex chemistry.

Throughout the laboratory work, the investigative approach is stressed as important in maintaining student interest and in providing training for more advanced study. Each exercise is considered as part of an investigative effort, and not as a routine operation. Challenge is offered through the analyses of "unknown" samples. Constant attention to cleanliness, orderliness, correctness of manipulation, and the written record reflect our considered belief that the sooner these habits are developed, the more successful the student will be in his or her study of chemistry.

This book is based upon the senior author's more comprehensive "Qualitative Analysis." Its compilation reflects the experience of the users of that volume and, thus, the assistance given, consciously or unconsciously, by our colleagues and students over an extended period. We are deeply grateful for all the suggestions and criticisms that have come to us; without this help the current effort would have been impossible. It is our hope that those who use this textbook will again give freely of their opinions and suggestions. We shall be grateful for all comments.

Therald Moeller
Rod O'Connor

Ions
in
Aqueous Systems

part one

Principles of Solution Chemistry

1
Some Characteristics of Aqueous Solutions

The reactions characteristic of the anions and cations usually take place in solution. Since water is the solvent most commonly used to dissolve inorganic substances, the following discussion is restricted to aqueous systems.

WATER AS A SOLVENT

The solvent properties of water are commonly considered to be unique. Strictly speaking, such a view is incorrect, for it is possible to duplicate in other solvents (e.g., liquid ammonia, liquid sulfur dioxide, liquid hydrogen fluoride) most of the phenomena described as characteristic of aqueous solutions. However, it is undeniably true that water does dissolve more substances and substances that differ from each other more extensively in properties than does any other solvent. The convenient liquid range of water and its lack of objectionable properties make it particularly useful from the point of view of ease of handling. Since water is the most common natural liquid and plays many vital roles in living systems, the properties of water have always been of great interest.

The versatility of water as a solvent can be related to its structure. In each individual water molecule the difference in electronegativities* renders each hydrogen-oxygen bond markedly polar in character, with the electron pair of the bond displaced appreciably toward the oxygen atom. In addition, the electron density of the oxygen atom is increased by the presence of two unshared pairs of valence electrons. The polarity of the hydrogen-oxygen bond is often represented as:

$$\overset{\delta+}{\mathrm{H}}\text{—}\overset{\delta-}{\mathrm{O}} \quad \text{or} \quad \overset{+\!\!\longrightarrow}{\mathrm{H—O}}$$

If the water molecule were linear, the two polar bonds would be exactly oppositely directed, and the water molecule as a whole would be nonpolar. However, it can be easily demonstrated that water molecules are polar. For example, a stream of water from a buret is deflected toward either a positively or negatively charged object. Indeed, experimental evidence from such different procedures as infrared spectroscopy and neutron diffraction indicates a structure for the water molecule consistent with its polar character.

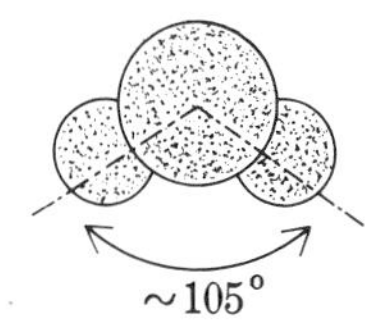

This structure suggests that the two polar bonds are not mutually neutralized. The significant difference between the electron density on the oxygen atom and that on each hydrogen atom then results in a molecule having an unsymmetrical charge distribution. Each water molecule thus constitutes an electric *dipole*.

δ⁻

H H H H −

δ⁺ +

In any collection of water molecules (ice, liquid water, or molecular clusters in water vapor) each "positive" hydrogen atom may be attracted electrostatically by a "negative" oxygen atom in another water molecule.

* The electronegativity of an atom is a measure of the tendency of the atom to attract the electron pair of a covalent bond in which the atom is involved.

This attraction results in a relatively weak bond, the *hydrogen bond*, in which a hydrogen atom functions as a bridge between two oxygen atoms. Since the electron density of the oxygen atom is concentrated in the two unshared pairs of valence electrons, each oxygen atom can bond to hydrogen atoms of two neighboring water molecules. In the resulting collection, each "positive" hydrogen atom tends, by mutual electrostatic repulsion, to assume an equilibrium position as far from the others as possible. If all four bonds to the central oxygen atom were equivalent, the optimum geometry would be described as a tetrahedral configuration around the central oxygen atom.

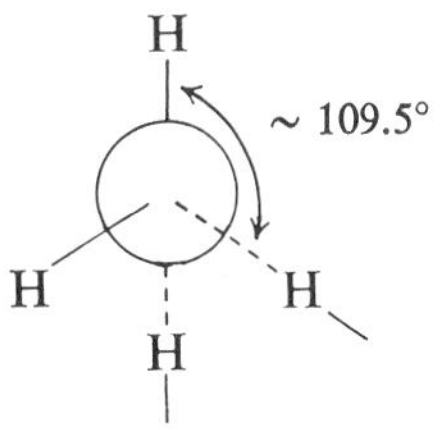

Since the hydrogen bond is somewhat longer (and weaker) than the "original" covalent bond between an oxygen atom and a hydrogen atom in the free water molecule, the hydrogen bond is slightly different in polarity. Thus, mutual hydrogen-hydrogen repulsions are not exactly equivalent, and the resulting configuration is a bit distorted from a true tetrahedral geometry. In ice, complete hydrogen bonding links all water molecules together, giving a tetrahedron of hydrogen atoms surrounding each oxygen atom. This rather open structure persists in liquid water to about 4°C. Here, a re-arrangement to a somewhat collapsed and less open structure involving fewer hydrogen bonds takes place. As the temperature rises, more and more hydrogen bonds are ruptured by thermal motions.

Thus, liquid water contains aggregated groups of the type $(H_2O)_n$. These aggregates influence any property of liquid water related to molecular motion because of additional energy requirements for the destruction of the aggregates. The comparatively high freezing and boiling points, low vapor pressure, high heat capacity, and large heats of fusion and vaporization of liquid water are all related to the presence of these hydrogen-bonded aggregates.

The hydrogen bonds in water are readily broken. If any substance that contains ions or centers of electric charge (from polar bonds) is brought into contact with water, sufficient electrical disturbance results to rupture the hydrogen bonds. The released water dipoles are then attracted to these charge centers. If these attractions are sufficiently large to overcome the interionic or intermolecular attractions initially present in the material, the particles of that substance will be pulled away from the positions they originally occupied in their own aggregations, and the substance will dissolve.

BEHAVIOR OF SOLUTES IN CONTACT WITH WATER

It is useful to consider solubility in water in terms of familiar concepts of electrostatics. The force between two regions of opposite charge is a function of the distance between the charge centers and of the magnitude of the charge density of the regions, i.e., the amount of charge and the effective volume in which it is concentrated. Just as it requires energy to separate two oppositely charged regions, energy is released when such regions approach each other. The electrostatic considerations of solutes in contact with water may then be summarized as *energy-releasing processes*

1. Orientation and approach of water dipoles to charged or polar regions of the solute
2. Orientation and approach of water dipoles to each other
3. Orientation and approach of solute particles (of opposite charge or oppositely arranged dipoles) to each other

and *energy-requiring processes*

1. Destruction of hydrogen bonds as water molecule clusters break apart
2. Separation of oppositely charged (or oppositely oriented dipoles) of solute particles as they leave their own aggregates

Obviously then, there are difficulties in predicting the solubility of substances in water since the relative magnitudes of opposing effects must be assessed, even for this simplified consideration. More rigorous treatment of solubility in later studies of physical chemistry will reveal other necessary considerations that are best discussed in thermodynamic terms. Some generalizations may be stated on the basis of the simple electrostatic considerations outlined here.

Ionic compounds Crystals of ionic compounds amount to orderly arrangements of ions and owe their stabilities to the electrostatic attractions existing between ions of opposite charge. If an ionic crystal is brought into contact with water, each ion on the surface of the crystal and water dipoles from the solvent will attract each other, the negative ends of the water dipoles lining up with the cations and the positive ends with the anions. As a result of these attractions, bonds holding the ions into the crystal will be weakened, and if the interionic forces are overcome, the ions of the crystal will be pulled into solution, each ion being surrounded by an atmosphere of water molecules. This process is depicted for a general case in Fig. 1-1.

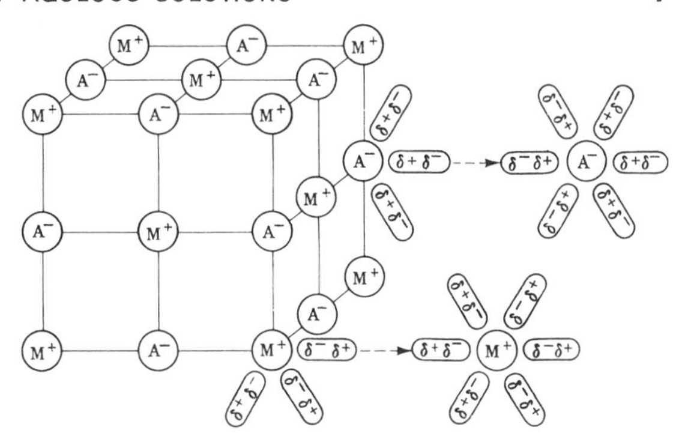

Fig. 1-1 Dissolution of an ionic crystal in water.

It is important to realize that when an ionic substance is dissolved in water, ions are not *created* in the process. In effect, the sole function of the water is to separate existing ions from their fixed positions in the crystal. It was a lack of understanding of this situation which led to much early confusion in the interpretation of aqueous solutions of electrolytes (ions).

Interionic forces, which favor retention of ions in the crystal lattice, increase with increasing charge density of the ions (net charge per unit volume). However, the same effect is noted for the attraction between ions and polar water molecules. Other factors, such as the nature of the bonding between adjacent particles in crystals, the characteristics of particular lattice geometries, and the cation-anion ratios involved, also affect the water solubility of ionic substances in ways that are difficult to estimate accurately. In the absence of simple principles permitting good predictions, experimental determinations of solubility must be employed (Table 1-1).

For practical purposes, a general set of so-called "solubility rules" is often extremely valuable in studying solution chemistry via qualitative analysis. Typical rules are as follows:

1. All chlorides are soluble in water except those of silver, mercury(I), and lead(II). The same is generally true of bromides and iodides, except that mercury(II) bromide is only moderately soluble, and iodides such as the mercury(II), bismuth(III), and tin(IV) compounds are insoluble. In general, all fluorides are insoluble in water except silver fluoride, ammonium fluoride, and the alkali-metal fluorides. Lithium fluoride is only moderately soluble.
2. All oxides are insoluble in water except those of the alkali metals, barium, and strontium. Calcium oxide is moderately soluble in water. Since, when dissolved in water, oxide ion is converted completely to hydroxide ion, the same general considerations apply to hydroxides.

Table 1-1 Approximate water solubilities of some typical compounds at 20°C

Approximate solubility: moles/1,000 g (as power of 10)*

Ion		F^-	Cl^-	Br^-	I^-	OH^-	NO_3^-	S^{2-}	SO_4^{2-}	CO_3^{2-}	PO_4^{3-}
	Radius, Å	1.36	1.81	1.95	2.16			1.84			
Na^+	0.97	0	1	1	1	1	1	0	0	0	0
Ag^+	1.26	1	−5	−6	−8		1	−17	−2	−4	−7
K^+	1.33	1	0	1	1	1	0		0	1	1
Mg^{2+}	0.65	−3	1	1	1	−4	1		0	−2	−5
Zn^{2+}	0.74	−1	1	1	1	−6	1	−12	0	−5	
Cu^{2+}	0.72		1			−7	1	−17	0	−5	
Ni^{2+}	0.72		1	1	0	−6	1	−10	0	−4	
Fe^{2+}	0.74		1	1		−5	1	−9	0	−6	
Ca^{2+}	0.99	−4	1	1	1	−2	1		−2	−4	−9
Sn^{2+}	1.12		1		−2	−9		−13	0		
Hg^{2+}	1.10		−1	−2	−9	−8		−27			
Pb^{2+}	1.20	−3	−3	−3	−4	−5	0	−14	−4	−7	−11
Ba^{2+}	1.34	−2	0	0	1	−1	−1		−5	−5	
Al^{3+}	0.51	−1	1			−11			0		−3
Cr^{3+}	0.64					−8					
Fe^{3+}	0.64		0			−10	1	−18			

* Number is exponent of 10 for closest solubility value, for example, $-5 = 10^{-5}$ mole/1,000 g H_2O.

3. All sulfides, selenides, and tellurides are insoluble in water except those of ammonium ion, the alkali metals, and the alkaline earth metals.
4. All simple acetates, chlorates, nitrates, nitrites, perchlorates, and permanganates are soluble in water. Potassium and ammonium perchlorates are only moderately soluble. Silver nitrite and acetate are moderately soluble.
5. All sulfates are soluble in water except those of lead(II), barium, and strontium. Calcium and silver sulfates are only moderately soluble.
6. All arsenates, arsenites, borates, carbonates, cyanides, ferricyanides, ferrocyanides, oxalates, phosphates, and sulfites are insoluble in water except those of ammonium ion and the alkali metals. Except for a few cyanides, ferricyanides, and ferrocyanides, these compounds dissolve in dilute acids.
7. Most chromates, except those of ammonium ion and the alkali metals, are insoluble in water. Important exceptions are the calcium, copper(II), and magnesium compounds. All chromates are soluble in acids.
8. Most fluorosilicates, thiocyanates, and thiosulfates are soluble in water. Important exceptions are barium and the alkali metal fluorosilicates; mercury, lead, and silver thiocyanates; and lead and silver thiosulfates.

9. All silicates are insoluble in water except the simple alkali-metal compounds. All silicates yield insoluble silicon dioxide when treated with acids.
10. Many compounds containing highly acidic cations or highly basic anions undergo hydrolysis in contact with water and, thereby, produce insoluble substances. This situation is particularly true of antimony and bismuth compounds, and of many sulfides and cyanides.

It is strongly recommended that these generalizations be memorized.

Nonionic compounds Aggregations of uncharged molecules are held together by intermolecular forces that are generally much weaker than those in ionic crystals, although the electrostatic character of the forces is similar. The interionic forces in ionic crystals result from interactions of regions of high charge density, whereas intermolecular forces depend on regions of much lower charge density, as determined by the relative electronegativities of bonded atoms (bond polarity) and molecular geometry.

The strongest intermolecular forces are those referred to as hydrogen bonds (described earlier for water molecules), since the covalently bonded hydrogen atom is uniquely small and, when present in a typical polar bond, results in a region of uniquely high charge density. Some common molecular substances in which the principal intermolecular forces are hydrogen bonds include alcohols, phenols, ammonia and amines, amides, carboxylic acids (Table 1-2), and hydrogen fluoride.

In general, hydrogen bonds are formed when a hydrogen atom covalently bonded to a small, highly electronegative atom is attracted to the

Table 1-2 Some molecular substances capable of forming hydrogen bonds

Class	*Examples*
Alcohols	Methanol (wood alcohol), CH_3OH; ethanol (grain alcohol), CH_3CH_2OH; glycerine, $HOCH_2CH(OH)\cdot CH_2OH$; glucose, $C_6H_{12}O_6$
Phenols	Phenol (carbolic acid), C_6H_5OH; hexachlorophene, $C_{13}H_6O_2Cl_6$
Ammonia and amines	Ammonia, NH_3; methylamine, CH_3NH_2; aniline, $C_6H_5NH_2$; ethylenediamine, $H_2NCH_2CH_2NH_2$
Amides	Acetamide, CH_3CONH_2; thioacetamide, CH_3CSNH_2
Carboxylic acids	Formic acid, HCOOH; acetic acid, CH_3COOH; benzoic acid, C_6H_5COOH; oxalic acid, HOOCCOOH

negative region of an unshared valence-electron pair on some other small, highly electronegative atom. Such situations most commonly involve oxygen, nitrogen, or fluorine. Molecules such as H_2S or HBr do not form effective hydrogen bonds with similar molecules because the electronegative atoms involved are so large that their valence-electron regions repel each other before the molecules can approach closely enough for hydrogen bonding and, perhaps more important, because the bonds involving hydrogen are considerably less polar than in molecules like H_2O, NH_3, and HF. Although the carbon atom is small, its electronegativity is so close to that of hydrogen that the carbon-hydrogen bond is not sufficiently polar to induce strong intermolecular attractions (except in those few cases where other atoms bonded to carbon strongly affect the C—H bond polarity, e.g., in the molecule $CHCl_3$).

A variety of molecules have intermolecular forces similar to, but weaker than, hydrogen bonds. Such dipole-dipole interactions result anytime the molecule contains polar bonds arranged in a geometry such that the overall molecule is polar. For example, formaldehyde,

```
H
 \
  C=O
 /
H
```

and acetone,

```
          O
          ‖
H   H     C    H   H
  \ | /     \  | /
    C          C
    |          |
    H          H
```

are polar compounds; whereas carbon dioxide, O=C=O, the polar bonds of which are mutually neutralized because of the linear shape of the molecule, is nonpolar.

Many molecules, or regions within molecules, are so electrically symmetrical that only very weak intermolecular forces (van der Waals' forces) are active. Such nonpolar molecules or regions occur in substances like methane,

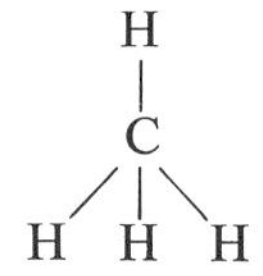

or in the hydrocarbon portions of most organic compounds, e.g., 1-butanol.

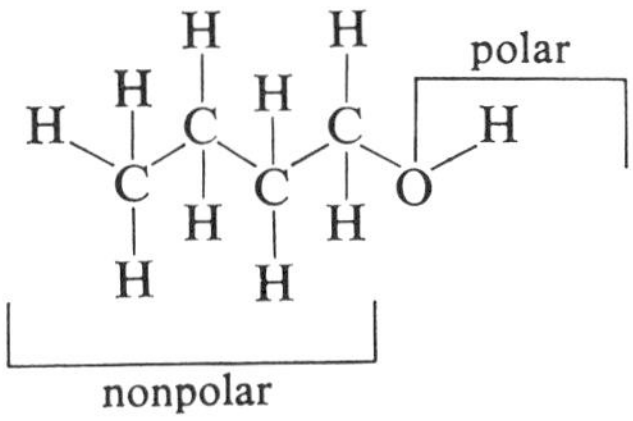

If nonionic compounds are to dissolve in water, they must contain regions (functional groups) that can form hydrogen bonds with water molecules. These functional groups may act as hydrogen acceptors, for example,

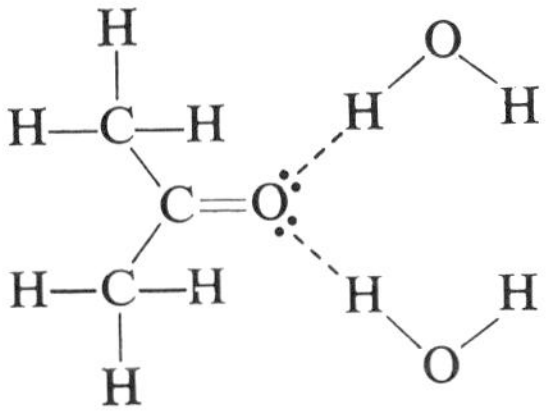

or, as is the case among water molecules themselves, as both hydrogen donors and acceptors, for example,

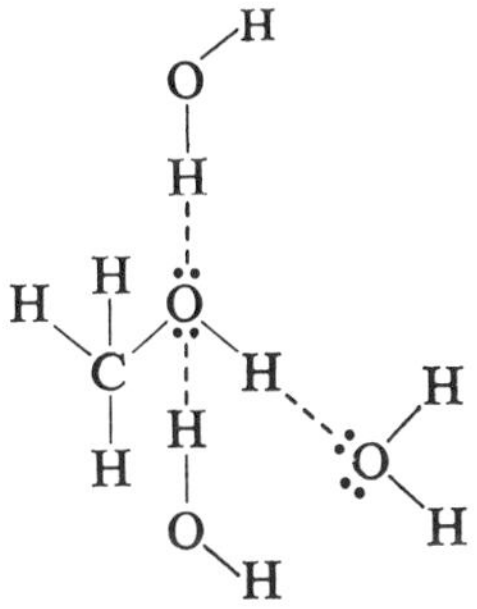

Molecules incapable of hydrogen bonding with water are not appreciably water-soluble (even though the molecules may not be strongly attracted to others of their own kind), because they cannot disrupt the hydrogen-bonded clusters of water molecules and attain the molecular dispersion necessary for true solution. An excellent analog of this situation can be seen if a collection of marbles is shaken with a collection of magnets. Although the marbles do not attract each other, they are prevented from uniformly mixing with the magnets because of the strong forces among the latter. Molecules that contain both nonpolar regions and functional groups capable of hydrogen bonding dissolve in water only if the nonpolar character does not predominate. Thus, methanol (CH_3OH) and ethanol (CH_3CH_2OH) are completely miscible with water (mix in any proportions), but alcohols like 1-butanol ($CH_3CH_2CH_2CH_2OH$) are not appreciably water-soluble. In general, a nonpolar region equivalent to or greater than that of 1-butanol will predominate over the effect of a single hydrogen-bonding group to the extent of preventing significant solubility in water. A detailed treatment of this subject would require a discussion of molecular substances which is beyond the scope of this book.

Soluble, nonionic compounds most commonly dissolve in water in the molecular form to give nonconducting solutions containing only molecular species. If, however, the polar groups are very highly polar in nature, attractions exerted by the water molecules may be sufficiently large to rupture bonds in molecules of the solute to create ions. In practice, this circumstance is limited to the removal of hydrogen ions (i.e., protons) by water, and is characteristic of the common hydrogen acids. These compounds are covalent in the pure state, but contain such strongly polarized bonds that water causes rupture of the bonds holding hydrogen atoms, giving solutions that contain the hydrated proton. This process can be illustrated for hydrogen chloride by the equation

$$\mathrm{H}\overset{\cdot}{\times}\underset{\cdot\cdot}{\ddot{\mathrm{Cl}}}: + \mathrm{H}\overset{\cdot}{\times}\underset{\underset{\mathrm{H}}{\cdot\times}}{\ddot{\mathrm{O}}}: \rightleftharpoons \mathrm{H}\overset{\cdot}{\times}\overset{\mathrm{H}^+}{\underset{\underset{\mathrm{H}}{\cdot\times}}{\ddot{\mathrm{O}}}}: + \overset{\cdot}{\times}\underset{\cdot\cdot}{\ddot{\mathrm{Cl}}}:^-$$

and in less detail for other substances by the equations

$$HNO_3 + H_2O \rightleftharpoons H_3O^+ + NO_3^-$$

$$H_2SO_4 + H_2O \rightleftharpoons H_3O^+ + HSO_4^-$$

$$HSO_4^- + H_2O \rightleftharpoons H_3O^+ + SO_4^{2-}$$

$$CH_3CO_2H + H_2O \rightleftharpoons H_3O^+ + CH_3CO_2^-$$

On the other hand, ions can also be created through the rupture of water molecules by other polar linkages, notably those containing basic nitrogen atoms. Thus, dissolution of ammonia gas in water can be represented by the equation

$$\mathrm{H}\overset{\cdot}{\times}\overset{\overset{\mathrm{H}}{\cdot\times}}{\underset{\underset{\mathrm{H}}{\cdot\times}}{\mathrm{N}}}: + \mathrm{H}\overset{\cdot}{\times}\underset{\underset{\mathrm{H}}{\cdot\times}}{\ddot{\mathrm{O}}}: \rightleftharpoons \mathrm{H}\overset{\cdot}{\times}\overset{\overset{\mathrm{H}}{\cdot\times}}{\underset{\underset{\mathrm{H}}{\cdot\times}}{\mathrm{N}}}:\mathrm{H}^+ + \overset{\cdot}{\times}\underset{\cdot\cdot}{\ddot{\mathrm{O}}}\overset{\cdot}{\times}\mathrm{H}^-$$

and the same behavior is characteristic of the soluble organic derivatives of ammonia (amines).

AQUEOUS SOLUTIONS OF ELECTROLYTES

All substances that are ionic in the pure state, and all nonionic substances that dissolve in water to give solutions that contain ions, are called electrolytes. Electrolytes may be of the following general kinds*:

Salts Compounds that amount in the pure state to aggregations of ions; for example, Na^+Cl^-, K^+OH^-, $Ca^{2+}O^{2-}$, $Fe^{2+}SO_4^{2-}$.

Acids Covalent hydrogen compounds that in contact with water yield hydronium ions; for example, HCl, H_2SO_4, HSO_4^-, CH_3CO_2H.

Bases Covalent compounds that attract protons so strongly that they yield hydroxide ions in contact with water; for example, NH_3, CH_3NH_2, $(CH_3)_2NH$, $(CH_3)_3N$.

Properties of aqueous solutions of electrolytes Solutions of electrolytes are characterized by physical properties that are determined by the presence of ions. Inasmuch as these properties have been considered in detail in introductory courses, only a brief review is essential here.

Conductance All these solutions conduct electric current, the ease with which the current flows being determined by the nature of the solute, the concentration of the solution, and the temperature. The conductance of an electrolyte is often expressed numerically per gram equivalent of that substance. At constant temperature, the equivalent conductance always increases as solute concentration decreases. The limiting conductance, which is approached at zero concentration, is known as the conductance at infinite dilution and is useful in the experimental determination of ionization constants.

Vapor pressure The vapor pressure of an aqueous solution of a nonvolatile electrolyte at a particular temperature is always less than that of a similar solution of a nonvolatile nonelectrolyte of the same molality because of the presence of more dissolved particles.† The vapor pressure of any such solution is always less than that of pure water at any given temperature.

Boiling point and freezing point As a consequence of vapor-pressure lowering, the boiling point of an aqueous solution containing a nonvolatile

* These classifications are special cases for the water system, based on the following general definitions which are discussed in greater detail in a later section:

Salt Any compound existing in the pure state as an aggregation of ions, even though one or more of those ions may be an acid or a base

Acid Any substance acting to donate one or more protons to another substance

Base Any substance acting to accept one or more protons from another substance

† Properties, the numerical magnitudes of which depend solely on the number of foreign particles (atoms, molecules, or ions) present, are called "colligative properties."

solute is higher than that of pure water, and the freezing point of any aqueous solution is lower than that of pure water. For nonelectrolytes, the molal boiling-point and molal freezing-point constants average about 0.513 and 1.87°C, respectively. For electrolytes yielding two ions per solute particle, the effects approach twice these values; for those giving three ions, three times these values; etc. That these effects are never exactly two times, three times, etc., is due mainly to the influence of interionic attraction.

Osmotic pressure The tendency of a solution, separated from the pure solvent by only a porous membrane, to dilute itself by passage of solvent molecules through the membrane is called "osmosis." The extent to which this dilution occurs can be measured by the hydrostatic pressure, the osmotic pressure, developed within the solution. Osmotic pressure is dependent in magnitude upon the concentration of the solution involved, and is thus affected by the number of solute particles present in a given quantity of solvent. Solutions of electrolytes, because they contain more particles per mole of solute, give greater osmotic effects than do those of nonelectrolytes of the same molarity (page 17).

Degree of electrolytic dissociation Quantitatively, the "strength" of an electrolyte can be expressed in terms of its degree of electrolytic dissociation or ionization, α, that is, in terms of the fraction of the material that appears to be in the form of ions. This fraction can be determined by comparing measured conductance with that extrapolated for "infinite" dilution, or by comparing observed freezing- and boiling-point effects with those assumed for completely ionic substances. The values given in Table 1-3 indicate both how electrolytes vary among themselves and how degree of dissociation depends upon concentration.

The original theory of electrolytic dissociation, as propounded by Svante Arrhenius in 1887, assumed that when a solute is placed in contact with water, an equilibrium is established between solute molecules and ions in solution. An equilibrium of this type might be formulated as

$$\text{Solute molecules} + H_2O \rightleftharpoons \text{solute ions} \qquad (1\text{-}1)$$

It is apparent that for this equilibrium the degree of dissociation has a real meaning since it is actually a measure of the extent to which ions have been formed.

However, since it has been determined that salts in the pure state are actually aggregations of ions, the fallacy of treating solutions of all electrolytes in this way is obvious, since with these ionic materials water cannot "create" ions. The term "degree of dissociation or ionization" has no significance when applied, as such, to aqueous solutions of salts. Yet these solutions do not behave as if all the solute material is present in the ionic

Table 1-3 Apparent degrees of electrolytic dissociation in aqueous solutions at about 25°C

Solute	*Concentration, g equiv liter^{-1}*	α	*Solute*	*Concentration, g equiv liter^{-1}*	α
HCl	0.1	0.92	KCl*	0.1	0.86
HNO_3	0.1	0.92		0.05	0.89
H_2SO_4	0.1	0.92		0.01	0.94
HO_2CCO_2H	0.1	0.40	$NaNO_3$	0.1	0.83
H_3PO_4	0.1	0.27		0.01	0.93
CH_3CO_2H	0.1	0.0135	K_2SO_4†	0.1	0.72
H_2CO_3	0.1	0.0017	$BaCl_2$	0.1	0.76
H_2S	0.1	0.0007		0.01	0.88
HCN	0.1	0.0001	$MgSO_4$‡	0.1	0.45
$NaOH$	0.1	0.91		0.01	0.67
KOH	0.1	0.91	$CuSO_4$	0.1	0.40
$Ba(OH)_2$	0.1	0.77		0.01	0.63
NH_3	0.1	0.013	$HgCl_2$§	0.1	0.01

* In general, for M^+A^- salts in 0.1 *N* solution α = about 0.86.
† In general, for $M^{2+}(A^-)_2$ or $(M^+)_2A^{2-}$ salts in 0.1 *N* solution α = about 0.75.
‡ In general, for $M^{2+}A^{2-}$-type salts in 0.1 *N* solution α = about 0.40.
§ See later discussion of complex formation.

form, as has been indicated by the *apparent* degrees of dissociation listed in Table 1-3. How, then, can one reconcile these seemingly opposite statements?

The answer is provided by the theory of *interionic attraction*, which was proposed in its fundamental form by P. Debye and E. Hückel in 1923. Compounds ionic in the solid state should, of necessity, be completely ionic when dissolved. However, because of forces of electrostatic attraction, each cation in solution will, in effect, surround itself with an "atmosphere" of anions, and each anion will surround itself with an "atmosphere" of cations. This situation can be illustrated for sodium chloride, for example, without indicating ion hydration, as

$$\begin{array}{ccccccc} & Cl^- & & & Na^+ & & \\ Cl^- & Na^+ & Cl^- & \quad Na^+ & Cl^- & Na^+ \\ & Cl^- & & & Na^+ & & \end{array}$$

it being understood that the actual situation is three-dimensional and need not involve any absolutely fixed number of associated ions. These forces of interionic attraction then prevent each ion from moving independently and, thereby, render it less effective in altering the properties of the solvent or solution. Thus, when an electric current is applied to such a solution, each ion is held back by its associated ions of opposite charge ("drag" effect) and is not "free" to move in the field. A decrease in observed conductance is the net result. The same situation pertains in solutions of the strong acids, since these compounds dissociate essentially completely in water.

Aqueous solutions of strong electrolytes must be regarded, therefore, as being completely ionic, and any observed modifications in properties are due to the effects of interionic attraction. With weak electrolytes, however, equilibria between undissociated molecules and ions do exist, and treatment of the Arrhenius type is completely correct. This distinction should be kept in mind at all times.

Hydration of ions In aqueous solutions, ions attract polar water molecules as well as other ions, and these attractions are sufficient so that it can be assumed that each ion carries with it a certain number of water molecules. A case in point is the proton, or hydrogen ion, which is sometimes depicted as the l-hydrate, $H^+ \cdot H_2O$ or H_3O^+, the *hydronium* ion.* Unfortunately, it is very difficult to determine with any certainty the number of water molecules associated with a given ion in solution. Rather than write only approximate formulas, or those containing indefinite numbers of water molecules, for example, $Al(H_2O)_x{}^{3+}$ or $F(H_2O)_y{}^-$, most chemists have found it convenient to ignore water of hydration with ionic species in writing equations, except in the very few instances where it is important. This practice is employed in this book. Because it seems inconsistent, then, to hydrate the proton and no other ion, the unsolvated proton is used throughout most of the discussions that follow. Actually, the inclusion of water of hydration does little more to a chemical equation than add water which is then carried along unchanged.

CONCENTRATIONS OF SOLUTIONS

Solution concentrations can be expressed either in physical or chemical units. Physical units express the composition of the solution in terms of weight of solute present in a given weight or volume of solution or solvent. Chemical units express the composition of the solution in terms of mole, g ion, or equivalent quantities of solute present in a given quantity of solution or solvent. Chemical units are directly applicable to quantitative interpretations of chemical reactions. Physical units are only indirectly applicable. Some common methods of expressing concentration are the following:

1. Weight of solute per unit volume of solution
2. Percentage of solute by weight
3. Molarity

* There is evidence that more extensively hydrated species, for example, $H_9O_4{}^+$, $[H(H_2O)_4]^+$, are present also. A more nearly correct formulation is probably $H^+(aq)$, where *aq* refers to a number of water molecules that may vary.

4. Normality
5. Formality
6. Mole fraction
7. Molality

Each of these is now reviewed in detail.

Weight of solute per unit volume of solution When concentration is so expressed, a measured volume of solution gives a definitely known weight of solute. The concentration may be given in terms of grams of solute per liter of solution (g liter^{-1} or g/liter), in grams per milliliter (g ml^{-1} or g/ml), etc. For solutions conveniently employed in qualitative analysis, concentrations in milligrams per milliliter (mg ml^{-1} or mg/ml) are particularly useful. For micro work, micrograms per milliliter (μg ml^{-1} or μg/ml) is a useful unit. Any such solution is prepared by dissolving a weighed quantity of solute in a quantity of solvent, and then diluting with solvent to the appropriate volume.

For electrolytes, it is often desirable to prepare solutions containing given quantities of particular ions in unit volumes from weighed amounts of salts containing those ions. Thus, the weight of silver nitrate required to prepare 125 ml of a solution containing 10 mg of Ag^+ ml^{-1} can be obtained as follows, using the dimensional approach (Appendix A).

$$\frac{\text{Wt } Ag^+}{125 \text{ ml}} = \frac{10 \cancel{\text{mg}}}{1 \cancel{\text{ml}}} \times 125 \cancel{\text{ml}} \times \frac{1 \text{ g}}{1{,}000 \cancel{\text{mg}}} = 1.25 \text{ g}$$

$$\text{Wt } AgNO_3 \text{ needed} = 1.25 \cancel{\text{g } Ag^+} \times \frac{1 \cancel{\text{mole } Ag^+}^{*}}{107.9 \cancel{\text{g } Ag^+}} \times \frac{169.9 \text{ g } AgNO_3}{1 \cancel{\text{mole } Ag^+}}$$
$$= 1.97 \text{ g}$$

Percentage of solute by weight The solute concentration may be expressed as a percentage of the total weight of the solution. Thus, a 10 percent solution of sodium chloride would contain 10 g of the salt in 100 g of solution. Because of the impracticality of preparing solutions by weight alone, it is customary to employ volumes with the conversion factor of density of solution. Thus, the weight of solute present in 500 ml of a sulfuric acid solution that contains 62 percent H_2SO_4 by weight and has a density of 1.52 g ml^{-1} can be obtained as follows:

$$\text{Wt } H_2SO_4 = 500 \cancel{\text{ml}} \times \frac{1.52 \text{ g}}{\cancel{\text{ml}}} \times \frac{62}{100} = 471 \text{ g}$$

Molarity The molarity or molar concentration, M, of a solution amounts to the number of moles of solute present in a liter of solution. Thus, a

* For the sake of uniformity, the term "mole" will be routinely employed to represent the quantity of any species (atoms, ions, molecules, or electrons) associated with Avogadro's number (6.023×10^{23}) of unit particles of the species.

1 M solution of sulfuric acid contains 1 mole of H_2SO_4 per liter of solution. Since the molecular weight of this compound is 98.08, this solution contains 98.08 g of solute per liter. By the same token, an $M/10$, or 0.1 M, sulfuric acid solution contains 0.1 mole, or 9.808 g of solute per liter, etc. The molarity of a solution is given by the ratio of the number of grams of solute present in 1 liter of solution to the *gram-molecular weight* (mole) of that solute. Thus, for the sulfuric acid solution discussed in the preceding section,

$$\begin{aligned} \text{Concentration} &= \frac{471\ \cancel{\text{g H}_2\text{SO}_4}}{500\ \cancel{\text{ml}}} \times \frac{1{,}000\ \cancel{\text{ml}}}{1\ \text{liter}} \times \frac{1\ \text{mole H}_2\text{SO}_4}{98.08\ \cancel{\text{g H}_2\text{SO}_4}} \\ &= 9.61\ \text{moles H}_2\text{SO}_4/\text{liter} \\ &= 9.61\ M \end{aligned}$$

A solution of a given molarity is prepared by dissolving the appropriate quantity of solute in a suitable quantity of solvent and then diluting with solvent to the required volume.

Normality The normality or normal concentration, N, of a solution amounts to the number of gram equivalents of solute present in a liter of solution. Thus, a 1 N solution of sulfuric acid contains one gram equivalent of H_2SO_4 per liter of solution. The number of gram equivalents is determined by the ratio of the number of grams of solute present to the gram-equivalent weight of that solute.

The gram-equivalent weight of a substance is defined as the weight in grams of that material which is chemically equivalent to 1.008 g of hydrogen or to 8.000 g of oxygen. In any chemical process where oxidation-reduction is not involved, the gram-equivalent weight is the ratio of the gram-formula weight of the substance in question to the total net charge of either the positive or negative radical, without regard to algebraic sign. In a process where oxidation-reduction is involved, the gram-equivalent weight is the ratio of the gram-formula weight of the substance in question to the number of moles of electrons* gained or lost in the process. Some examples illustrating the evaluation of gram-equivalent weights are given in Table 1-4.

* A mole of electrons is 6.023×10^{23} electrons. Its significance is apparent if one considers that for the oxidation

$$K \rightarrow K^+ + e^-$$

each gram atom (6.023×10^{23} atoms) of potassium (39.102 g) loses 6.023×10^{23} electrons. Or, for the reduction

$$MnO_4^- + 8H^+ + 5e^- \rightarrow Mn^{2+} + 4H_2O$$

each mole (6.023×10^{23} ions) of permanganate (118.94 g) gains $5 \times 6.023 \times 10^{23}$ electrons. The gram-equivalent weight of a substance is, thus, exactly defined as the quantity of that substance (in grams) the formation or reaction of which involves the loss, gain, or sharing of 1 mole of electrons.

Table 1-4 Gram-equivalent weights of electrolytes

Reaction type	Substance involved	Gram-equivalent weight	
		Method of calculation	*Examples*
Metathesis	Acid	$\frac{\text{Gram-formula weight}}{\text{No. of acidic protons}}$	$\frac{HCl}{1} = \frac{36.46\ g}{1} = 36.46\ g$
			$\frac{H_2SO_4}{2} = \frac{98.08\ g}{2} = 49.04\ g$
	Base	$\frac{\text{Gram-formula weight}}{\text{No. of proton-acceptor sites}}$	$\frac{NaOH}{1} = \frac{40.00\ g}{1} = 40.00\ g$
			$\frac{Ba(OH)_2}{2} = \frac{171.4\ g}{2} = 85.7\ g$
	Salt	$\frac{\text{Gram-formula weight}}{\text{Total net charge of either ionic species}}$	$\frac{NaCl}{1} = \frac{58.45\ g}{1} = 58.45\ g$
			$\frac{(NH_4)_2SO_4}{2} = \frac{132.14\ g}{2} = 66.07\ g$
			$\frac{Al_2(SO_4)_3}{6} = \frac{342.16\ g}{6} = 57.03\ g$
Oxidation-reduction	Oxidizing agent	$\frac{\text{Gram-formula weight}}{\text{No. of moles of electrons gained}}$	$\frac{HNO_3}{3} = \frac{63.02\ g}{3} = 21.01\ g$
			$(HNO_3 + 3e^- \rightarrow NO)$
	Reducing agent	$\frac{\text{Gram-formula weight}}{\text{No. of moles of electrons lost}}$	$\frac{H_2S}{2} = \frac{34.08\ g}{2} = 17.04\ g$
			$(H_2S \rightarrow S + 2e^-)$

Normality is a concentration unit of particularly great utility in the study of chemical reactions in solution. Inasmuch as solutions of equal normality contain the same number of gram equivalents in a given volume, equal volumes of such solutions react exactly with each other. By the same token, if the normalities are unequal, the volumes reacting vary correspondingly. Since solutions are commonly measured by volume in the laboratory, knowledge of normalities permits one to carry out reactions by merely measuring the volumes of the reacting solutions. It is apparent that here one is merely equating numbers of chemical equivalents; that is, for two materials A and B

$$\text{Equivalents of A} = \text{equivalents of B} \tag{1-2}$$

where the equivalents are usually given as gram equivalents or milligram equivalents. The number of equivalents is related to the normality and the volume as

$$\text{Number of equivalents} = \text{volume} \times \text{normality} \tag{1-3}$$

Therefore, one may say

$$\text{Volume}_A \times \text{normality}_A = \text{volume}_B \times \text{normality}_B \tag{1-4}$$

for solutions of A and B. If the volume is given in liters, the equivalents become gram equivalents*; if the volume is given in milliliters, the equivalents become milligram equivalents (or milliequivalents, meq). The latter unit is most useful for laboratory operations. It is apparent that both volumes must be expressed in the same units.

Formality The terms "molarity" and "mole" are most exactly restricted to compounds that are molecular in character, since it is only for these that the terms "molecule" and "molecular weight" have true significance. For an ionic compound such as sodium chloride, the terms mole and molecular weight have no real meaning. To say, for example, that 58.45 g of sodium chloride represent 1 mole of the compound is not rigorously correct. It would be better to speak of *gram-formula weights* in discussing solutes that are ionic, and to employ the term "formality," F, rather than molarity in describing the concentrations of their solutions. Thus, a 1 F solution of sodium chloride would contain 58.45 g of the solute per liter of solution. Formality is a better term for all types of solutes than molarity. *However, molarity has been so commonly employed for all types of materials that continued usage has made it generally acceptable.*

In many instances the concentrations of individual ions are of greater importance than the bulk concentrations of solutes as such. These ion

* If not otherwise indicated, the term "equivalents" will be used to indicate *gram equivalents*.

concentrations can be expressed in terms of the number of gram ions (g ion) per liter of solution, a notation comparable with molarity or formality. Thus, a solution containing 58.45 g of sodium chloride per liter contains 1 g ion of sodium ion and 1 g ion of chloride ion per liter. On the other hand, in a solution containing 1 gram-formula weight of calcium chloride per liter, the calcium-ion concentration is 1 g ion liter^{-1}, but the chloride-ion concentration is 2 g ion liter^{-1}. Other cases are treated similarly.

Mole fraction The mole fraction of solute B, N_B, in a solution containing solvent A is given by the relationship

$$\text{Mole fraction of B} = N_B = \frac{\text{No. moles of B}}{\text{No. moles of B} + \text{No. moles of A}} \tag{1-5}$$

Thus, for a solution containing 25 g of ethyl alcohol dissolved in 100 g of water, the mole fraction of ethyl alcohol is obtained as

$$\text{Mole fraction of } C_2H_5OH = \frac{\dfrac{25 \text{ g } C_2H_5OH}{46 \text{ g/mole}}}{\dfrac{25 \text{ g } C_2H_5OH}{46 \text{ g/mole}} + \dfrac{100 \text{ g } H_2O}{18 \text{ g/mole}}} = 0.089$$

Mole fractions are seldom used in the qualitative study of chemical reactions in solution.

Molality The molality, m, of a solution amounts to the number of moles of solute present in 1,000 g of *solvent*. Thus, a 1 m aqueous solution of sulfuric acid contains 98.08 g of H_2SO_4 in 1,000 g of water. It is apparent that molality and molarity are not exactly the same. Molality is the concentration unit employed in studying the physical chemistry of solutions. It has been mentioned in connection with vapor-pressure, boiling-point, and freezing-point effects, and it is also used in defining electrode potentials (Chap. 7).

Problems involving the use of solution concentrations in describing the quantitative aspects of chemical reactions appear in the next chapter.

ILLUSTRATIVE EXAMPLES

Example 1-1 The specific gravity of a phosphoric acid solution which is 18.0 percent H_3PO_4 by weight is 1.10. Calculate (*a*) the weight of H_3PO_4 in 500 ml of this solution, and (*b*) the molarity and normality of the solution.

Solution (*a*) The specific gravity is the ratio of the density of the substance to that of pure water. It is assumed that water has a density of 1.00 g ml^{-1} under the conditions involved. Thus, the density of the phosphoric acid solution is 1.10 g ml^{-1}. Since the phosphoric acid is 18.0 percent of the total solution weight,

$$\frac{\text{Wt of } H_3PO_4}{500 \text{ ml}} = \frac{1.10 \text{ g}}{1 \cancel{\text{ml}}} \times 500 \cancel{\text{ml}} \times \frac{18}{100} = 99.0 \text{ g}$$

(*b*) Since both the molarity and normality relate to the quantity of solute per liter of solution,

$$\frac{\text{Wt of } H_3PO_4}{\text{liter}} = \frac{\text{wt of } H_3PO_4}{500 \cancel{\text{ml}}} \times \frac{1{,}000 \cancel{\text{ml}}}{1 \text{ liter}}$$

$$= \frac{99.0 \text{ g}}{500 \cancel{\text{ml}}} \times \frac{1{,}000 \cancel{\text{ml}}}{1 \text{ liter}}$$

$$= 198 \text{ g liter}^{-1}$$

Then, since the formula weight of H_3PO_4 = 98.0 (g/mole),

$$\text{Molarity of } H_3PO_4 \text{ solution} = \frac{198 \cancel{\text{g}}}{1 \text{ liter}} \times \frac{1 \text{ mole}}{98.0 \cancel{\text{g}}} = 2.02 \text{ mole liter}^{-1}$$

$$= 2.02\ M$$

Since the equivalent weight is one-third the molecular weight (Table 1-3), then

$$\text{Normality of } H_3PO_4 = \frac{198 \cancel{\text{g}}}{1 \text{ liter}} \times \frac{3 \text{ g-eq wts}}{98.0 \cancel{\text{g}}}$$

$$= 6.06 \text{ g eq liter}^{-1} = 6.06\ N$$

Example 1-2 What volume of 0.0100 *M* $Pb(NO_3)_2$ solution could be prepared from 5.00 g of lead(II) nitrate?

Solution Since the formula weight for lead nitrate is 331 g/mole,

$$\text{Volume of solution obtainable} = \frac{1 \text{ liter}}{0.0100 \cancel{\text{mole}}} \times \frac{1 \cancel{\text{mole}}}{331 \cancel{\text{g}}} \times 5.00 \cancel{\text{g}}$$

$$= 1.51 \text{ liter}$$

Example 1-3 What volume of 16.0 *N* nitric acid is needed to prepare 50.0 ml of 6.00 *N* nitric acid?

Solution

$$\text{Volume needed} = 50.0 \cancel{\text{ml}} \times \frac{6.00 \cancel{\text{meq}}}{1 \cancel{\text{ml}}} \times \frac{1 \text{ ml}}{16.0 \cancel{\text{meq}}}$$

$$= 18.8 \text{ ml}$$

Example 1-4 What weight of copper(II) sulfate 5-hydrate must be used to prepare 500 ml of a solution containing 0.0200 mole of Cu^{2+} per liter?

Solution Since each gram-formula weight (mole) of $CuSO_4 \cdot 5H_2O$ contains 1 mole of Cu^{2+}, and the formula weight is 249.7, then

$$\text{Wt. of compound needed} = \frac{249.7 \text{ g}}{1 \cancel{\text{mole}}} \times \frac{0.0200 \cancel{\text{mole}}}{\cancel{\text{liter}}} \times 0.500 \cancel{\text{liter}}$$

$$= 2.50 \text{ g}$$

Example 1-5 A solution contains 16.0 g of methanol and 250 g of water. Determine (*a*) the molality of this solution, and (*b*) the mole fraction of methanol present.

Solution (*a*) Since the molecular weight of methanol (CH_3OH) is 32.0, then

$$\text{Molality of methanol} = \frac{\text{No. moles solute}}{\text{No. kg solvent}}$$

$$= \frac{16.0\ \text{g}}{0.250\ \text{kg}} \times \frac{1\ \text{mole}}{32.0\ \text{g}}$$

$$= 2.00\ \text{moles/kg}\ H_2O\ (2.00\ m)$$

$$(b)\ \text{Mole fraction of methanol} = \frac{\text{No. moles methanol}}{\text{No. moles methanol} + \text{No. moles water}}$$

$$= \frac{\dfrac{16.0\ \text{g}}{32.0\ \text{g/mole}}}{\dfrac{16.0\ \text{g}}{32.0\ \text{g/mole}} + \dfrac{250\ \text{g}}{18.0\ \text{g/mole}}}$$

$$= \frac{0.5}{0.5 + 13.9} = \frac{0.5}{14.4}$$

$$= 0.035$$

EXERCISES

1-1. Draw structural representations to indicate what results, on the molecular level, when (*a*) a small amount of formaldehyde (H_2CO) is added to a large volume of water; (*b*) a few grains of table salt (NaCl) are added to a cup of water; and (*c*) a drop of 1-decanol ($CH_3CH_2CH_2CH_2CH_2CH_2CH_2CH_2CH_2CH_2OH$) is added to the surface of a large pan of water.

1-2. Describe experiments that could be performed with equipment found in the average home to demonstrate that the water molecule is a dipole.

1-3. Compare two or more physical properties of (*a*) CO_2 and CS_2, and (*b*) H_2O and H_2S. What does this evidence suggest about the nature of intermolecular forces in water?

1-4. Which member of each of the following pairs of substances would you expect to be the more soluble in water? Explain each answer. Are there any pairs for which no simple prediction is satisfactory? Why?

(*a*) Calcium sulfate or calcium chloride
(*b*) Sodium chloride or silver chloride
(*c*) Glycerine or 1-butanol
(*d*) Ethylenediamine or aniline
(*e*) Ammonia gas or hydrogen gas
(*f*) Potassium chloride or sodium bromide

1-5. Classify each of the following as a strong electrolyte, a weak electrolyte, or a non-electrolyte. Explain each classification.

(*a*) Methanol
(*b*) Sodium acetate
(*c*) Ammonia
(*d*) Acetic acid
(*e*) Ammonium acetate

1-6. An organic compound prepared from ammonia and cyanic acid (HCNO) has the molecular formula CH_4N_2O. The compound is very soluble in water, but the resulting solution does not conduct an electric current significantly better than does pure water. A solution made from 6 g of the compound and 100 g of water freezes at about $-1.86°C$.

(*a*) What do you conclude about the nature of the chemical bonds in this compound?

(*b*) Suggest a possible structural formula for the compound.

1-7. Why are freezing-point depression and boiling-point elevation for a particular solvent functions of the relative *numbers* of solute and solvent particles present?

1-8. Although molecular weights of many nonionic compounds of small size and molecular weight, such as formaldehyde and ethanol, can be determined fairly closely by measurement of the freezing-point depressions of their aqueous solutions, corresponding measurements of boiling-point elevation give very inaccurate data. Suggest a reasonable explanation.

1-9. One can write a functional definition for molality as

$$m = \frac{\text{No. g-formula wts. of solute}}{\text{No. kg of solvent}}$$

Write similar definitions for:

(*a*) Molarity
(*b*) Formality
(*c*) Normality

1-10. One can write a simple formula for calculations involving dilution of a concentrated solution as

$$M_C \times V_C = M_D \times V_D$$

where M denotes molarity, V denotes volume of solution, C refers to "concentrated" solution, and D refers to "dilute" solution.

(*a*) Explain why this expression is correct.

(*b*) Such an equation cannot be used if the concentration is expressed as molality, although it is accurate for F or N. Explain.

1-11. Calculate the percent by weight of ethanol in an ethanol-water solution in which the mole fraction of ethanol is 0.25.

1-12. The specific gravity of a sulfuric acid solution which is 20 percent H_2SO_4 by weight is 1.1394. Calculate the molarity, normality, and molality of the solution.

1-13. Calculate the weight of aluminum chloride 6-hydrate needed to prepare 2.5 liters of a solution containing 0.15 mole of Al^{3+} per liter.

1-14. It is desired to prepare 500.0 ml of 0.1000 N HCl from a stock solution that is 6.000 N.

(*a*) What volume of the stock solution is needed?

(*b*) Could the solution be prepared to the desired accuracy by measuring separately the necessary quantity of stock solution and mixing it with 500.0 ml (stock-solution volume) of distilled water? Why or why not?

1-15. It has recently been suggested that the number of deaths by drowning in swimming pools might be significantly reduced if the pool water contained the same concentration of sodium chloride as that of the blood stream, thus reducing osmotic effects from the water entering the lungs. If the concentration of sodium chloride in the blood is about 0.86 percent by weight, what quantity of the salt would be needed for a swimming pool containing 4,000 gallons of pure water?

1-16. A 0.010 *N* solution of potassium permanganate is sometimes used as an antidote for oxalic acid poisoning. How would you prepare such a solution from a stock solution labeled 5.00 *F* $KMnO_4$ if the reaction involves conversion of MnO_4^- to Mn^{2+}?

1-17. Determine the equivalent weight of potassium dichromate when used according to each of the following processes:

(*a*) $Cr_2O_7^{2-} + 2Pb^{2+} + H_2O \rightarrow 2PbCrO_4 + 2H^+$

(*b*) $Cr_2O_7^{2-} + 14H^+ + 6e^- \rightarrow 2Cr^{3+} + 7H_2O$

Do you foresee any problems in using as a routine laboratory stock solution a bottle labeled 1.00 *N* $K_2Cr_2O_7$? Explain.

1-18. What volume of ethylene glycol (a common "antifreeze") should be mixed with 3.0 gal of water to prepare a solution that will not freeze until it reaches −40°F? The density of ethylene glycol is approximately 1.1088 g/ml. Assume that the freezing point depression for water remains constant over the temperature range involved.

1-19. A solution containing 5.85 g of sodium chloride and 100 g of water begins to freeze between −2.5 and −3.0°C. Explain.

1-20. A 2.00-g sample of an unknown organic compound is mixed with 10.00 g of camphor, and the melting point of the mixture is found to be 17.5°C lower than that of pure camphor. What is the approximate molecular weight of the sample if the molal freezing-point constant for camphor is −37.7?

RECOMMENDED SUPPLEMENTARY READING

Clapp, L. B.: "The Chemistry of the OH Group," Prentice Hall, Englewood Cliffs, N.J., 1967 (paperback).

Dreisbach, D.: "Liquids and Solutions," Houghton Mifflin, Boston, 1966 (paperback).

Lagowski, J. J.: "The Chemical Bond," Houghton Mifflin, Boston, 1966 (paperback).

2
Ionic Reactions in Aqueous Solution

Chemical reactions in aqueous solutions may involve both charged (ionic) and neutral (molecular or atomic) species. Reactions among molecules are of major interest in the study of organic chemistry. These reactions are quite slow and usually inefficient, resulting in appreciably less than 100 percent conversion of reactants to products. Chemical changes depending on simple collision of oppositely charged ions, on the other hand, are typically rapid and usually efficient. These reactions, and others involving ions (and sometimes molecules) in more complex processes, form the basis for inorganic analytical procedures.

IONIC EQUATIONS

Every chemical reaction can be represented by an equation. A chemical equation should indicate as nearly accurately as possible the exact reacting materials and products. Although it is still common practice in courses of elementary chemistry to utilize equations in which reactants and products are represented by "molecular" formulas,* it is apparent that this practice should usually be reserved for those cases where true molecular substances are involved. For reactions that take place in aqueous solution, such a practice is appropriate in only a limited number of cases and fails to show that the majority of the chemical changes involved are based upon the behavior of ions. For these cases, the use of net ionic equations is more desirable.

In describing a chemical reaction by means of a net ionic equation, only those species that are actually reactants or products in the *observed chemical change* are included. Other materials (ions or molecules) are omitted as contributing nothing to the observed reaction. For example, it can be shown that treating a silver salt solution (for example, $AgNO_3$, Ag_2SO_4, $AgClO_4$) with a solution of any soluble chloride (for example, HCl, $NaCl$, $CaCl_2$, $AlCl_3$) produces a white precipitate of silver chloride. It may be concluded that this reaction is a general one involving only silver and chloride ions. The process can be described by the ionic equation

$$Ag^+(aq) + Cl^-(aq) \rightarrow \underset{\text{White}}{\mathbf{AgCl}(s)}$$

This equation clearly describes any of the precipitations carried out, and may be so written because it is observed that neither the anion from the silver salt solution nor any of the added cations influence the observed reaction. A single net ionic equation thus describes accurately a type of reaction that might be indicated much less clearly by dozens of molecular equations.

In writing net ionic equations a number of generally accepted conventions are used regarding which materials are to be written in the molecular form and which are to be written in the ionic form. These conventions are summarized as follows:

1. Write as ions all strong electrolytes in solution; e.g., hydrochloric acid as $H^+ + Cl^-$, potassium hydroxide as $K^+ + OH^-$, and sodium sulfate as $Na^+ + SO_4^{2-}$.
2. Write as molecular formulas:
   a. All solids, whether present initially as such or formed as precipitates in the reaction; e.g., insoluble sulfides such as CuS and ZnS, and insoluble hydroxides such as $Al(OH)_3$ and $Cu(OH)_2$. This convention is admittedly only a convenience since many solid substances are either completely ionic or predominantly so (Chap. 1). However, equations are concerned more with ions in solution, and for this reason the convention is acceptable. In this book, the practice

* "Molecular" formula refers to the use of a *complete formula* for any chemical substance, for example, H_2, $NaCl$, H_2SO_4, $Ca(OH)_2$.

of printing the formulas of solid substances in equations in **boldface** type is employed. This serves both to point out such substances more clearly and to fix the insoluble materials in the mind of the student. Colors of precipitates are frequently written under their formulas.

b. All gases, whether used as reactants or formed as products in the reaction; e.g., hydrogen sulfide as H_2S or sulfur dioxide as SO_2. The condition of a species is often indicated by appropriate letters, such as (*s*) for solid, (*g*) for gas, (*l*) for liquid, or (*aq*) for aqueous.

$$\mathbf{Zn}(s) + 2H^+(aq) \rightarrow Zn^{2+}(aq) + H_2(g)$$

c. All soluble weak electrolytes; e.g., hydrosulfuric acid as H_2S, hydrocyanic acid as HCN, or water as H_2O.

These conventions are illustrated by the following examples:

1. Acids react with solutions of metal hydroxides to give salts and water. When the resulting salts are soluble, these reactions can be described by the general equation

$$H^+ + OH^- \rightarrow H_2O$$

If, however, the salt is insoluble, both the reacting ions and the product salt must be shown. For example, in the reaction between solutions of sulfuric acid and barium hydroxide all products and ions are shown.

$$2H^+ + SO_4{}^{2-} + Ba^{2+} + 2OH^- \rightarrow \underset{\text{White}}{\mathbf{BaSO_4}(s)} + 2H_2O$$

2. Insoluble salts are precipitated when their components are brought together in aqueous solutions. Only the ions actually involved are included in the equations. From the solubility rules (Chap. 1) can be deduced such equations as the following:

$$Ag^+ + Br^- \rightarrow \underset{\text{Cream}}{\mathbf{AgBr}(s)}$$

$$Ni^{2+} + 2OH^- \rightarrow \underset{\text{Pale green}}{\mathbf{Ni(OH)_2}(s)}$$

$$Ba^{2+} + CrO_4{}^{2-} \rightarrow \underset{\text{Yellow}}{\mathbf{BaCrO_4}(s)}$$

$$Mn^{2+} + S^{2-} \rightarrow \underset{\text{Pink}}{\mathbf{MnS}(s)}$$

3. Many metals react with dilute nitric acid to give solutions of the corresponding nitrates, liberating nitric oxide gas. Using the letter M to

represent various metals; e.g., Cu, Zn, Cd, and Pb; a general equation including only the essential materials may be written as

$$3M(s) + 8H^+ + 2NO_3^- \rightarrow 2NO(g) + 4H_2O + 3M^{2+}$$

Other examples are considered later.

Ionic equations, like all other types of equations, are complete only if balanced. Here balancing involves not only atoms of the species involved, but also ionic charges. Although several combinations of numbers may correctly balance the atoms in a given ionic equation, only that combination that also balances the ionic charges is the correct one. Thus, in both the following equations the atoms are balanced, but only the first is correctly balanced in terms of ionic charges. It is an excellent practice always to check the equalities of both atoms and charges in balancing ionic equations. By convention, the smallest whole-number coefficients sufficient for a balanced equation are used.

$$As_2S_3(s) + 3S^{2-} + 2S(s) \rightarrow 2AsS_4^{3-}$$

$$As_2S_3(s) + 2S^{2-} + 3S(s) \rightarrow 2AsS_4^{3-}$$

TYPES OF REACTIONS

Chemical reactions are of two general types—those in which no changes in oxidation state occur, and those in which changes in oxidation state do occur (electron transfer). The first type represents the metathetical reactions, the second the oxidation-reduction (or redox) reactions. Examples of both types are common in analytical chemistry.

Metathetical reactions Ionic reactions in which there are no changes in oxidation state take place, in general, because of the removal of one or more products from the field of reaction. Removal of a product may involve tying up ions in some fashion; such as the liberation of a gas, formation of a precipitate, or formation of a slightly ionized substance. Each of these reactions may be illustrated as shown below.

Liberation of a gas Solutions of sulfides and carbonates react readily with strong acids because of the liberation of gases.

$$S^{2-} + 2H^+ \rightarrow H_2S(g)$$

$$CO_3^{2-} + 2H^+ \rightarrow CO_2(g) + H_2O$$

Formation of a precipitate Salt solutions can be mixed with each other without reaction (if no electron transfer is possible) unless there are two or more ions present that form an insoluble compound. Then reaction occurs because precipitation removes those ions from solution. Typical equations for processes of this tpye are given in the first two examples of the preceding section on ionic equations.

Formation of a slightly ionized substance Solutions of acids and soluble hydroxides react because the formation of water removes ions from solution. Other slightly ionized products are weak acids, weak bases, or complex ions (Chap. 6). Examples of slightly ionized substances are given in the following equations:

$$CN^- + H_2O \rightarrow OH^- + \underset{\text{Weak acid}}{HCN}$$

$$OH^- + NH_4^+ \rightarrow H_2O + \underset{\text{Weak base}}{NH_3}$$

$$\mathbf{Cu(OH)_2}(s) + 4NH_3 \rightarrow 2OH^- + \underset{\text{Complex ion}}{[Cu(NH_3)_4]^{2+}}$$

Oxidation-Reduction reactions These are reactions in which electrons are transferred from one substance to another. Oxidation is defined as a loss of electrons, and reduction is defined as a gain of electrons. A reagent that causes a material to lose electrons by gaining them itself is known as an oxidizing agent. A reagent that causes a material to gain electrons by supplying those electrons itself is known as a reducing agent. Inasmuch as an observable chemical reaction cannot occur by only a loss or only a gain of electrons, oxidation and reduction proceed simultaneously.

Balancing equations for oxidation-reduction reactions Oxidation-reduction reactions appear to take place by the transfer of electrons from the reducing agent to the oxidizing agent. It is usually possible to separate physically the oxidation and reduction portions of a given reaction; to connect the two externally by means of a wire, and internally by means of a porous diaphragm or salt bridge; and to observe the flow of electrons in the external wire of the resulting chemical cell. Viewing an oxidation-reduction process in terms of a transfer of electrons is thus quite generally correct and very useful. Inasmuch as all the electrons supplied by one material must be utilized by the other, the balancing of equations for oxidation-reduction reactions can be approached by trying to find some means of equalizing electron loss and gain.

Oxidation number

A reaction is defined as oxidation-reduction if a change in oxidation number (oxidation state) occurs. The oxidation number is useful in chemical "bookkeeping" and has no physical reality, i.e., it does not represent an actual charge on a particle except in the case of a free element or a monatomic ion. Oxidation numbers for certain species are assigned as indicated in Table 2-1.

Table 2-1 Some oxidation numbers

Type of substance	*Oxidation number assignment*	*Examples*
"Free" element	0	0 Na, 0 F_2, 0 As_4, 0 S_8
Monatomic ion	Same as the charge on the ion	+I H^+, +II Mg^{2+}, +III Al^{3+}, −I F^-, −II S^{2-}
Group IA elements in compounds	+I	+I NaCl, +I Na_3PO_4
Group IIA elements in compounds	+II	+II $CaSO_4$, +II $Ca_3(PO_4)_2$
Fluorine in compounds	−I	−I HF, −I AlF_3, −I CF_4
Oxygen in compounds or polyatomic ions	−II [except as −I in peroxides (O—O bond, as +II in OF_2)]	−II H_2O, −II CO_2, −II $PO_4{}^{3-}$; −I H_2O_2, −I Na_2O_2, +II OF_2
Hydrogen in compounds or polyatomic ions	+I (except as −I when combined with metals)	+I HCl, +I H_2O, +I NH_3, +I OH^-; −I CaH_2, −I $LiAlH_4$

There are two ways of determining oxidation numbers to be assigned to other elements in compounds and polyatomic ions.

1. When the element is combined only with other elements of known oxidation number, the algebraic sum of all the oxidation numbers equals zero for compounds and the charge on the ion for polyatomic ions.

$$\sum (\text{Oxidation No. per symbol} \times \text{No. of times symbol appears}) = \text{net charge} \qquad (2\text{-}1)$$

2. When more than one element of unknown oxidation number is present, the electron-dot formula for the substance is written. Both shared electrons in a bond are assigned to the more electronegative atom in the bond. Unshared electrons are assigned to their "parent" atom. The oxidation number is then the kernel charge (atomic number − electrons not represented in dot formulas) minus the number of dots assigned to the atom.

The following examples illustrate the assignment of oxidation numbers.

Example 2-1 Assign oxidation numbers to each element in calcium carbonate.

Solution Calcium carbonate is formulated as $CaCO_3$. According to Table 2-1, Ca is +II and O is −II in this compound. Let N be the oxidation number for carbon. Then, according to Eq. 2-1:

$$+\mathrm{II} + N + (3 \times -\mathrm{II}) = 0$$

From which

$$N = +\mathrm{IV}$$

(Note that this does not indicate an actual *charge* of +4 on carbon.)

Example 2-2 Assign oxidation numbers to each element in potassium dichromate.

Solution The formula of potassium dichromate is $K_2Cr_2O_7$. According to Table 2-1, K is +I and O is −II in this compound. Let N be the oxidation number for chromium. Then, according to Eq. 2-1:

$$(2 \times +\mathrm{I}) + (2N) + (7 \times -\mathrm{II}) = 0$$

From which

$$N = +\mathrm{VI}$$

Example 2-3 Assign an oxidation number for sulfur in the $S_4O_6^{2-}$ ion.

Solution It is assumed that some indication will be given if a substance is a peroxide. In the absence of other evidence, then, the oxidation number −II is assigned to oxygen. Let N be the oxidation number for sulfur. According to Eq. 2-1:

$$(4N) + (6 \times -\mathrm{II}) = -2$$

From which

$$N = +\mathrm{II}\tfrac{1}{2}*$$

Example 2-4 Methyl amine (CH_3NH_2) is an organic base quite similar to ammonia. Assign oxidation numbers to the elements in this compound.

Solution The electron-dot formula, using the four valence electrons of the carbon atom, the five of nitrogen, and the one of hydrogen, is

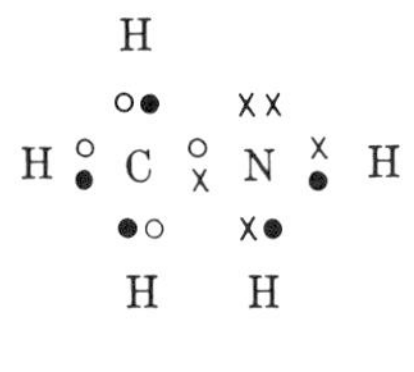

```
      H
      o●     xx
  H o● C ox N x● H
      ●o     x●
      H      H
```

Note: There is no distinction between the "different" electrons in the molecule. The use of ○, ×, and ● is just another "bookkeeping" simplification.

* Unfortunately the Roman system does not express fractions, so the rather unusual form shown here is used.

The relative order of electronegativities is N > C > H, so assigning bond-pair electrons to the more electronegative atom in each bond gives

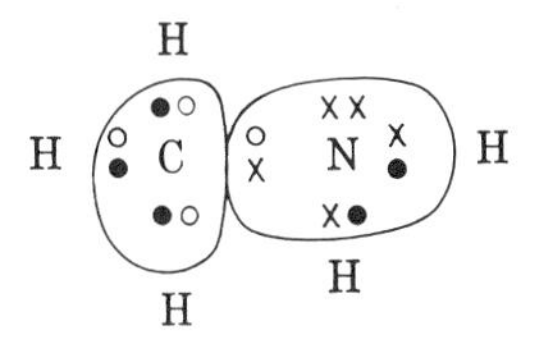

Remember, oxidation number equals kernel charge minus number of assigned dots (electrons). Thus,

for H, $1 - 0 = +\mathrm{I}$

for C, $4 - 6 = -\mathrm{II}$

and

for N, $5 - 8 = -\mathrm{III}$

Example 2-5 Which of the reactions, described by the following equations, is classified as an oxidation-reduction reaction?

$$CO_2 + H_2O \rightarrow HCO_3^- + H^+$$

$$CO_2 + 3H_2 \rightarrow CH_3OH + H_2O$$

Solution For the first equation, both hydrogen and oxygen have the same oxidation numbers (+I and −II, respectively) in both reactants and products. Since something must be reduced whenever something else is oxidized, the carbon atom must also retain its oxidation number (+IV). Hence, the first equation represents a metathetical reaction.

For the second equation, hydrogen has an oxidation number of 0 in the reactants and +I in the products. Thus, this equation does represent oxidation-reduction.

Ion-electron equations

When it is observed that a chemical reaction involves changes in oxidation numbers, the reaction must be classified as oxidation-reduction. In many cases equations representing these reactions may be balanced in a simple fashion, similar to that commonly used for metathetical equations; i.e., adjusting numerical coefficients to balance both elementary symbols and charges. Frequently, however, oxidation-reduction equations do not readily yield to a simple trial-and-error approach. There are various systematic ways of balancing oxidation-reduction equations. One method, which is particularly useful in electrochemical studies, uses *ion-electron equations* (*half-reactions*).*

* Half-reactions are considered in terms of potential values and equilibrium concepts in Chap. 7.

An oxidation-reduction reaction may be considered to be the summation of two so-called "half-reactions," one describing the loss of electrons by the reducing agent and the other the gain of electrons by the oxidizing agent. Each such half-reaction can then be formulated in an equation that includes the ions and molecules involved and sufficient electrons to account for the observed oxidation or reduction. Such an equation is an ion-electron equation, and the process described is an ion-electron half-reaction. To illustrate, consider the process described by the skeleton equation

$$Sn^{2+} + Fe^{3+} \rightarrow Fe^{2+} + Sn^{4+}$$

This may be regarded as the result of summation of oxidation and reduction processes described by the ion-electron equations

$$Sn^{2+} \rightarrow Sn^{4+} + 2e^- \qquad \text{(oxidation)}$$

$$Fe^{3+} + e^- \rightarrow Fe^{2+} \qquad \text{(reduction)}$$

Or, in a more complicated case, the reaction described by the equation

$$H_2S(g) + H^+ + NO_3^- \rightarrow NO(g) + H_2O + S(s)$$

may be represented by the ion-electron equations

$$H_2S(g) \rightarrow S(s) + 2H^+ + 2e^- \qquad \text{(oxidation)}$$

and

$$NO_3^- + 4H^+ + 3e^- \rightarrow NO(g) + 2H_2O \qquad \text{(reduction)}$$

Similarly, the equation

$$Br^- + H^+ + Cr_2O_7^{2-} \rightarrow Cr^{3+} + H_2O + Br_2$$

amounts to the ion-electron equations

$$2Br^- \rightarrow Br_2 + 2e^- \qquad \text{(oxidation)}$$

$$Cr_2O_7^{2-} + 14H^+ + 6e^- \rightarrow 2Cr^{3+} + 7H_2O \qquad \text{(reduction)}$$

Each ion-electron equation describes an observable chemical process, and indicates clearly the materials involved and the conditions of acidity necessary. Inasmuch as the reactions considered occur in aqueous solution, water may appear as either a reactant or a product, and may be included in an equation wherever necessary. Reactions taking place in acidic solution ($C_{H^+} > 10^{-7}M$) include sufficient hydrogen ions to convert oxygen in oxoanions to water. Reactions taking place in alkaline solution ($C_{H^+} < 10^{-7}M$) include hydroxide ion in sufficient quantity to form oxoions or insoluble hydroxides. Examples illustrating behaviors in acidic and alkaline solutions are given in Table 2-2. Sufficient electrons are always included

Table 2-2 Ion-electron equations in acidic and alkaline solutions

Couple	*Acidic solution*	*Alkaline solution*
Na(0)–Na(I)	$Na(s) \rightleftharpoons Na^+ + e^-$	$Na(s) \rightleftharpoons Na^+ + e^-$
Fe(0)–Fe(II)	$Fe(s) \rightleftharpoons Fe^{2+} + 2e^-$	$Fe(s) + 2OH^- \rightleftharpoons Fe(OH)_2(s) + 2e^-$
Fe(II)–Fe(III)	$Fe^{2+} \rightleftharpoons Fe^{3+} + e^-$	$Fe(OH)_2(s) + OH^- \rightleftharpoons Fe(OH)_3(s) + e^-$
N(II)–N(V)	$NO + 2H_2O \rightleftharpoons NO_3^- + 4H^+ + 3e^-$	$NO + 4OH^- \rightleftharpoons NO_3^- + 2H_2O + 3e^-$
Cr(III)–Cr(VI)	$2Cr^{3+} + 7H_2O \rightleftharpoons Cr_2O_7^{2-} + 14H^+ + 6e^-$	$CrO_2^- + 4OH^- \rightleftharpoons CrO_4^{2-} + 2H_2O + 3e^-$

to take care of the observed change in oxidation number. A system described by an ion-electron equation is sometimes referred to as a *couple*. Thus, the N(V)–N(II) couple would refer to the NO_3^-–NO reaction; the Cr(VI)–Cr(III) couple to the $Cr_2O_7^{2-}$–Cr^{3+} or CrO_4^{2-}–CrO_2^- reaction; etc. (Roman numerals indicate the oxidation state in each case).*

The principle of balancing oxidation-reduction equations by equalizing electrons amounts here merely to adjusting the two ion-electron equations involved to contain the same number of electrons, and then adding the two algebraically as simultaneous equations to cancel out the electrons.

Systematic procedure for balancing oxidation-reduction equations by the method of half-reactions

Step 1 Consider the reaction† as two half-reactions, one for oxidation and the other for reduction. Include only species involved in electron transfer as they appear in the overall equation.

Step 2 Use appropriate numerical coefficients to balance all symbols except H and O in each half-reaction equation.

Step 3 Add numbers of water molecules as needed to balance O symbols in each half-reaction equation.

Step 4 Add numbers of H^+ ions as needed to balance H symbols in each half-reaction equation.

Step 5 Add numbers of electrons as needed to balance net reactant and product charge in each half-reaction equation.

Step 6 Multiply all coefficients in each half-reaction equation by the smallest whole number required to make the electron loss equal the electron gain.

Step 7 Add the two half-reaction equations resulting from Step 6, subtracting or combining numbers of duplicated species that appear in

* Zero is used to indicate zero oxidation state, for example, Na(0)–Na(I).

† If the original equation given is not a "net" equation, write it in that form; then change the species in the final balanced equation as necessary to return it to its original form.

both reactants and products so that each species appears only in one place in the total equation.

Step 8 **Check.** The same number of each elementary symbol should appear in both reactants and products. The *net* charge should be the same for both reactants and products. The numerical coefficients should be the smallest whole numbers required for balancing. No electrons should appear in the final equation.

Step 9 (*For reactions in alkaline solution only*) change all H^+ ions in the balanced equation to H_2O and add the same number of OH^- ions to the other side of the equation. If H_2O now appears on both sides of the equation, subtract the smaller number of H_2O units from each side.

Example 2-6 Balance (in acidic solution):

$$CH_3CH_2OH + Cr_2O_7^{2-} \rightarrow CO_2 + H_2O + Cr^{3+}$$

Solution

Step 1

$$CH_3CH_2OH \rightarrow CO_2$$

$$Cr_2O_7^{2-} \rightarrow Cr^{3+}$$

Note that H_2O need not be included at this step since neither oxygen nor hydrogen changes oxidation state. However, no harm would be done if it were included.

Step 2

$$CH_3CH_2OH \rightarrow \underline{2}CO_2$$

$$Cr_2O_7^{2-} \rightarrow \underline{2}Cr^{3+}$$

Step 3

$$\underline{3H_2O} + CH_3CH_2OH \rightarrow 2CO_2$$

$$Cr_2O_7^{2-} \rightarrow 2Cr^{3+} + \underline{7H_2O}$$

Step 4

$$3H_2O + CH_3CH_2OH \rightarrow 2CO_2 + \underline{12H^+}$$

$$\underline{14H^+} + Cr_2O_7^{2-} \rightarrow 2Cr^{3+} + 7H_2O$$

Step 5

$$3H_2O + CH_3CH_2OH \rightarrow 2CO_2 + 12H^+ + \underline{12e^-}$$

$$[0 = 12(+) + 12(-)]$$

$$\underline{6e^-} + 14H^+ + Cr_2O_7^{2-} \rightarrow 2Cr^{3+} + 7H_2O$$

$$[6(-) + 14(+) + 2(-) = 2 \times 3(+)]$$

Step 6 To get electron loss equal to electron gain, multiply the lower half-reaction equation (reduction) in Step 5 by 2.

$$3H_2O + CH_3CH_2OH \rightarrow 2CO_2 + 12H^+ + 12e^-$$

$$12e^- + 28H^+ + 2Cr_2O_7^{2-} \rightarrow 4Cr^{3+} + 14H_2O$$

Step 7

$$\cancel{3H_2O} + CH_3CH_2OH + \cancel{12e^-} + \cancel{28H^+} + 2Cr_2O_7^{2-} \rightarrow 2CO_2 + 4Cr^{3+} + \cancel{14H_2O} + \cancel{12e^-} + \cancel{12H^+}$$

Balanced equation

$$CH_3CH_2OH + 16H^+ + 2Cr_2O_7^{2-} \rightarrow 2CO_2 + 4Cr^{3+} + 11H_2O$$

Step 8

$$2C,\ 22H,\ 15\ O,\ 4Cr = 2C,\ 4Cr,\ 15O,\ 22H;\ 16^+ + [2 \times 2(-)] = 4 \times 3(+)$$

Example 2-7 Balance (in alkaline solution):

$$H_2C{=}CH_2 + MnO_4^- \rightarrow HOCH_2CH_2OH + MnO_2$$

Solution

Step 1

$$H_2C{=}CH_2 \rightarrow HOCH_2CH_2OH$$

$$MnO_4^- \rightarrow MnO_2$$

Step 2 Omit, since C's and Mn's are already balanced.

Step 3

$$\underline{2H_2O} + H_2C{=}CH_2 \rightarrow HOCH_2CH_2OH$$

$$MnO_4^- \rightarrow MnO_2 + \underline{2H_2O}$$

Step 4

$$2H_2O + H_2C{=}CH_2 \rightarrow HOCH_2CH_2OH + \underline{2H^+}$$

$$\underline{4H^+} + MnO_4^- \rightarrow MnO_2 + 2H_2O$$

Step 5

$$2H_2O + H_2C{=}CH_2 \rightarrow HOCH_2CH_2OH + 2H^+ + \underline{2e^-} \qquad [0 = 2(+) + 2(-)]$$

$$\underline{3e^-} + 4H^+ + MnO_4^- \rightarrow MnO_2 + 2H_2O \qquad [3(-) + 4(+) + 1(-) = 0]$$

Step 6 To get electron loss equal to electron gain, note that the smallest number containing both 2 and 3 is 6, so multiply the oxidation half-reaction equation by 3 and the reduction half-reaction equation by 2.

$$6H_2O + 3H_2C{=}CH_2 \rightarrow 3HOCH_2CH_2OH + 6H^+ + 6e^-$$

$$6e^- + 8H^+ + 2MnO_4^- \rightarrow 2MnO_2 + 4H_2O$$

Step 7

$$\cancel{6H_2O} + 3H_2C{=}CH_2 + \cancel{6e^-} + \cancel{8H^+} + 2MnO_4^- \rightarrow 3HOCH_2CH_2OH + \cancel{6H^+} + \cancel{6e^-} + \cancel{4H_2O} + 2MnO_2$$

$$2H_2O + 3H_2C{=}CH_2 + 2H^+ + 2MnO_4^- \rightarrow 3HOCH_2CH_2OH + 2MnO_2$$

Step 8

$$18H, 6C, 2Mn, 10O{=}2Mn, 10O, 18H, 6C;\ 2(+) + 2(-) = 0$$

Step 9

$$2H_2O + 3H_2C{=}CH_2 + 2H_2O + 2MnO_4^- \rightarrow 2OH^- + 3HOCH_2CH_2OH + 2MnO_2$$

or, combining H_2O's

$$4H_2O + 3H_2C{=}CH_2 + 2MnO_4^- \rightarrow 2OH^- + 3HOCH_2CH_2OH + 2MnO_2$$

Example 2-8 Balance:

$$Cu(s) + HNO_3(aq) \rightarrow Cu(NO_3)_2(aq) + NO_2(g)$$

Solution First, write the equation in a form showing the main species involved:

$$Cu + H^+ + NO_3^- \rightarrow Cu^{2+} + NO_3^- + NO_2$$

Then write an unbalanced net equation showing only those species involved in electron transfer:

$$Cu + NO_3^- \rightarrow Cu^{2+} + NO_2$$

Note that H^+ need not be included and that NO_3^- is omitted from the products since it represents residual, unchanged reactant NO_3^-.

Step 1

$$Cu \rightarrow Cu^{2+}$$

$$NO_3^- \rightarrow NO_2$$

Step 2 Omit, since Cu's and N's are already balanced.

Step 3

$$Cu \rightarrow Cu^{2+}$$

$$NO_3^- \rightarrow NO_2 + \underline{H_2O}$$

Step 4

$$Cu \rightarrow Cu^{2+}$$

$$\underline{2H^+} + NO_3^- \rightarrow NO_2 + H_2O$$

Step 5

$$Cu \rightarrow Cu^{2+} + \underline{2e^-} \qquad [O = 2(+) + 2(-)]$$

$$\underline{e^-} + 2H^+ + NO_3^- \rightarrow NO_2 + H_2O \qquad [(-) + 2(+) + (-) = O]$$

Step 6 To get electron loss equal to electron gain, multiply the reduction half-reaction equation by 2.

$$Cu \rightarrow Cu^{2+} + 2e^-$$

$$2e^- + 4H^+ + 2NO_3^- \rightarrow 2NO_2 + 2H_2O$$

Step 7

$$Cu + \cancel{2e^-} + 4H^+ + 2NO_3^- \rightarrow 2H_2O + 2NO_2 + Cu^{2+} + \cancel{2e^-}$$

$$Cu + 4H^+ + 2NO_3^- \rightarrow Cu^{2+} + 2NO_2 + 2H_2O$$

Step 8

$$1Cu, 4H, 2N, 6O = 1Cu, 2N, 6O, 4H; 4(+) + 2(-) = 2(+)$$

Then, to return the equation to its original form, add $2NO_3^-$ to each side and combine the H^+'s with NO_3^- in the reactants and Cu^{2+} with the NO_3^-'s in the products.

$$Cu + 4HNO_3 \rightarrow Cu(NO_3)_2 + 2NO_2 + 2H_2O$$

Acid-Base reactions Acids are often considered to be compounds that give hydrogen ions in aqueous solutions and bases to be compounds that give hydroxide ions in aqueous solutions. These definitions are credited to Svante Arrhenius (about 1887) and have been used widely for many years. However, it has become apparent that they are somewhat restrictive, particularly as regards bases. A desire to broaden the concept to include other materials that react in identical fashions prompted Brønsted and Lowry in 1923 to define acids as substances capable of giving up protons (i.e., proton donors) and bases as substances capable of adding protons (i.e., proton acceptors). The most obvious advantage of these definitions is the extension of the term base to a number of materials containing other than the hydroxide ion.

Acids may be of the following types:

1. Molecular; for example; HCl, HNO_3, H_2SO_4, CH_3COOH, H_2O
2. Anionic; for example; HSO_4^-, $H_2PO_4^-$, HPO_4^{2-}
3. Cationic; for example; NH_4^+ and hydrated cations such as $Al(H_2O)_4^{3+}$

Correspondingly, bases may be of the following types:

1. Molecular; for example; NH_3, CH_3NH_2, $(CH_3)_2NH$, $(CH_3)_3N$, H_2O
2. Anionic; for example; O^{2-}, Cl^-, OH^-, NO_2^-, CN^-, CO_3^{2-}, HSO_4^- (any anion)
3. Cationic; for example; $Al(OH)^{2+}$ or $Al(OH)(H_2O)_3^{2+}$

The most common acids are molecular in character; the most common bases, anionic. It is important to realize that salts, which are aggregations of ions, are neither acids nor bases as such, but contain acids and/or bases. The most common cases in point are the metal hydroxides. These compounds are not bases as such, but they do contain the basic hydroxide ion. Extreme care in terminology is necessary here.

Substances can be inherently acidic but behave as acids only if bases are present to accept protons. Similarly, substances can be inherently basic but behave as bases only if acids are present to provide protons. Acidic and basic properties are of particular importance in aqueous solutions, for water can behave as either a proton acceptor or donor. These behaviors are exemplified by the equations*

$$\underset{\text{Acid}}{CH_3CO_2H} + \underset{\text{Base}}{H_2O} \rightleftharpoons \underset{\text{Acid}}{H_3O^+} + \underset{\text{Base}}{CH_3CO_2^-}$$

$$\underset{\text{Acid}}{H_2O} + \underset{\text{Base}}{NH_3} \rightleftharpoons \underset{\text{Acid}}{NH_4^+} + \underset{\text{Base}}{OH^-}$$

These equations are illustrative of what might be called "general acid-base equilibria" of the type

$$\text{Acid}_1 + \text{base}_2 \rightleftharpoons \text{acid}_2 + \text{base}_1 \tag{2-1}$$

where base_1, which is derived from acid_1 by the loss of a proton, is called the conjugate base of acid_1; and base_2, by the same token, is called the conjugate base of acid_2. It is apparent that the strengths of acids and bases in aqueous solution are measured by the extents to which reactions with water occur. This topic is discussed quantitatively in Chap. 4. The reaction of an acid with a base is called "neutralization." The most common neutralization in aqueous solution involves the hydroxide ion.

An even more general definition of acid-base behavior was suggested in 1923 by G. N. Lewis. Careful study of the substances classified as bases by the Brønsted-Lowry system reveals that they all react by makng available an unshared electron pair for the formation of a covalent bond with a proton. Lewis recognized that many other substances react as does the proton, by seeking an electron pair for bond formation. These substances are now referred to as Lewis acids. An acid-base reaction can then be considered as competition for an electron pair.

Examples 2-9

$$PbCl_2(s) + Cl^- \rightleftharpoons PbCl_3^-$$

```
 ..  ..  ..          .. -         ..  ..  ..
:Cl:Pb:Cl:  +      :Cl:     ⇌   :Cl:Pb:Cl: -
 ..      ..          ..           ..  ..  ..
                                     :Cl:
                                      ..
Lewis acid       Lewis base       Complex ion
```

$$PbCl_3^- + Cl^- \rightleftharpoons PbCl_4^{2-}$$

```
                                    Cl   Cl 2-
 ..  ..  .. -        .. -            ...
:Cl:Pb:Cl:  +      :Cl:  ⇌          :Pb:
 ..  ..  ..          ..              . .
    :Cl:                            Cl   Cl
     ..
Lewis acid       Lewis base       Complex ion
```

*The double arrow convention ($\rightleftharpoons$) indicates a condition in which reactants and products coexist in a condition of dynamic equilibrium (Chap. 3).

$$\begin{array}{ccccc} H^+ & + & F^- & \rightleftharpoons & HF \\ H^+ & + & :\ddot{\underset{\cdot\cdot}{F}}:^- & \rightleftharpoons & H:\ddot{\underset{\cdot\cdot}{F}}: \\ \text{Lewis acid} & & \text{Lewis base} & & \text{Molecular compound} \end{array}$$

$$\begin{array}{ccccccc} FeCl_3 & + & Cl_2 & \rightleftharpoons & FeCl_4^- & + & Cl^+ \\ :\ddot{\underset{\cdot\cdot}{Cl}}:\underset{:\ddot{\underset{\cdot\cdot}{Cl}}:}{Fe}:\ddot{\underset{\cdot\cdot}{Cl}}: & + & :\ddot{\underset{\cdot\cdot}{Cl}}:\ddot{\underset{\cdot\cdot}{Cl}}: & \rightleftharpoons & :\ddot{\underset{\cdot\cdot}{Cl}}:\overset{:\ddot{Cl}:}{\underset{:\ddot{\underset{\cdot\cdot}{Cl}}:}{Fe}}:\ddot{\underset{\cdot\cdot}{Cl}}:^- & + & :\underset{\cdot\cdot}{Cl}:^+ \\ \text{Lewis acid}_1 & & \text{Lewis base}_2 & & \text{Lewis base}_1 & & \text{Lewis acid}_2 \\ & & & & \text{(Complex ion)} & & \end{array}$$

For purposes of studying most simple reactions in aqueous solution, the Brønsted-Lowry terminology is used. Specific reference to Lewis acids is made in discussions of many important reactions in organic chemistry and in some aspects of the discussion of complex ions (Chap. 6).

QUANTITATIVE ASPECTS OF REACTIONS IN SOLUTION

Chemical reactions in solution are subject to the same quantitative treatment as reactions that take place under other conditions. In addition to situations involving only weight or mole quantities, however, those concerning volumes of reacting solutions must be considered. Calculations for these cases commonly require knowing the molarity (or formality) and normality units.

Problems involving weight-weight relationships only These problems are concerned with the quantities of materials that react or result in particular chemical reactions. Quantities may be expressed in weight or mole units. Consider, for example, the reaction expressed by the equation

$$3\mathbf{Cu}(s) + 8H^+ + 2NO_3^- \rightarrow 2NO(g) + 4H_2O + 3Cu^{2+}$$

It is apparent that for every 3 moles of copper consumed, 3 moles of Cu^{2+} and 2 moles of nitrogen(II) oxide are produced. A mole (gram atom) of copper or a mole (gram ion) of Cu^{2+} represents 63.5 g, whereas a mole of nitrogen(II) oxide represents 30.0 g. If, therefore, it is desired to calculate the actual weight of nitrogen(II) oxide obtained from 100 g of copper, the operation may be carried out as follows:

$$\begin{aligned} \text{Wt of NO} &= 100\,\cancel{\text{g Cu}} \times \frac{1\,\cancel{\text{mole Cu}}}{63.5\,\cancel{\text{g Cu}}} \times \frac{2\,\cancel{\text{moles NO}}}{3\,\cancel{\text{moles Cu}}} \times \frac{30.0\text{ g NO}}{1\,\cancel{\text{mole NO}}} \\ &= 31.5\text{ g} \end{aligned}$$

Other problems of this general type are handled similarly.

Problems involving mole quantities and solution volumes Whenever solution concentrations are expressed in terms of moles or gram-formula weights per liter (or similarly with other volumes), the relative volumes of solutions involved in a particular reaction can be determined only if the equation is known for the reaction in question. Consider, for example, the reaction expressed by the equation

$$Al^{3+} + 3OH^- \rightarrow \mathbf{Al(OH)_3}(s)$$

It is apparent that for each mole of Al^{3+} present, 3 moles of OH^- are required. Thus, the volume of 2 M sodium hydroxide solution required to react exactly with 250 ml of 3.5 M aluminum nitrate solution can be calculated as

$$\text{Vol of NaOH solution} = 250\text{ ml} \times \frac{3.5\ \cancel{\text{moles Al}^{3+}}}{1\ \cancel{\text{liter}}} \times \frac{3\ \cancel{\text{moles OH}^-}}{1\ \cancel{\text{mole Al}^{3+}}} \times \frac{1\ \cancel{\text{liter}}}{2\ \cancel{\text{moles OH}^-}} = \sim 1{,}300\text{ ml}$$

A number of variations of this type of calculation are obviously possible.

Problems involving equivalent quantities and solution volumes Whereas a solution can have only a single molarity or formality, it is possible for it to be described by more than one normality since the normality is determined by the type of reaction the solute undergoes (Chap. 1, Exercise 1-17). In terms of the discussion in Chap. 1 it is apparent that, if solution concentrations are expressed in equivalents (normalities), the volumes involved in reactions are related as

$$\text{Volume}_A \times \text{normality}_A = \text{volume}_B \times \text{normality}_B \tag{2-2}$$

This relation holds irrespective of the equation for the reaction or whether the reaction is of the metathetical or oxidation-reduction type, provided only that the reaction is of the *type* for which both normality_A and normality_B were calculated.

For example, 100 ml of 0.10 N oxalic acid (N based on its action as a proton donor) is exactly equivalent to 100 ml of a 0.10 N solution of *any base* (the normality of which is determined by its action as a proton acceptor). However, if the same solution of oxalic acid is used with a reducing agent that converts it to ethylene glycol, its normality is not 0.10 for this reaction, and calculations must be changed accordingly. The ion-electron equation for the latter reaction is

$$\underset{\text{Oxalic acid}}{H_2C_2O_4} + 8H^+ + 8e^- \rightarrow \underset{\text{Ethylene glycol}}{HOCH_2CH_2OH} + 2H_2O$$

from which, N (of oxalic acid) $= 8 \times M$ (of oxalic acid). When acting as an acid capable of furnishing two protons, N (of oxalic acid) $= 2 \times M$ (of oxalic acid).

ILLUSTRATIVE EXAMPLES

Example 2-10 Determine the weight of ammonia obtainable from the reduction of excess nitrate ion by 25 g of zinc according to the equation

$$4Zn(s) + 7OH^- + NO_3^- \rightarrow NH_3(g) + 2H_2O + 4ZnO_2^{2-}$$

Solution According to the equation, 1 mole of ammonia (17.0 g) is obtained for every 4 moles of zinc (4 × 65.4 g) consumed. Hence,

$$\text{Wt of } NH_3 = 25 \text{ g Zn} \times \frac{1 \text{ mole Zn}}{65.4 \text{ g Zn}} \times \frac{1 \text{ mole } NH_3}{4 \text{ moles Zn}} \times \frac{17.0 \text{ g } NH_3}{1 \text{ mole } NH_3}$$
$$= 1.63 \text{ g}$$

Example 2-11 Determine the volume of concentrated hydrochloric acid solution (sp gr = 1.201, 40.0 percent HCl by weight) needed to dissolve 10.0 g of iron(III) oxide according to the equation

$$Fe_2O_3(s) + 6H^+ \rightarrow 2Fe^{3+} + 3H_2O$$

Solution The actual quantity of HCl may be calculated first. Since, according to the equation, each gram-formula weight, or mole, of iron(III) oxide (159.7 g) requires 6 moles of hydrogen chloride, (6 × 36.46 g),

$$\text{Wt of HCl} = 10.0 \text{ g } Fe_2O_3 \times \frac{1 \text{ mole } Fe_2O_3}{159.7 \text{ g } Fe_2O_3} \times \frac{6 \text{ moles HCl}}{1 \text{ mole } Fe_2O_3} \times \frac{36.46 \text{ g HCl}}{1 \text{ mole HCl}}$$
$$= 13.7 \text{ g}$$

The volume of solution containing this quantity of hydrogen chloride is then obtained as

$$\text{Vol of hydrochloric acid soln.} = 13.7 \text{ g HCl} \times \frac{1 \text{ ml soln.}}{1.201 \text{ g soln.}} \times \frac{100 \text{ g soln.}}{40.0 \text{ g HCl}}$$
$$= 28.5 \text{ ml}$$

Example 2-12 Calculate the quantity of silver phosphate that can be precipitated from 500 ml of a solution containing 10 mg of silver ion per milliliter. The equation for the reaction is assumed to be

$$3Ag^+ + PO_4^{3-} \rightleftharpoons Ag_3PO_4(s)$$

Solution The equation tells us that 3 moles of silver ion (3 × 108 g) will yield 1 mole of silver phosphate (419 g). The total quantity of silver ion present is calculated as

$$\text{Wt of } Ag^+ = 500 \text{ ml} \times \frac{10 \text{ mg}}{1 \text{ ml}} \times \frac{1 \text{ g}}{1{,}000 \text{ mg}}$$
$$= 5.0 \text{ g}$$

Hence,

$$\text{Wt of } Ag_3PO_4 = 5.0 \text{ g } Ag^+ \times \frac{1 \text{ mole } Ag^+}{108 \text{ g } Ag^+} \times \frac{1 \text{ mole } Ag_3PO_4}{3 \text{ moles } Ag^+} \times \frac{419 \text{ g } Ag_3PO_4}{1 \text{ mole } Ag_3PO_4}$$

$$= 6.5 \text{ g}$$

Example 2-13 Calculate the quantity of chromium(III) hydroxide that can be dissolved by 500 ml of 0.10 N sulfuric acid solution.

Solution The number of milliequivalents of chromium(III) hydroxide dissolved must equal the number of milliequivalents of sulfuric acid used. These quantities are:

$$\text{meq of } Cr(OH)_3 = x\text{g } Cr(OH)_3 \times \frac{1 \text{ mole } Cr(OH)_3}{103 \text{ g } Cr(OH)_3} \times \frac{3 \text{ g equiv } Cr(OH)_3}{1 \text{ mole } Cr(OH)_3} \times \frac{1{,}000 \text{ meq}}{1 \text{ g equiv } Cr(OH)_3}$$

and

$$\text{meq of } H_2SO_4 = 500 \text{ ml } H_2SO_4 \times \frac{0.10 \text{ meq}}{1 \text{ ml } H_2SO_4}$$

Equating and simplifying gives

$$\text{Wt of } Cr(OH)_3 = \frac{103 \text{ g } Cr(OH)_3}{1 \text{ mole } Cr(OH)_3} \times \frac{1 \text{ mole } Cr(OH)_3}{3 \text{ g equiv } Cr(OH)_3} \times \frac{1 \text{ g equiv } Cr(OH)_3}{1{,}000 \text{ meq}} \times 500 \text{ ml } H_2SO_4 \times \frac{0.10 \text{ meq}}{1 \text{ ml } H_2SO_4}$$

$$= 1.7 \text{ g}$$

Example 2-14 Calculate the quantity of hydrogen sulfide that can be oxidized to sulfur by 100 ml of 0.050 N nitric acid. The concentration of the nitric acid is based upon reduction to nitrogen(II) oxide.

$$3H_2S + 2H^+ + 2NO_3^- \rightarrow 2NO(g) + 4H_2O + 3S(s)$$

Solution Again milliequivalents may be equated. The number of milliequivalents of hydrogen sulfide is determined by the two-electron change in the ion-electron equation

$$H_2S \rightarrow S + 2H^+ + 2e^-$$

Thus

$$\text{meq of } H_2S = \text{wt of } H_2S \times \frac{1 \text{ mole } H_2S}{34.1 \text{ g } H_2S} \times \frac{2 \text{ g equiv } H_2S}{1 \text{ mole } H_2S} \times \frac{1{,}000 \text{ meq}}{1 \text{ g equiv } H_2S}$$

Correspondingly,

$$\text{meq of } HNO_3 = 100 \text{ ml } HNO_3 \times \frac{0.050 \text{ meq}}{1 \text{ ml } HNO_3}$$

Equating and simplifying then gives

$$\text{Wt of } H_2S = \frac{34.1 \text{ g } H_2S}{1 \text{ mole } H_2S} \times \frac{1 \text{ mole } H_2S}{2 \text{ g equiv } H_2S} \times \frac{1 \text{ g equiv } H_2S}{1{,}000 \text{ meq}} \times 100 \text{ ml } HNO_3 \times \frac{0.050 \text{ meq}}{1 \text{ ml } HNO_3}$$

$$= 0.085 \text{ g}$$

EXERCISES

2-1. Convert each of the following skeleton molecular equations into a balanced net ionic equation:

(*a*) $Ba(NO_3)_2 + Al_2(SO_4)_3 \rightarrow \mathbf{BaSO_4}(s) + Al(NO_3)_3$
(*b*) $\mathbf{CuCO_3}(s) + HCl \rightarrow CuCl_2 + H_2O + CO_2(g)$
(*c*) $FeCl_3 + NaOH \rightarrow \mathbf{Fe(OH)_3}(s) + NaCl$
(*d*) $FeSO_4 + HNO_3 \rightarrow NO(g) + H_2O + Fe_2(SO_4)_3$
(*e*) $Ba(NO_3)_2 + K_2Cr_2O_7 + H_2O \rightarrow \mathbf{BaCrO_4}(s) + KNO_3 + HNO_3$

2-2. From outside readings in scientific periodicals or popular magazines find one example of each of the following that is of significance in the fields of medicine, home economics, agriculture, or environmental control. Cite the reference used.

(*a*) A metathetical reaction involving Mg^{2+} or Al^{3+} ion.
(*b*) An oxidation-reduction reaction involving copper or manganese.
(*c*) A solid oxidizing agent.
(*d*) An aqueous reducing system.
(*e*) A half-reaction useful in an electrochemical cell.
(*f*) A reaction involving a Lewis acid in a nonaqueous solvent.

2-3. Balance each of the following equations.

(*a*) $\mathbf{Zn}(s) + H^+ \rightarrow H_2(g) + Zn^{2+}$
(*b*) $Fe^{2+} + H^+ + Cr_2O_7^{2-} \rightarrow Cr^{3+} + H_2O + Fe^{3+}$
(*c*) $Cl^- + H^+ + NO_3^- \rightarrow NO(g) + H_2O + Cl_2(g)$
(*d*) $\mathbf{Al}(s) + H_2O + OH^- + NO_2^- \rightarrow NH_3 + Al(OH)_4^-$
(*e*) $Cr(OH)_4^- + OH^- + ClO^- \rightarrow Cl^- + H_2O + CrO_4^{2-}$
(*f*) $H_2S(g) + Fe^{3+} \rightarrow Fe^{2+} + H^+ + \mathbf{S}(s)$
(*g*) $\mathbf{Cu}(s) + H^+ + SO_4^{2-} \rightarrow SO_2(g) + H_2O + Cu^{2+}$

2-4. Balance each of the following equations.

(*a*) $I^- + MnO_4^- \rightarrow I_3^- + Mn^{2+}$ (in acidic solution)
(*b*) $HSO_3^- + Fe^{3+} \rightarrow HSO_4^- + Fe^{2+}$ (in acidic solution)
(*c*) $Mn^{2+} + \mathbf{NaBiO_3}(s) \rightarrow MnO_4^- + Bi^{3+} + Na^+$ (in acidic solution)
(*d*) H_2CO (formaldehyde) $+ Cr_2O_7^{2-} \rightarrow HCO_2H$ (formic acid) $+ Cr^{3+}$ (in acidic solution)
(*e*) $CH_3CH_2OH + MnO_4^- \rightarrow CO_3^{2-} + \mathbf{MnO_2}(s)$ (in alkaline solution)
(*f*) $\mathbf{Cr(OH)_3}(s) + H_2O_2 \rightarrow CrO_4^{2-}$ (in alkaline solution)
(*g*) $\mathbf{Cu_2O}(s) + CrO_4^{2-} \rightarrow \mathbf{Cu(OH)_2}(s) + \mathbf{Cr(OH)_3}(s)$ (in alkaline solution)

2-5. Which of the following reactants acts, in each case, as a Brønsted-Lowry base?

(*a*) $HCl(g) + NH_3(g) \rightarrow \mathbf{NH_4Cl}(s)$
(*b*) $HCN + H_2O \rightarrow CN^- + H_3O^+$
(*c*) $CH_3NH_2 + H_2O \rightarrow CH_3NH_3^+ + OH^-$

2-6. Calculate the weight of calcium carbonate needed to react completely with 100 ml of 6 *N* hydrochloric acid solution.

2-7. Calculate the volume of 0.50 *M* barium hydroxide solution required to precipitate all the sulfate ion obtainable from 100 g of iron(II) sulfate 7-hydrate.

2-8. Determine the volume of 0.30 *N* hydrobromic acid solution required to neutralize 40 g of solid sodium hydroxide.

2-9. Calculate the weight of elemental sulfur obtainable by oxidation of 200 liters [at standard conditions (STP)] of hydrogen sulfide gas.

2-10. Calculate the weight of aluminum oxide needed to react completely with 100 ml of 20.0 percent (by weight) nitric acid solution of specific gravity 1.1150.

2-11. Determine the volume of 0.10 *N* potassium permanganate solution needed to oxidize hydrochloric acid solution to 100 ml of elemental chlorine (measured at STP).

2-12. Calculate the quantity of iron(II) hydroxide that can be precipitated from 750 ml of a solution containing 15 mg of iron(II) ion liter^{-1}.

2-13. Determine the concentration of a barium nitrate solution, 100 ml of which will precipitate all the sulfate ion in 50 ml of a solution containing 0.02 mole of SO_4^{2-} ion liter^{-1}. Express your answer in milligrams of Ba^{2+} ion ml^{-1}.

2-14. An unknown solution of sodium hydroxide was standardized against a solution of perchloric acid labeled 0.1028 *N* $HClO_4$.

(*a*) If 50.00 ml of the sodium hydroxide solution was neutralized by 38.22 ml of the acid, what normality would be calculated for the base?

(*b*) It was later learned that the perchloric acid used had actually been standardized for an oxidation-reduction reaction in which perchlorate ion was reduced to chloride ion. What should the label for the sodium hydroxide solution read? Why?

RECOMMENDED SUPPLEMENTARY READING

Campbell, J. A.: "Why Do Chemical Reactions Occur?," Prentice Hall, Englewood Cliffs, N.J., 1965 (paperback).

Vanderwerf, C. A.: "Acids, Bases, and the Chemistry of the Covalent Bond," Reinhold, New York, 1961 (paperback).

Kieffer, W. F.: "The Mole Concept in Chemistry," Reinhold, New York, 1963 (paperback).

Sienko, M. J.: "Stoichiometry and Structure," Benjamin, New York, 1964 (paperback).

Sisler, H. H.: "Chemistry in Non-Aqueous Solvents," Reinhold, New York, 1961 (paperback).

3
Reaction Rates, Mechanisms, and Chemical Equilibria

It is logical to believe that chemical reactions occur only as a result of collisions between atoms, molecules, or ions of the reacting substances. On the other hand, it is experimentally true that collisions may occur without causing reactions to take place. Thus, one can mix hydrogen and chlorine gases in the dark or in subdued light and preserve the resulting mixtures indefinitely under these conditions without the formation of hydrogen chloride. Certainly the principles of kinetic theory would require large numbers of collisions between hydrogen and chlorine molecules under these circumstances. On the other hand, increasing the temperature of this mixture or exposing it to light of appropriate wavelength causes an immediate and very rapid reaction to occur, initiated by thermal or photochemical conversion of chlorine molecules to reactive chlorine atoms.

These and other similar observations are perhaps best explained by stating that although specific chemical substances may have a tendency to react, they will do so only when a certain limiting quantity of energy is available to initiate the reaction. This energy is called the "energy of activation." Each reaction is characterized by its own particular energy of activation. It follows, therefore, that a collision involving atoms, molecules, or ions can result in reaction only if it is sufficiently energetic. In other words, although reactions occur because of collisions between particles of the reacting substances, only certain collisions are actually effective. The rate of a reaction (i.e., the speed at which reactants are converted to products) is then determined by the frequency of effective collisions. Any factor affecting the number of effective collisions taking place in a given interval of time will affect correspondingly the rate of the reaction.

FACTORS AFFECTING REACTION RATES

The frequencies of effective collisions are determined by the following factors:

1. Temperature of the system
2. Contact between reacting substances
3. Presence or absence of a catalyst
4. Concentrations of reacting substances

Each of these may be considered in more detail.

Temperature of the system Temperature is a measure of the average kinetic energy of a system. For a system made up of a collection of particles, such as a given volume of gas or a fixed quantity of a solution, the temperature is related to the *average* kinetic energy of all the particles present. At an increased temperature some of the particles may have a lower kinetic energy, but the number increasing in energy will outweigh the number decreasing. The opportunity for effective collisions (i.e., for collisions of sufficient energy to initiate a chemical change) is, therefore, better at higher temperatures than at lower temperatures. In addition, since the average velocity of a particle is greater at higher temperatures, collisions between particles will occur more often in a given time interval at elevated temperatures than at lower temperatures. Experimentally, the majority of chemical reactions do proceed more rapidly at higher temperatures. A few reactions, however, proceed more slowly at elevated temperatures. For those reactions that are speeded up, a careful analysis of rate data usually shows that the rate is increased more than could be expected simply on the basis of more frequent collisions. Increased collision frequency, then, must be only one of the factors determining the temperature dependence of reaction rates.

In order for reactions to occur when chemical bonds must be broken, the collisions must occur with sufficient energy to break these bonds. The number of collisions per unit time may, therefore, be less important than the number of *high-energy* collisions per unit time. The Maxwell-Boltzmann energy distribution (Fig. 3-1) suggests that the frequency of high-energy

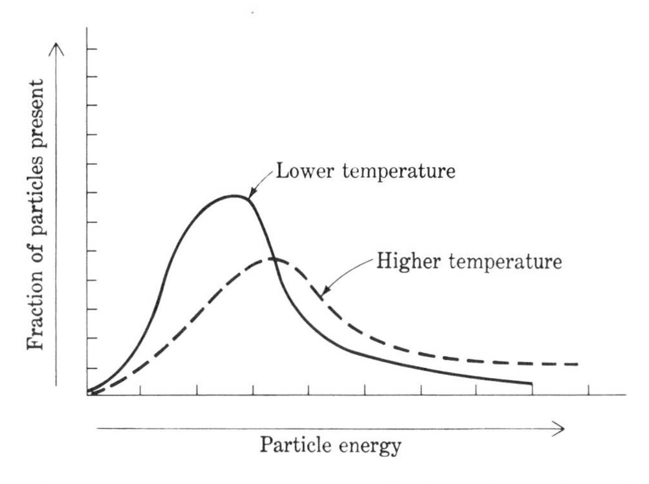

Fig. 3-1 Maxwell-Boltzmann distribution of energies of particles.

collisions increases rapidly with increasing temperature. This suggestion is supported by experimental measurements of velocity distribution in molecular beams. This effect is a major factor in temperature dependence of reaction rates.

Contact between reacting substances The intimacy with which particles of the reacting substances contact each other markedly affects the frequencies of effective collision among these particles. Thus, massive metals such as iron or lead may oxidize slowly upon exposure to air, but they do not burn. On the other hand, these same metals when sufficiently finely divided ignite and burn upon contact with air. Dust explosions involving finely divided substances are not uncommon. A readily demonstrated reaction involves mercury(II) nitrate and potassium iodide, both white solids yielding colorless solutions, to give red mercury(II) iodide. If solid mercury(II) nitrate and solid potassium iodide are shaken together, the red color of the reaction product develops only very slowly. If the two solids are ground together in a mortar, thereby increasing the intimacy of contact, the red product forms with obviously greater rapidity. If the two solids are dissolved separately in water and the solutions mixed, precipitation of the red product is instantaneous. Reactions in solution provide maximum intimacy of contact and are, therefore, of particular utility.

Contact between oppositely charged ions in solution is readily effected. Monatomic ions, even when shielded by oriented clusters of water molecules, are essentially spherical, and collisions can be effective without regard to geometry. Polyatomic ions, especially large ions, may require certain spatial orientations for effective collisions, but the interactions of the charged regions of the approaching species will greatly facilitate necessary orientations.

Reactions between uncharged molecules are generally much slower than those involving ions. Usually only a fraction of the possible collision geometries are effective; i.e., lead to reaction; as illustrated by the following examples:

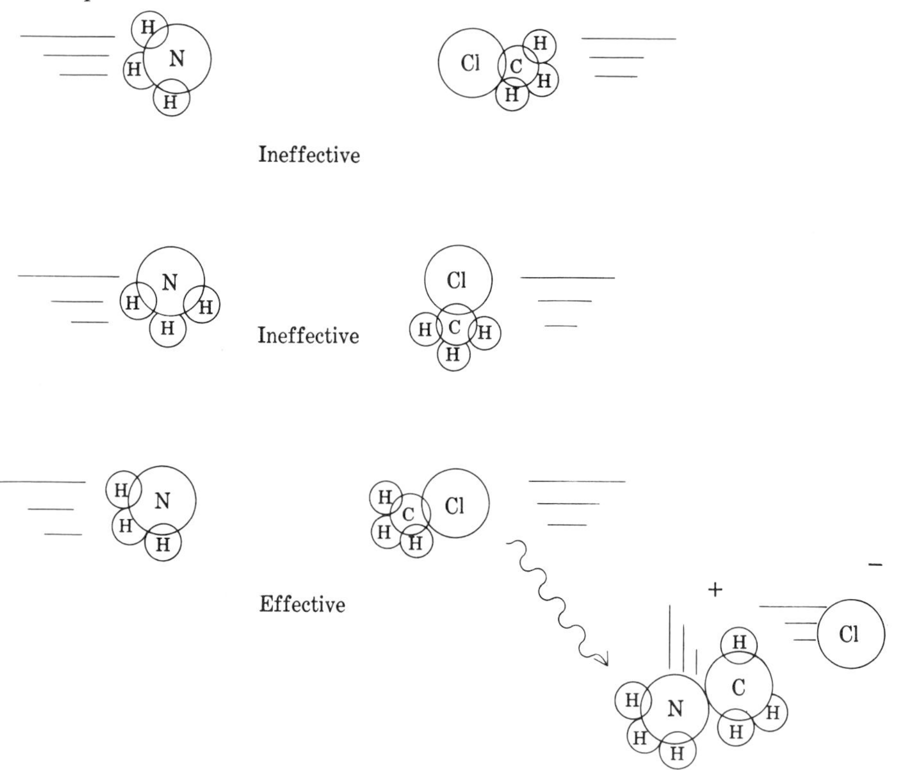

Not only does increasing temperature affect translational motion of particles, but it also leads to changes in internal vibrations and rotations of polyatomic particles. The latter effects may introduce complications in collision geometry, sometimes to such an extent that over certain temperature ranges reaction rate actually decreases with increasing temperature. Temperature effects on reaction rates will be large when the primary consideration is the increase in frequency of effective collisions, but small when most collisions are generally effective (oppositely charged ions) or when complications of collision geometry are predominant.

Presence or absence of a catalyst By definition a catalyst is a substance that only alters the rate of a chemical reaction. A catalyst is incapable of initiating a reaction that cannot inherently take place in the absence of the catalyst, nor can it influence the total quantity of material that can be produced in an equilibrium system—it can only alter the time necessary for a given reaction to take place.

A catalyst apparently alters the frequency of effective collisions by altering the energy of activation for the reaction in question. This it may do by providing either a new path for the reaction or an active surface on which the reaction can occur. There is evidence that a so-called *homogeneous* catalyst provides a new reaction pathway by entering into the reaction and forming an intermediate substance. This intermediate then regenerates the catalyst in its original form. A reaction can, in fact, generate its own catalyst and be, therefore, *autocatalytic* in nature. Thus, when permanganate ion is reduced to manganese(II) ion by oxalate ion in acidic solution, the reaction is observably slow until sufficient manganese(II) ion is formed to catalyze the process. A *surface*, or *contact*, catalyst appears to provide points at which the energy of activation is changed, sometimes by the formation of strong bonds to adsorbed reactant particles so that electron distribution in the particles is appreciably altered.

Perhaps the most useful catalysts increase the rates of chemical reactions. Numerous examples of such positive catalysts have been given in beginning chemistry courses. The importance of many newly developed catalysts is their ability to direct reactions to yield specific products. Other substances, usually called "inhibitors," decrease reaction rates. Examples of these "negative catalysts" are substances used to decrease the rate of decomposition of hydrogen peroxide solutions, to inhibit the oxidation of rubber, and to inhibit the corrosion of iron.

Concentrations of reacting substances It is an experimentally observed fact that reaction rates usually increase as the concentrations of the reacting substances increase. One has only to recall how much more rapidly materials burn in pure oxygen than in the diluted oxygen of the atmosphere to realize that this is true. An increase in the concentration of a reacting substance means an increase in the total number of collisions involving particles of that substance in a given interval of time. This circumstance must give, inevitably, an increase in the frequency of effective collisions and, thus, an increase in reaction rate. When the reaction concerned takes place in solution, concentration has its usual meaning. When a reaction involves a gas, concentration is usually expressed in terms of pressure ($P = nRT/V$).

Concentration is a factor of considerable importance in determining reaction rates for reactions taking place in solution since it is the one factor that is, perhaps, most easily altered or controlled. In a general way in laboratory practice, such as qualitative analysis, the factors of temperature, contact, and catalyst remain reasonably constant and, except in special cases, only concentrations are altered. It is important, therefore, to examine in more detail the effects of concentration changes, other factors remaining constant.

EFFECTS OF CONCENTRATION ON REACTION RATES: THE MECHANISM OF A REACTION

To begin the study of the effects of concentration changes on reaction rates consider first a simple reaction, the collision of hydrogen iodide molecules in the gas phase to form hydrogen and iodine.

$$2HI(g) \rightarrow H_2(g) + I_2(g)$$

In a simple experiment, 0.100 ml of pure, liquid hydrogen iodide is introduced into an evacuated 4.00-liter vessel maintained at 100°C. The liquid volatilizes immediately, and the resulting gas is observed until sufficient color is noted to match a standard comparison vessel containing iodine vapor at known concentration. The time elapsed between introduction of the sample and formation of the necessary concentration of iodine is recorded. The experiment is repeated, using, successively, 2.00-liter and 1.00-liter vessels. The concentrations in the three cases will be inversely proportional to the volumes of the containers since each vessel contained initally the same number of molecules of hydrogen iodide. Observations show that the reaction is four times as rapid in the 2.00-liter vessel as in the 4.00-liter one, and sixteen times as fast in the 1.00-liter vessel. These observations suggest that the rate of the reaction is proportional to the square of the molecular concentration of the hydrogen iodide. Can a simple *mechanism* (i.e., description of effective collisions occurring) account for this?

Let us assume that this reaction will proceed only if two molecules of HI collide with each other. If, now, there are six molecules of HI present, the total number of possible collisions, as determined by the representation,

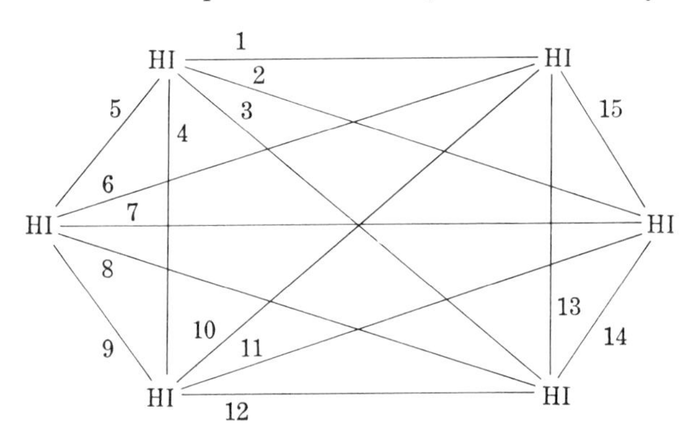

is fifteen. A similar approach shows that if eight molecules of HI are present, there are twenty-eight possible collisions; if there are ten molecules, forty-five possible collisions; if there are twelve molecules, sixty-six possible collisions; etc. Mathematically this reduces to

$$\text{Total possible collisions} = \frac{N_{HI}(N_{HI} - 1)}{2} \tag{3-1}$$

where N_{HI} represents the number of HI molecules initially present. Since, in any observable reaction, the total number of particles in a given volume is almost inconceivably large (1 mole is 6.02×10^{23} particles), no detectable practical difference exists between N_{HI} and $N_{HI} - 1$. Therefore, it is essentially correct to write for this reaction

$$\text{Total possible collisions} = \frac{N_{HI} \times N_{HI}}{2} = \frac{N_{HI}^2}{2} \tag{3-2}$$

and to say that

$$\text{Total possible collisions} \propto N_{HI}^2 \tag{3-3}$$

where $\propto$ represents proportionality.

Since the number of effective collisions is proportional to the number of total collisions, and reaction rate is proportional to the former, one may conclude that

$$\text{Effective collisions} \propto N_{HI}^2 \tag{3-4}$$

and

$$\text{Reaction rate} \propto N_{HI}^2 \tag{3-5}$$

In a practical sense, one does not measure the actual number of atoms, ions, or molecules present for a given reaction. However, one does measure concentration in terms of moles or gram-formula weights of substance per liter. Since it is apparent that these units are proportional to the actual numbers of particles present in a given volume, it follows that for the system described

$$\text{Reaction rate} \propto C_{HI}^2 \tag{3-6}$$

where C_{HI} represents the concentration of HI in moles (or gram-formula weights) per liter.* Equation (3-6) can be made an equality by including a proportionality constant k.

$$\text{Reaction rate} = kC_{HI}^2 \tag{3-7}$$

* Many prefer the older convention of using square brackets to represent concentrations in these units. Thus, C_A and $[A]$ would have the same meaning. However, since square brackets are also used in another sense in describing complex ions, concentrations are expressed uniformly in this book as C_A, C_{Ag^+}, C_{NH_3}, etc.

Here k is known as a *rate constant* and has the correct magnitude and dimensions to convert concentrations to observed reaction rate.

Very few chemical reactions are believed to proceed according to as simple a mechanism as that suggested for the hydrogen iodide case. The study of concentration effects on reaction rates is a most important field of chemistry. When coupled with modern instrumental investigations, this study can frequently lead to a knowledge of the details of reactions proceeding in a number of successive collision stages so that these reactions can be better controlled for more economical routes to desired products.

It should be understood that, since direct observation of reacting molecules and ions is not possible even with the most modern instrumentation, the mechanisms proposed for reactions are simply the best interpretations of data available. Frequently, new experiments or improved data require modifications in the description of reaction mechanisms.

CHEMICAL EQUILIBRIA

Irreversible and reversible reactions In certain chemical reactions the products appear to be incapable of reacting to give the initial substances except under conditions so drastic that they cannot be realized in ordinary laboratory practice. A case in point is the reaction of calcium with water.

$$\mathbf{Ca}(s) + 2H_2O(l) \rightarrow H_2(g) + \mathbf{Ca(OH)_2}(s)$$

It can be shown that appreciable reduction of calcium hydroxide (or oxide) with hydrogen would require hydrogen pressures so large as to be unattainable on a practical basis. Such reactions are said to be *irreversible*, although in an exact sense no chemical reaction is truly irreversible if sufficiently rigorous conditions exist.

In other cases, the products in a given reaction when brought in contact with each other can react to give the initial substances. These reactions are said to be *reversible*. Thus, iron liberates hydrogen from water vapor at elevated temperatures according to the equation

$$3\mathbf{Fe}(s) + 4H_2O(g) \rightarrow 4H_2(g) + \mathbf{Fe_3O_4}(s)$$

In the other direction, if magnetic iron oxide is heated with hydrogen gas, iron and water vapor are regenerated as shown by

$$4H_2(g) + \mathbf{Fe_3O_4}(s) \rightarrow 3\mathbf{Fe}(s) + 4H_2O(g)$$

Each reaction is the exact reverse of the other. Each reaction proceeds in the direction indicated, because as it is carried out experimentally, one product is allowed to escape as a gas and is no longer in contact with the other product.

Suppose now that iron and water vapor are heated together in a closed vessel. Magnetic iron oxide and hydrogen are again produced, but this time they remain in contact with each other and, therefore, can react to produce the initial substances. At first, this "reverse" reaction is not particularly significant because the quantities of iron oxide and hydrogen present are small, but as these quantities increase, their interaction becomes more and more pronounced. Eventually, an apparently static situation is attained where all four substances are present in constant quantities. The same condition is attained if the starting materials are hydrogen and magnetic iron oxide, assuming temperature and other conditions to be the same. Although the condition is seemingly static, it is actually one in which both reactions are taking place simultaneously, but at the same rate. This is a condition of *chemical equilibrium* and is represented by the equation

$$3\mathbf{Fe}(s) + 4H_2O(g) \rightleftharpoons 4H_2(g) + \mathbf{Fe_3O_4}(s)$$

Chemical equilibrium results whenever two exactly opposite chemical reactions take place at the same time and the same rate.

Characteristics of chemical equilibria Equilibrium reactions may involve solids, liquids, gases, or any combination thereof, and can be established in any of these three phases or in systems involving more than one phase. In all instances they are dynamic in character, all materials involved being altered at all times. However, under a given set of conditions, a particular system at equilibrium will contain definite quantities of all the substances involved. Thus, it is possible to say that for that system some particular fraction (usually on a mole basis) of the reactants has been converted into products. Thus, when acetic acid is dissolved in water to give a 0.1 M solution, about 1.35 percent of the 0.1 mole of acetic acid present per liter of solution is converted into ions, as indicated by the equation

$$CH_3CO_2H + H_2O \rightleftharpoons CH_3CO_2^- + H_3O^+$$

In other terms, equilibrium is attained in a 0.1 M aqueous solution of acetic acid when 1.35 percent of the acid has reacted with water.

It should be emphasized that equilibria involving all degrees of conversion of initial reactants are known. The double arrow in an equation should never be interpreted as meaning that 50 percent conversion has occurred. It is sometimes useful, though, to indicate qualitatively the relative amounts of the species in an equilibrium mixture by using arrows of unequal lengths in the equation written. This has been done in the equation in the preceding paragraph.

Principle of Le Châtelier It is often desirable to alter the quantities of materials present in a particular equilibrium system. This circumstance is especially true in systems of industrial importance where "unfavorable" equilibria can often be altered sufficiently to provide increased quantities of desired products. It is also true in many systems studied in the laboratory, in particular where alteration and control of ion concentrations are desired. Changes in the composition of an equilibrium system can be effected by altering the conditions under which the equilibrium was established. The nature of this displacement can be ascertained qualitatively by applying the *principle of Le Châtelier* which states: *If some stress is applied to a system in equilibrium, the reaction which tends to relieve that stress will be favored and the equilibrium reestablished accordingly.* Stresses involve alteration in temperature, gas pressure, or concentration.

The principle of Le Châtelier can be illustrated using the synthesis of ammonia, which is described by the equation

$$N_2(g) + 3H_2(g) \rightleftharpoons 2NH_3(g) + 24.4 \text{ kcal}^*$$

Since the decomposition of ammonia absorbs heat, it follows that the stress induced by an increase in the temperature of the entire system can be relieved in this way, and the equilibrium concentration of ammonia decreased with increasing temperature. Since the formation of ammonia involves a decrease in volume (4 moles of gas giving 2 moles), it follows that the stress induced by an increase in total pressure of the system can be relieved by the conversion of nitrogen and hydrogen to ammonia, the equilibrium concentration of ammonia thus increasing with increasing pressure. That these conclusions are true is shown by the data in Table 3-1. If, on the other hand, the concentration of nitrogen is increased (by adding more of it while maintaining the concentration of hydrogen), a greater equilibrium concentration of ammonia again results because its formation relieves the stress induced by the increased number of collisions between nitrogen and hydrogen

Table 3-1 Effects of temperature and pressure on formation of ammonia

Pressure atm	*Volume percent of NH_3 at equilibrium*		
	200°C	*400°C*	*600°C*
1	15.3	0.44	0.05
200	85.8	36.3	8.25
1,000	98.3	80.0	31.5

* That is, 12.2 kcal of heat are evolved for each mole of ammonia formed. In thermodynamic terminology, for this reaction $\Delta H = -12.2$ kcal mole^{-1}.

molecules.* Alterations in concentrations are of great significance with reactions that take place in solution.

Concentration effects in chemical equilibrium: A treatment independent of reaction mechanism Since the late nineteenth century, chemists have used the reaction of hydrogen and iodine to form hydrogen iodide as an example of a simple collision mechanism. Recent evidence (*Chem. Eng. News*, pp. 40–41, January 16, 1967) suggests that the actual mechanism is much more complicated. This situation can serve as a useful example of the way in which concentrations are treated in equilibrium systems so that the actual mechanism by which the equilibrium is established need not be known.

Assume that the recent data were not yet available, so that the best evidence suggests that the reaction

$$H_2(g) + I_2(g) \rightleftharpoons 2HI(g)$$

proceeds by a simple bimolecular process in which a hydrogen molecule and an iodine molecule colliding with sufficient energy and proper geometry will react to form two molecules of hydrogen iodide. Then, as discussed earlier, the rate of this reaction (the "forward" reaction) can be shown as:

$$\text{Rate}_{\text{forward}} = k \times C_{H_2} \times C_{I_2} \tag{3-8}$$

As the concentration of hydrogen iodide molecules begins to increase, these molecules will occasionally collide in such a fashion as to re-form free hydrogen and free iodine. The rate of this "reverse" reaction will then be

$$\text{Rate}_{\text{reverse}} = k' \times C_{HI}^2 \tag{3-9}$$

where k' is some proportionality constant probably different from k in Eq. 3-8.

The conditions for chemical equilibrium will be attained when, in the reaction vessel, both reactions proceed simultaneously at the same rate. Under these conditions no further change in concentrations of any species involved would be detectable. Thus, *at equilibrium*,

$$H_2(g) + I_2(g) \rightleftharpoons 2HI$$

$$\text{Rate}_{\text{forward}} = \text{rate}_{\text{reverse}} \tag{3-10}$$

and

$$k \times C_{H_2} \times C_{I_2} = k' \times C_{HI}^2 \tag{3-11}$$

* The student is counseled to avoid the common practice of saying that an equilibrium is "shifted to the right" or "shifted to the left." Since the terms "right" and "left" have significance only in describing a particular equation as it happens to be written, it is better to avoid their use. It is more exact to speak of an equilibrium as being temporarily disturbed, then reestablished.

Rearranging the terms in Eq. 3-11 gives

$$\frac{C_{HI}^2}{C_{H_2} \times C_{I_2}} = \frac{k}{k'} \tag{3-12}$$

and, since both k and k' are constants, $k/k' = K_{equil}$, where K_{equil} is called the "equilibrium constant" for the reaction.

$$K_{equil} = \frac{C_{HI}^2}{C_{H_2} \times C_{I_2}} \tag{3-13}$$

Any chemical equilibrium can be treated in this fashion to yield an equilibrium-constant expression. It is conventional and convenient to write such expressions with the concentrations of the products, in the equation describing the process in question, in the numerator of the fraction. If this is done, the magnitude of the equilibrium constant is then a direct measure of the concentrations of the products relative to those of the reactants. It is not incorrect, of course, to write equilibrium-constant expressions in the inverted form, and sometimes it is convenient to do this. This latter procedure gives a constant which is the reciprocal of the usual value.

Although the reaction of hydrogen iodide molecules to form hydrogen and iodine is still believed to be a simple bimolecular process following the rate law shown in Eq. 3-9, modern experiments indicate that the forward reaction; that is, $H_2 + I_2 \rightarrow 2HI$; probably* proceeds by successive reaction steps as:

$I_2(g) \rightleftharpoons 2I(g)$ rapidly established

$2I(g) + H_2(g) \rightarrow 2HI(g)$ slow: rate-determining

Thus, the rate law for the reaction should be

$$\text{Rate} = k'' \times C_I^2 \times C_{H_2} \tag{3-14}$$

rather than that shown in Eq. 3-8. This would, at first glance, appear to invalidate the derivation of the equilibrium equation as shown by Eq. 3-13. Is such, indeed, the case?

Actually, once the equilibrium $I_2(g) \rightleftharpoons 2I(g)$ is attained, a constant ratio of iodine atoms to iodine molecules is established and, by reasoning similar to that employed in arriving at Eq. 3-13,

$$K = \frac{C_I^2}{C_{I_2}} \tag{3-15}$$

* At least one alternative mechanism is also considered feasible.

Then

$$C_I^2 = KC_{I_2} \tag{3-16}$$

and, substituting in Eq. 3-14,

$$\text{Rate} = k''KC_{I_2} \times C_{H_2} \tag{3-17}$$

Evaluation of the constants involved show that

$$\underset{\text{Eq. 3-17}}{k''K} = \underset{\text{Eq. 3-8}}{k}$$

Thus, *even though originally based on an incorrect reaction mechanism*, Eq. 3-13 describing the final equilibrium state is valid.

Similar treatment* of any other equilibrium system will also show that the equilibrium state can be described in terms of appropriate concentration factors for initial reactants and final products, independent of the mechanism by which the equilibrium condition is attained.

The equilibrium constant used for a particular chemical reaction is unaffected by reasonable changes in concentration, provided all other conditions (particularly temperature) remain fixed. Thus, the equilibrium

$$CH_3CO_2H + H_2O \rightleftharpoons H_3O^+ + CH_3CO_2^-$$

is described by a constant that at 25°C has a value of 1.80×10^{-5} mole liter^{-1} over a range of acetic acid concentrations of at least 0.00015 to 0.1 mole liter^{-1}. In numerical magnitudes, equilibrium constants may vary from very large values, say $>10^{10}$, to very small values, say $<10^{-50}$, depending upon the reaction in question. Some typical equilibria and their constants are summarized in Table 3-2.

When equilibrium constants are either rather large or rather small, they are expressed numerically as powers of 10 (see Appendix A). Because equilibrium constants, as normally used, are based upon actual concentrations of chemical species, they should be expressed in appropriate units. If molar concentrations are employed, the unit is moles liter^{-1} or some power thereof, depending upon the numerical magnitudes of the coefficients in the equation describing the process. It is common practice to recognize this, but to write the numerical values of the constants without units.

Characterization of an equilibrium reaction by an equilibrium constant clearly describes the expected effects of concentration changes upon such a system. Thus, if one considers Eq. (3-13), it is apparent that if more molecular hydrogen is added to the system described, constancy in K can result only if there is a simultaneous decrease in the concentration of iodine and an increase in the concentration of hydrogen iodide. This, of course, can

* For a more detailed discussion see K. J. Mysels, *J. Chem. Ed.*, **33**:86 (1956).

Table 3-2 Some examples of chemical equilibrium at 20 to 25°C

Equation	*Equilibrium constant*
$2SO_2(g) + O_2(g) \rightleftharpoons 2SO_3(g)$	$K = \dfrac{C^2_{SO_3}}{C^2_{SO_2} \times C_{O_2}} = 800$ liter mole^{-1} (at 527°C)
$CH_3CO_2H(aq) + H_2O(l) \rightleftharpoons H_3O^+(aq) + CH_3CO_2^-(aq)$	$K^* = \dfrac{C_{H_3O^+} \times C_{C_2H_3O_2^-}}{C_{HC_2H_3O_2}} = 1.80 \times 10^{-5}$ mole liter^{-1}
$CN^-(aq) + H_2O(l) \rightleftharpoons HCN(aq) + OH^-(aq)$	$K^* = \dfrac{C_{HCN} \times C_{OH^-}}{C_{CN^-}} = 4.8 \times 10^{-6}$ mole liter^{-1}
$HgS(s) \rightleftharpoons Hg^{2+}(aq) + S^{2-}(aq)$	$K^* = C_{Hg^{2+}} \times C_{S^{2-}} = 1.6 \times 10^{-54}$ mole^2 liter^{-2}
$[Cu(NH_3)_4]^{2+}(aq) \rightleftharpoons Cu^{2+}(aq) + 4NH_3(aq)$	$K = \dfrac{C_{Cu^{2+}} \times C^4_{NH_3}}{C_{[Cu(NH_3)_4]^{2+}}} = 4.7 \times 10^{-15}$ mole^4 liter^{-4}
$2Cl^-(aq) \rightleftharpoons Cl_2(g) + 2e^-$	$K = \dfrac{C_{Cl_2}}{C^2_{Cl^-}} = 1 \times 10^{-46}$ liter mole^{-1}

* The concentration of water is considered essentially constant and is included in the value of K (Chap. 4). The concentration of a solid is taken as unity (Chap. 5).

occur only if the equilibrium is reestablished when H_2 is added. It follows that the extent of the displacement required is determined by the constancy in K. It is apparent, too, that if the magnitude of K is known, the extent of the displacement can be calculated. Examples appear in subsequent chapters.

Types of chemical equilibria Among those equilibria which are of interest while studying solution chemistry and qualitative analysis are the following:

1. Ionic equilibria involving weak acids and bases (Chap. 4)
2. Equilibria involving water and its ions (Chap. 4)
3. Equilibria involving amphoteric substances (Chap. 4)
4. Equilibria involving precipitates (Chap. 5)
5. Equilibria involving complex ions (Chap. 6)
6. Equilibria involving oxidation-reduction systems (Chap. 7)

Approximations Equilibrium situations generally considered in analysis contain a number of approximations. The most fundamental of these is the assumption that the equilibrium "constant" does not vary with changes in concentration. As has been mentioned, this assumption is valid over a fairly wide concentration range, *if the solutions are sufficiently dilute.* In more concentrated solutions, aggregates of ions, molecules, or molecules with ions are formed, and their behavior is sufficiently different from that of the simpler species that detectable variations in equilibrium constants occur. Where highly accurate measurements and calculations are essential, experimentally determined values of *activities* (related to apparent concentration behavior) are employed. These refinements are not usually needed for equilibrium calculations in qualitative analysis; simple *molar* concentration terms are commonly adequate.

Additional approximations concerned with the concentration of water in aqueous solutions and the "concentration" of solids in contact with their saturated solutions are discussed in later chapters.

ILLUSTRATIVE EXAMPLES

Example 3-1 At elevated temperatures hydrogen iodide gas is in equilibrium with hydrogen and iodine as

$$2HI \rightleftharpoons H_2 + I_2$$

If the equilibrium concentrations of a particular mixture of hydrogen iodide, hydrogen, and iodine are found to be, respectively, 1.56×10^{-2} mole liter^{-1}, 3.56×10^{-3} mole liter^{-1}, and 1.25×10^{-3} mole liter^{-1} at 425.5°C, calculate the equilibrium constant at this temperature.

Solution The equilibrium-constant expression is

$$K = \frac{C_{H_2} \times C_{I_2}}{C_{HI}^2}$$

so direct substitution gives

$$K = \frac{(3.56 \times 10^{-3} \cancel{\text{mole liter}^{-1}}) \times (1.25 \times 10^{-3} \cancel{\text{mole liter}^{-1}})}{(1.56 \times 10^{-2} \cancel{\text{mole liter}^{-1}})^2}$$

$$= 1.83 \times 10^{-2}$$

Example 3-2 Using the data in Example 3-1, determine the equilibrium concentrations of hydrogen and iodine if after pure hydrogen iodide has been heated in a closed flask at 425.5°C the equilibrium concentration of hydrogen iodide remaining is 1.20×10^{-2} mole liter^{-1}.

Solution From the original equilibrium equation, it follows that at equilibrium

$$C_{H_2} = C_{I_2}$$

since for each molecule of hydrogen produced a molecule of iodine is also produced. Substitution in the equilibrium-constant expression then gives

$$\frac{C_{H_2}^2}{(1.20 \times 10^{-2} \text{ mole liter}^{-1})^2} = 1.83 \times 10^{-2}$$

or

$$C_{H_2} = \sqrt{1.83 \times 10^{-2} \times 1.44 \times 10^{-4} \text{ mole}^2 \text{ liter}^{-2}}$$

$$= 1.62 \times 10^{-3} \text{ mole liter}^{-1}$$

The concentration of iodine is then the same.

Example 3-3 Using the data in Example 3-1, calculate the equilibrium concentrations of hydrogen iodide, hydrogen, and iodine if 2.00 moles of hydrogen iodide are equilibrated in a 1-liter closed flask at 425.5°C.

Solution The equilibrium concentrations of the three components must be evaluated first. If

$$x = \text{moles of HI dissociated}$$

then at equilibrium

$$C_{\mathrm{HI}} = 2.00 - x$$

and since 2 moles of HI is required to give 1 mole of H_2 or I_2,

$$C_{\mathrm{H_2}} = C_{\mathrm{I_2}} = \frac{x}{2}$$

Substitution in the equilibrium-constant expression then gives

$$\frac{(x/2)(x/2)}{(2.00 - x)^2} = 1.83 \times 10^{-2}$$

$$\frac{x^2}{4(2.00 - x)^2} = 1.83 \times 10^{-2}$$

Taking the square root of both sides of this equation gives

$$\frac{x}{2(2.00 - x)} = 1.35 \times 10^{-1}$$

from which

$$x = 0.425 \text{ mole liter}^{-1}$$

Hence, at equilibrium

$$C_{\mathrm{HI}} = 1.575 \text{ moles liter}^{-1}$$

$$C_{\mathrm{H_2}} = C_{\mathrm{I_2}} = 0.213 \text{ mole liter}^{-1}$$

EXERCISES

3-1. What is meant by the energy of activation of a chemical reaction? Of what importance is it?

3-2. For each of the following pairs of reactions, indicate which would be expected to show the greater increase in rate with a temperature increase from 30 to 50°C. Explain your choices.

(*a*) $Ag^+ + Cl^- \rightarrow$ **AgCl(*s*)**

or

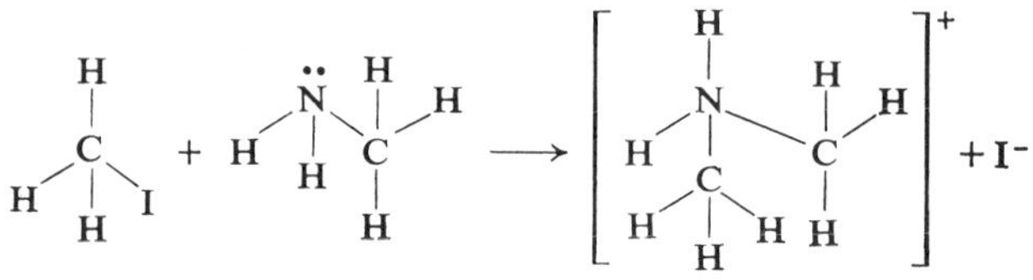

```
                                                 H               +
    H             ••  H                      [   |               ]
    |             N   |    H                     N     H     H
    C      +  H   |   C           --->      H    |     C          + I-
 H  |  I          H   |                          C     |
    H                 H                     H    |  H H
                                                 H
```

(*b*) $H_2 + Cl_2 \rightarrow 2HCl$

or

```
      H              H         H
      |  H   ..   H  |         |  H
      C  |   N    |  C         C  |  I
    /  \ |  / \   | / \       / \ | /
  H  |   C     \  C  |  H + H  |  C       ⟶
     H   |     C  |  H         H  |
         H  H/ | \H H             H
            H  |  H
             \ | /
               C
               |
               H
```

```
  ┌           H             ┐+
  |        H  |  H          |
  |         \ | /           |
  |           C             |
  |        H  |  H          |
  |         \ | /           |
  |    H      C       H     |
  |    |  H   |   H   |     |
  |    C  |   N   |   C     |  + I⁻
  |  /  \ |  / \  |  /  \   |
  | H  |  C   |   C  |   H  |
  |    H  |   |   |  H      |
  |       H   C   H         |
  |         / | \           |
  |       H   C   H         |
  |         / | \           |
  |       H   H   H         |
  └                         ┘
```

3-3. Large sheets of magnesium slowly form white surface coatings of magnesium oxide when in contact with air. Finely powdered magnesium ignites spontaneously when shaken with pure oxygen. Explain.

3-4. The reaction described by the equation $PCl_3(g) + Cl_2(g) \rightarrow PCl_5(g)$ is allowed to occur by mixing 10^{-3} mole of each gas in a 1.00-liter vessel, and the speed of the reaction is monitored by the rate of disappearance of $Cl_2(g)$. Assume that the reaction takes place by a mechanism the slowest step of which involves collision of a Cl_2 molecule with a PCl_3 molecule. If the initial rate is n mole liter^{-1},

(*a*) What would be the initial rate if $C_{Cl_2} = 10^{-3}$ mole liter^{-1} and $C_{PCl_3} = 3 \times 10^{-3}$ mole liter^{-1}?

(*b*) What would be the initial rate if 10^{-3} mole of each reactant was mixed in a 0.10-liter vessel?

3-5. Some texts state that "the speed of a chemical reaction is doubled for every 10°C rise in temperature." If this were true, would equilibrium constants be independent of temperature? Explain. Find some experimental data from an outside reference book that show that the statement is not accurate. Cite the reference used.

3-6. Why might it be important to determine the mechanism by which a chemical equilibrium is established? Can accurate calculations be made on relative concentrations in a system in chemical equilibrium if the mechanism by which it is established is not known? Explain.

3-7. An equilibrium system

$$2SO_2(g) + O_2(g) \rightleftharpoons 2SO_3(g)$$

is found to contain 2.0 moles liter^{-1} each of $SO_2(g)$ and $O_2(g)$ at 527°C. What is the concentration of $SO_3(g)$?

3-8. The system described in Exercise 3-7 is contained in a vessel equipped with a movable piston. The piston is pushed in until the volume of the gaseous mixture is reduced to half its original volume, while the temperature is maintained at 527°C.

(*a*) Describe chemical changes which result from this change of volume.

(*b*) Determine the concentrations of reactants and product when equilibrium is reestablished.

3-9. Give one example of an industrial or biological system that involves:

(*a*) Heterogeneous catalysis
(*b*) Homogeneous catalysis
(*c*) Use of an inhibitor (negative catalyst)
(*d*) An irreversible reaction
(*e*) A homogeneous equilibrium
(*f*) A heterogeneous equilibrium
(*g*) Enzymatic catalysis

3-10. Given an equilibrium system represented by the expression

$$2SO_2(g) + O_2(g) \rightleftharpoons 2SO_3(g) + 44.6 \text{ kcal}$$

indicate clearly how the equilibrium percentage of sulfur trioxide is affected by

(*a*) Decreasing the temperature
(*b*) Increasing the total pressure of the system
(*c*) Decreasing the partial pressure of sulfur dioxide
(*d*) Introducing more oxygen into the system

Explain in each case.

3-11. Given an equilibrium represented by the equation

$$\underset{\text{Colorless}}{N_2O_4(g)} + 14.7 \text{ kcal} \rightleftharpoons \underset{\text{Brown}}{2NO_2(g)}$$

what effect upon the color of the system would be produced by increasing the temperature? By increasing the total pressure? Explain.

3-12. One of the most important examples of industrial manipulation of a chemical equilibrium is the manufacture of ammonia by the process

$$N_2 + 3H_2 \rightleftharpoons 2NH_3 + 22.4 \text{ kcal}$$

(*a*) How is Le Châtelier's principle employed in increasing the efficiency of this process?

(*b*) Is a catalyst used to increase the ratio of product to reactants at equilibrium? If not, why is a catalyst employed?

3-13. Why are calculations on equilibrium systems correctly regarded as approximations when concentration terms such as moles liter^{-1} are used?

RECOMMENDED SUPPLEMENTARY READING

Barrow, Kenney, Lassila, Litle, and Thompson: "Understanding Chemistry," Vol. IV: "Chemical Equilibria," Benjamin, New York, 1967 (paperback) [a programmed approach].

Eyring, H. and Eyring, E. M.: "Modern Chemical Kinetics," Reinhold, New York, 1963 (paperback).

Nyman, C. J. and Hamm, R. E.: "Chemical Equilibrium," Raytheon Education Company, Boston, 1968 (paperback).

"Theory and Problems of College Chemistry," 5th ed., Schaum, New York, 1966 (paperback).

4
Acid-Base Equilibria in Aqueous Solutions

Chemical equilibria are said to be *homogeneous* if all the materials involved are in the same physical state (gas, liquid or solution, solid) or to be *heterogeneous* if not all the materials involved are in the same physical state. Although the principles of equilibrium as developed in the preceding chapter were applied there largely to homogeneous systems, they may be applied with equal exactness to heterogeneous systems as well. Solution chemistry involves both homogeneous and heterogeneous equilibria. Although it is difficult to say which of the two types is the more important, it is probable that more examples of homogeneous systems are encountered than of heterogeneous systems. Among the most common of these are the equilibria involving acids or bases.

It has been pointed out that reactions between acids and bases are invariably equilibria. As such, these reactions are described by equilibrium-constant expressions. Most important of these acid-base equilibria, however, are those which involve water. These may be formulated* as

$$HA(aq) + H_2O(l) \rightleftharpoons H_3O^+(aq) + A^-(aq) \tag{4-1}$$

* Without indicating hydration of the species involved (except for H_3O^+).

for any acid HA, and as

$$H_2O(l) + B(aq) \rightleftharpoons BH^+(aq) + OH^-(aq) \tag{4-2}$$

for any base. These reactions are of extreme importance not only for the laboratory study of reactions in solution, but also for many technological processes and in innumerable biological systems.

THE FUNCTION OF WATER

When clusters of water molecules collide with sufficient energy and proper orientation, the relatively weak hydrogen-oxygen bonds may break, and the resulting protons may, in turn, bond to unshared electron pairs of new oxygen atoms—frequently those already associated with the hydrogen atoms through hydrogen bonding. The reaction may be described as

$$(H_2O)_x + (H_2O)_y \rightarrow (OH)(H_2O)_{x-1}^- + H(H_2O)_y^+$$

The reverse of this reaction will be recognized as the general reaction between protons and hydroxide ions in aqueous solution, the most common example of *neutralization*. When the rates of the two opposing reactions become equal, a condition of dynamic chemical equilibrium is established which, for simplicity, may be shown as

$$H_2O(l) \rightleftharpoons H^+(aq) + OH^-(aq)$$

According to the procedure developed in Chap. 3, the equilibrium constant for this reaction is

$$K_{\text{equil}}{}^* = \frac{C_{H^+} \times C_{OH^-}}{C_{H_2O}}$$

Experiments show that the concentrations of hydrated protons and hydroxide ions in liquid water are very small compared to the concentration of undissociated water molecules. The latter may then be considered as essentially unchanged by this reaction, and a good approximation of the ion product for water is given by

$$K_w = K_{\text{equil}} \times C_{H_2O} = C_{H^+} \times C_{OH^-} \tag{4-3}$$

where K_w, the ion product, has the approximate value 1.0×10^{-14} at 20°C.

STRONG VERSUS WEAK ACIDS AND BASES

Any acid-base reaction, under the Brønsted-Lowry descriptions, involves a competition for protons. Acids, the conjugate bases of which (Chap. 2) are so much less effective than water as proton acceptors that the reaction of

* Remember that such an expression is exact only if the *activities* of the species are used. The use of molar concentrations involves an approximation.

the conjugate base with hydrated proton is negligible in aqueous solution, are referred to as *strong acids*. Examples of these acids are:

Hydrochloric acid	HCl
Hydrobromic acid	HBr
Perchloric acid	$HClO_4$
Sulfuric acid*	H_2SO_4

In a similar fashion, bases, the conjugate acids of which are significantly less effective than water as proton donors, are considered *strong bases*, as is the hydroxide ion itself. These bases are rarely encountered in aqueous systems.

There is no real justification, therefore, for attempting to assign ionization-constant values to strong acids and bases. Yet it must be admitted that interionic-attraction effects make these materials appear to be less than completely ionized in aqueous solution and, thus, permit determination of *apparent* degrees of dissociation. That such values are of no significance in giving ionization constants is shown by the data for hydrochloric acid solutions as given in Table 4-1. The dependence of the value of K_a upon concentration, and the wide variations in K_a values with changes in concentration are even more striking when Table 4-1 is compared with Table 4-3. Clearly, there is no point in trying to describe the strong acids or bases in terms of K_a or K_b values.

Weak acids and bases, then, are those, the behaviors of which as proton donors or acceptors, result in measurable equilibrium concentrations of conjugate pairs in aqueous solution.

Table 4-1 Apparent ionization constant of hydrochloric acid†

Concentration, mole liter^{-1}	K_a, *moles liter^{-1}*
0.200	1.56
0.100	1.05
0.050	0.73
0.010	0.32
0.001	0.12

† $K_a = \dfrac{C_{H^+} \times C_{Cl^-}}{C_{HCl}}$

where HCl is most probably closely associated ion pairs of partially hydrated H^+ and Cl^-.

* Note that H_2SO_4 is a strong acid in terms of loss of a single proton. Bisulfate ion, HSO_4^-, is not considered a strong acid by the criteria discussed here.

IONIZATION CONSTANTS OF WEAK ACIDS AND BASES

In terms of the discussion in Chap. 3, the general equilibrium expression shown by Eq. 4-1

$$HA(aq) + H_2O(l) \rightleftharpoons H_3O^+(aq) + A^-(aq)$$

describing the reaction taking place when an acid is placed in water, is characterized by the equilibrium-constant expression

$$K_{\text{equil}} = \frac{C_{H_3O^+} \times C_{A^-}}{C_{HA} \times C_{H_2O}} \tag{4-4}$$

Equilibrium problems that can be treated by the relatively simple procedures described in this text involve dilute solutions (usually $<1\ M$). Calculations on more concentrated solutions require corrections for non-ideal behavior, a procedure generally reserved for physical chemistry courses or situations requiring greater accuracy than is needed in simple analytical calculations.

In dilute aqueous solutions the concentration of water is always much larger than any other concentration involved (remember that a liter of water contains nearly 56 moles of H_2O), and the concentration of water is, therefore, essentially constant. Equation 4-4 may then be rewritten as

$$K_a = K_{\text{equil}} \times C_{H_2O} = \frac{C_{H_3O^+} \times C_{A^-}}{C_{HA}} \qquad \text{(4-5)*}$$

where K_a is the approximate ionization constant of the weak acid.

In the same fashion, the equilibrium expression for a weak base (Eq. 4-2)

$$H_2O(l) + B(aq) \rightleftharpoons BH^+(aq) + OH^-(aq)$$

is characterized by

$$K_b = \frac{C_{BH^+} \times C_{OH^-}}{C_B} \tag{4-6}$$

Ionization constants for some of the more common weak acids and bases are shown in Table 4-2. Additional values are shown in Appendixes B and C.

Numerical evaluations of ionization constants Ionization constants of weak acids and bases can be evaluated merely by substitution of appropriate concentration values in expressions such as Eqs. 4-5 or 4-6. For a given equilibrium, the individual concentrations can be calculated from the total concentration of the solution, provided the degree of dissociation α is known. As previously indicated, the degree of dissociation is obtained from measurement

* For simplicity, Eq. 4-5 is usually written as

$$K_a = \frac{C_{H^+} \times C_{A^-}}{C_{HA}}$$

Table 4-2 Ionization constants for typical weak acids and bases

Compound	*Equation*	*Equilibrium expression*	K_a *or* K_b
HF	$HF + H_2O \rightleftharpoons H_3O^+ + F^-$	$\frac{C_{H_3O^+} \times C_{F^-}}{C_{HF}}$	6.9×10^{-4}
CH_3CO_2H	$CH_3CO_2H + H_2O \rightleftharpoons H_3O^+ + CH_3CO_2^-$	$\frac{C_{H_3O^+} \times C_{CH_3CO_2^-}}{C_{CH_3CO_2H}}$	1.8×10^{-5}
H_2CO_3	$H_2CO_3 + H_2O \rightleftharpoons H_3O^+ + HCO_3^-$	$\frac{C_{H_3O^+} \times C_{HCO_3^-}}{C_{H_2CO_3}}$	4.2×10^{-7}
H_2S	$H_2S + H_2O \rightleftharpoons H_3O^+ + HS^-$	$\frac{C_{H_3O^+} \times C_{HS^-}}{C_{H_2S}}$	1.0×10^{-7}
HS^-	$HS^- + H_2O \rightleftharpoons H_3O^+ + S^{2-}$	$\frac{C_{H_3O^+} \times C_{S^{2-}}}{C_{HS^-}}$	1.3×10^{-13}
NH_3	$NH_3 + H_2O \rightleftharpoons NH_4^+ + OH^-$	$\frac{C_{NH_4^+} \times C_{OH^-}}{C_{NH_3}}$	1.8×10^{-5}
CH_3NH_2	$CH_3NH_2 + H_2O \rightleftharpoons CH_3NH_3^+ + OH^-$	$\frac{C_{CH_3NH_3^+} \times C_{OH^-}}{C_{CH_3NH_2}}$	5×10^{-4}

of certain physical properties of solutions, such as conductance. For a given weak acid or base, α has a particular numerical value for each overall concentration of the aqueous solution of the electrolyte, the values increasing with decreasing concentration. The calculations involved can be illustrated by a specific example.

For an aqueous solution of acetic acid, CH_3CO_2H, where the equilibrium

$$CH_3CO_2H(aq) + H_2O(l) \rightleftharpoons H_3O^+(aq) + CH_3CO_2^-(aq)$$

is established, the ionization constant is expressed as

$$K_a = \frac{C_{H_3O^+} \times C_{CH_3CO_2^-}}{C_{CH_3CO_2H}} \tag{4-7}$$

Values for the degree of dissociation at various concentrations at room temperature are given in Table 4-3.

Taking a representative value of α equal to 0.021 for a 0.040 M solution of acetic acid, a fraction, 0.021, of the 0.040 mole of acetic acid present exists as ions; whereas a fraction, 0.979, exists as molecules at equilibrium. It follows, then, that

$$C_{H_3O^+} = C_{H^+} = 0.021 \times 0.040 \text{ mole liter}^{-1} = 0.00084 \text{ mole liter}^{-1}$$

$$C_{CH_3CO_2^-} = C_{H^+} = 0.021 \times 0.040 \text{ mole liter}^{-1} = 0.00084 \text{ mole liter}^{-1}$$

$$C_{CH_3CO_2H} = 0.979 \times 0.040 \text{ mole liter}^{-1} = 0.0392 \text{ mole liter}^{-1}$$

Substitution in Eq. 4-7 gives

$$K_a = \frac{(0.00084) \times (0.00084)}{(0.0392)}$$

$$= 0.000018 \text{ mole liter}^{-1}$$

$$= 1.8 \times 10^{-5} \text{ mole liter}^{-1} \qquad \text{(or, by accepted practice, } 1.8 \times 10^{-5}\text{)}$$

Table 4-3 Ionization of acetic acid at 25°C

Concentration mole CH_3CO_2H liter^{-1}	*Degree of dissociation,* α	*Ionization constant* K_a, *mole liter*$^{-1}$
0.0500	0.0188	1.81×10^{-5}
0.0400	0.0210	1.80×10^{-5}
0.0128	0.0368	1.80×10^{-5}
0.00591	0.0537	1.80×10^{-5}
0.00241	0.0825	1.80×10^{-5}
0.00103	0.1232	1.78×10^{-5}
0.00015	0.2867	1.78×10^{-5}
		1.80×10^{-5} (Average)

In a similar fashion, the corresponding calculation can be made for any weak acid or base.* It should be apparent that, *if the degree of dissociation is very small* ($K_a < 10^{-4}$), *the concentration of unionized acid or base may be taken as the same as that of the solution without appreciable error.*

Constancy of the ionization constant For a concentration of 0.1 M or less, the ionization constant for a given weak acid or base is essentially independent of concentration provided the temperature remains constant. This circumstance is illustrated for acetic acid by the data in Table 4-3. Similar results might be recorded for other comparable acids or bases. As solution concentration increases, deviations from constancy become significantly large. Inasmuch as reactions in solution are conveniently studied in concentrations of approximately 0.01 M, errors introduced by variations in the magnitudes of ionization constants are too small to be significant.

THE $p(x)$ NOTATION

For solutions of weak acids and bases, both ion concentrations and ionization constants are of comparatively small numerical magnitudes. In order to increase the ease with which they can be handled, they are often expressed relative to powers of 10, as was done above. However, even this practice occasions some inconvenience, and an even shorter notation is desired. For this purpose, what might be called a p(x) notation is used.

The term p(x) is defined as

$$\text{p}(x) = -\log_{10} x = \log_{10} \frac{1}{x} \tag{4-8}$$†

where the quantity x may be the concentration of a given chemical species or an equilibrium constant. Thus, pH signifies the negative logarithm of the hydrogen-ion concentration, pCH_3CO_2 signifies the negative logarithm of the acetate-ion concentration, and $\text{p}K_a$ signifies the negative logarithm of the ionization constant of an acid. To illustrate, the $\text{p}K_a$ for acetic acid can be calculated from $K_a = 1.8 \times 10^{-5}$ as

$$\begin{aligned} \text{p}K_a &= -\log(1.8 \times 10^{-5}) = (-\log 1.8) + (-\log 10^{-5}) \\ &= (-0.26) + [-(-5)] = 4.74 \end{aligned}$$

Since logarithms are exponents, p(x) values have no units.

The advantage of expressing relatively small numbers in this logarithmic fashion is apparent. For convenience, negative rather than positive logarithms are employed to avoid giving p(x) values that are negative in as

* In these and other calculations, the student will find it convenient to express small concentration values and constants as powers of 10 rather than as decimals.

† In subsequent usage log will be employed to indicate $\log_{10}$.

many cases as possible. From which it follows that the p(x) value of any number greater than 1 is negative. It is relatively common to describe acids and bases in terms of their pK_a and pK_b values, respectively. This practice has been followed in the tabulations in Appendixes B and C. It is perhaps even more common to refer to hydrogen- and hydroxide-ion concentrations in aqueous solutions in terms of pH and pOH, respectively. This approach is discussed in the following section.

The properties of logarithms are such that the p(x) values in a multiplication are additive, and the p(x) values in a division are subtractive. Thus, if Eq. 4-7 for the ionization constant for acetic acid is considered, the negative logarithms of both sides of the expression may be taken to give

$$(-\log K_a) = (-\log C_{H^+}) + (-\log C_{CH_3CO_2^-}) - (-\log C_{CH_3CO_2H}) \quad (4\text{-}9)$$

Substitution of equivalent $p(x)$ values gives

$$pK_a = pH + pCH_3CO_2 - pCH_3CO_2H \quad (4\text{-}10)$$

This expression is convenient for evaluation of any of these quantities, or for relating each quantity to the others.

pH and pOH Inasmuch as the concentration of hydrogen or hydroxide ion is commonly comparatively small, it is often more convenient to express it in terms of the logarithmic p(x) type of notation previously discussed. As already indicated, then,

$$pH = -\log C_{H^+} = \log \frac{1}{C_{H^+}} \quad (4\text{-}11)$$

for hydrogen-ion concentrations and

$$pOH = -\log C_{OH^-} = \log \frac{1}{C_{OH^-}} \quad (4\text{-}12)$$

for hydroxide concentrations. The terms pH and pOH may be related to each other by taking the negative logarithm of both sides of Eq. 4-3, as

$$-\log K_w = (-\log C_{H^+}) + (-\log C_{OH^-})$$

This expression then reduces to

$$pK_w = pH + pOH \quad (4\text{-}13)$$

or, since $K_w = 10^{-14}$, it reduces further to

$$pH + pOH = 14 \quad (4\text{-}14)$$

By means of this relationship either pH or pOH can be determined if the other is known. Actually, pH is the more commonly employed term, regardless of whether the solution is acidic or alkaline, and solutions are often characterized in terms of pH values alone. These relationships are presented in Table 4-4.

For solutions of strong acids or soluble metal hydroxides, pH values are readily determined since these materials are completely ionic in solution. Thus, for a 0.02 *M* $HClO_4$ solution

$$C_{H^+} = 0.02 \text{ mole liter}^{-1}$$

and

$$pH = -\log 0.02 = -\log(2 \times 10^{-2}) = -\log 2 + (-\log 10^{-2})$$
$$= -0.30 + 2 = 1.70$$

Or, for a 0.05 *N* NaOH solution

$$C_{OH^-} = 0.05 \text{ mole liter}^{-1}$$

and

$$pOH = -\log 0.05 = -\log 5 + (-\log 10^{-2})$$
$$= -0.70 + 2 = 1.30$$

Table 4-4 Acidity, neutrality, and alkalinity* in aqueous solutions

C_{H^+}, mole liter^{-1}	C_{OH^-}, mole liter^{-1}	pH	pOH	*Nature*
1×10^{-14}	1×10^{0}	14	0	
1×10^{-13}	1×10^{-1}	13	1	
1×10^{-12}	1×10^{-2}	12	2	
1×10^{-11}	1×10^{-3}	11	3	Alkaline
1×10^{-10}	1×10^{-4}	10	4	
1×10^{-9}	1×10^{-5}	9	5	
1×10^{-8}	1×10^{-6}	8	6	
1×10^{-7}	1×10^{-7}	7	7	Neutral
1×10^{-6}	1×10^{-8}	6	8	
1×10^{-5}	1×10^{-9}	5	9	
1×10^{-4}	1×10^{-10}	4	10	
1×10^{-3}	1×10^{-11}	3	11	Acidic
1×10^{-2}	1×10^{-12}	2	12	
1×10^{-1}	1×10^{-13}	1	13	
1×10^{0}	1×10^{-14}	0	14	

* Although the term "basic" is also used in this same sense, alkaline is employed throughout this book as being a term more nearly free from ambiguity than "basic." Classically, the term "base" was associated with the hydroxide ion, but with the broadened use of the term, this association has disappeared. It would follow that all materials adding protons are basic, irrespective of the presence or absence of hydroxide ion.

therefore,

$$pH = 14.00 - 1.30 = 12.70$$

For solutions of weak acids and bases, however, account must be taken of the degree of dissociation. Thus, for a 0.040 M solution of acetic acid that is 2.1 percent ionized (Table 4-3)

$$C_{H^+} = 0.021 \times 0.04 = 0.00084 \text{ mole liter}^{-1}$$

and

$$pH = -\log(8.4 \times 10^{-4}) = -\log 8.4 + (-\log 10^{-4})$$
$$= -0.92 + 4 = 3.08$$

SOME APPLICATIONS OF IONIZATION CONSTANTS

Inasmuch as numerical values for ionization constants are available for many weak acids and bases, it is possible to calculate ion concentrations in solutions containing these substances under a variety of conditions. Ion concentrations are often necessary for an adequate understanding of why many laboratory operations in qualitative analysis, quantitative analysis, physical chemistry, and biochemistry are possible. Some types of calculations involving ionization constants are given in the following sections.

Concentration of individual ions in solutions of weak acids or bases The equilibrium established when gaseous ammonia is dissolved in water has been considered earlier, and the expression for K_b has been given in Eq. 4-6. At room temperature the numerical value of this K_b is 1.8×10^{-5} mole liter^{-1}, the same value, incidentally, as that for the K_a of acetic acid. Suppose that it is desired to determine the concentration of hydroxide ion in a 0.010 M solution of aqueous ammonia. Since the equilibrium expression shows that ammonium and hydroxide ions are present in equal quantities, it follows that

$$C_{OH^-} = C_{NH_4^+}$$

Furthermore, since the overall concentration must be corrected for the quantity of ammonia converted to ammonium and hydroxide ions, the equilibrium concentration of ammonia is

$$C_{NH_3} = (0.010 - C_{OH^-}) \qquad \text{mole liter}^{-1}$$

Substitution of these values into Eq. 4-6 then gives

$$1.8 \times 10^{-5} = \frac{C^2_{OH^-}}{0.010 - C_{OH^-}}$$

or

$$C^2_{OH^-} + (1.8 \times 10^{-5})C_{OH^-} = 1.8 \times 10^{-7}$$

Solution of this quadratic equation then leads to the desired answer,

$$C_{OH^-} = 4.2 \times 10^{-4} \text{ mole liter}^{-1}$$

This solution is an exact* one because it indicates correctly the concentrations of all species at equilibrium. However, it does involve a quadratic equation and is, therefore, somewhat cumbersome to handle. A simplification can be effected by pointing out that, since the concentration of hydroxide ion (4.2×10^{-4} mole liter^{-1}) is small compared with the overall concentration of the solution (10^{-2} mole liter^{-1}), the quantity $0.010 - C_{OH^-}$ will not differ significantly from 0.010. It may be said, then, as an approximation, that

$$C_{NH_3} = 0.010 \text{ mole liter}^{-1}$$

Substitution in Eq. 4-6 then gives

$$1.8 \times 10^{-5} = \frac{C^2_{OH^-}}{0.010}$$

and

$$C_{OH^-} = 4.2 \times 10^{-4} \text{ mole liter}^{-1}$$

It follows that the approximation is a valid one since the results calculated by the two procedures are essentially identical. These approximations are particularly useful, especially for materials which are even weaker bases or acids than ammonia or acetic acid.

Calculations such as these can be extended to evaluate the degree of dissociation. Thus, the degree of dissociation α amounts to the ratio of the equilibrium concentration of an ion present to the overall concentration of the solution, or, for the system just considered, to

$$\alpha = \frac{C_{OH^-}}{0.010 \text{ mole liter}^{-1}}$$

which gives

$$\alpha = 4.2 \times 10^{-2} = 0.042$$

Other situations are handled similarly.

Evaluation of K_a or K_b from constants for conjugate bases or acids† Salts consisting of ions of essentially equivalent tendency to donate or accept protons form aqueous solutions of pH ~7 (i.e., their

* The solution is mathematically exact, but it must be remembered that the use of concentration units, such as mole liter^{-1}, introduces an approximation due to nonideal behavior in solution.

† This situation has frequently been separately treated as "hydrolysis." Since all acid-base reactions in aqueous solution involve participation of water, the only unique characteristic of these situations is the necessity to calculate the K_a or K_b from data tabulated for the conjugate acid or base.

solutions are "neutral"). For example, ammonium acetate dissolves in water to form a solution of pH 7 because the ammonium ion has the same tendency to donate protons as the acetate ion does to accept them (K_a for $NH_4^+ = K_b$ for $CH_3CO_2^-$). Thus the concentration of hydrated protons is essentially the same as that in pure water, 10^{-7} M. Salts, the cations of which are alkali metal (Group I*A*) or alkaline earth (Group II*A*) ions and the anions of which are conjugate ions of simple strong acids (for example, Cl^-, Br^-, ClO_4^-), also form neutral aqueous solutions.

In cases where the K_a of the cation is not equal to the K_b of the anion, however, aqueous solutions will not have a pH of 7. Thus, salts such as sodium acetate or potassium cyanide form alkaline solutions in water (pH > 7), whereas salts like ammonium chloride form acidic solutions (pH < 7). Other cases of nonneutral salts are those in which the hydrated cation behaves essentially like water clusters, but the anion is an effective weak acid (HSO_4^-, $H_2PO_4^-$) or weak base (CO_3^{2-}, PO_4^{3-}) from a polyprotic system. Some hydrated cations, such as $Al(H_2O)_4^{3+}$, are themselves weak acids, and an aqueous solution of aluminum chloride is, therefore, acidic. The last case is discussed in greater detail in Chap. 6 in connection with complex ions.

Evaluation of acid-base behavior of salts in aqueous solution follows the same pattern as that developed for solutions of weak acids or bases. In some cases, however, K_a or K_b values involved are not conveniently available and must be calculated from tabulated constants for conjugate bases or acids. For example, suppose one desired to calculate the approximate pH of a 0.010 M solution of sodium acetate. First, it should be recognized that the hydrated sodium ion does not affect the pH. Acetate ion, however, is the conjugate base of acetic acid and, therefore, the following equilibrium is established in aqueous solution.

$$CH_3CO_2^-(aq) + H_2O(l) \rightleftharpoons CH_3CO_2H(aq) + OH^-(aq) \tag{4-15}$$

As with any weak base, the equilibrium-constant expression (compare Eq. 4-6) will be

$$K_b = \frac{C_{CH_3CO_2H} \times C_{OH^-}}{C_{CH_3CO_2^-}} \tag{4-16}$$

Before C_{OH^-} can be calculated, the K_b must be evaluated. This constant is seldom given in brief tables, but in Table 4-2 K_a for acetic acid is found to be 1.8×10^{-5}. Equation 4-16 contains some of the terms found in Eq. 4-7. In fact, division of Eq. 4-3 by Eq. 4-7 as follows

$$\frac{K_w}{K_a} = \frac{C_{H^+} \times C_{OH^-}}{\dfrac{C_{CH_3CO_2^-} \times C_{H^+}}{C_{CH_3CO_2H}}} = \frac{C_{CH_3CO_2H} \times C_{OH^-}}{C_{CH_3CO_2^-}}$$

shows that, by comparison with Eq. 4-16,

$$K_b \text{ (for } CH_3CO_2^-) = \frac{K_w}{K_a \text{ (for } CH_3CO_2H)} \tag{4-17}$$

and, thus, the K_b required can be evaluated as

$$K_b = \frac{1.0 \times 10^{-14}}{1.8 \times 10^{-5}} = 5.6 \times 10^{-10}$$

The remaining part of the original problem then becomes relatively simple. From Eq. 4-15

$$C_{CH_3CO_2H} = C_{OH^-}$$

Then Eq. 4-16 becomes

$$5.6 \times 10^{-10} = \frac{C^2_{OH^-}}{10^{-2}}$$

from which

$$C_{OH^-} = 2.4 \times 10^{-6} \text{ mole liter}^{-1}$$

and, from Eq. 4-12

$$\text{pOH} = -\log (2.4 \times 10^{-6}) = 5.62$$

Then, from Eq. 4-13

$$\text{pH} = 14.00 - 5.62 = 8.38$$

A more general treatment of this situation using Eqs. 4-5 and 4-6 gives

$$K_b = \frac{K_w}{K_{a,\text{ conj}}} \tag{4-18}$$

$$K_a = \frac{K_w}{K_{b,\text{ conj}}} \tag{4-19}$$

EFFECTS OF COMMON IONS

Earlier discussions suggest qualitatively that the addition of a common ion to a solution of a weak acid (hydrogen ion or acetate ion to acetic acid solution) or a weak base (ammonium ion or hydroxide ion to aqueous ammonia) will increase the quantity of undissociated acid or base. It follows that not only can the concentration of free acid or base be controlled by this *common-ion effect*, but also that the concentration of one ion can be controlled by varying the concentration of the other ion in the equilibrium. This is, of course, of particular importance for the separation of one ion from another using various laboratory operations. By the use of ionization constants, one can determine quantitatively the effects of common ions.

A calculation of this type may be illustrated for an aqueous ammonia solution containing an added ammonium salt. Inasmuch as the concentration of hydroxide ion in 0.010 *M* aqueous ammonia has been determined already, it is of interest to see how the value so calculated would compare with the hydroxide-ion concentration in a 0.010 *M* aqueous ammonia solution that is also 0.050 *M* in ammonium chloride. As before,

$$C_{NH_3} = (0.010 - C_{OH^-}) \qquad \text{mole liter}^{-1}$$

but this time there is additional ammonium ion present besides that from the ionization process, and therefore, $C_{NH_4^+}$ and C_{OH^-} are not equal. Since ammonium chloride is a salt, it yields 0.050 mole liter^{-1} of ammonium ion; so at equilibrium

$$C_{NH_4^+} = (0.050 + C_{OH^-}) \qquad \text{mole liter}^{-1}$$

C_{OH^-} being equal to the concentration of ammonium ion from the ionization itself. Substitution of these values into Eq. 4-6 then gives

$$1.8 \times 10^{-5} = \frac{(0.050 + C_{OH^-})(C_{OH^-})}{0.010 - C_{OH^-}}$$

or

$$C^2_{OH^-} + 0.050C_{OH^-} + 1.8 \times 10^{-5}C_{OH^-} = 1.8 \times 10^{-7}$$

This quadratic equation can be solved for C_{OH^-}, as was done before, to give

$$C_{OH^-} = 3.6 \times 10^{-6} \text{ mole liter}^{-1}$$

However, since C_{OH^-}, even in the absence of added ammonium salt, is small (4.2×10^{-4} mole liter^{-1}) in comparison with either 0.010 mole liter^{-1} or 0.050 mole liter^{-1} and since in the presence of ammonium ion it must be even smaller, it may be neglected in the expressions $0.010 - C_{OH^-}$ and $0.050 + C_{OH^-}$. Thus,

$$C_{NH_3} = 0.010 \text{ mole liter}^{-1}$$

$$C_{NH_4^+} = 0.050 \text{ mole liter}^{-1}$$

Then one has

$$1.8 \times 10^{-5} = \frac{0.050}{0.010} \times C_{OH^-}$$

or

$$C_{OH^-} = \tfrac{1}{5} \times 1.8 \times 10^{-5} = 3.6 \times 10^{-6} \text{ mole liter}^{-1}$$

The approximate solution is, thus, entirely as accurate as the exact solution. It follows, of course, that in any solution of aqueous ammonia where

$$\frac{C_{NH_4^+}}{C_{NH_3}} = 5$$

C_{OH^-} must be 3.6×10^{-6} mole liter^{-1}.

The effects of common ions are considered again in connection with buffer solutions and controlled precipitation processes.

SIMPLE BUFFERS: CONJUGATE-PAIR BUFFERS

Any solution that can accept added acid or base with a resulting smaller pH change than would have occurred with the same conditions for the acid or base added to pure water is said to be "buffered." Many different kinds of substances can act as buffers. Concentrated solutions of strong acids or bases, amphoteric substances such as amino acids, or mixtures of weak acids and weak bases are examples of buffers. They differ in their effectiveness and utility depending on many factors, some of which are considered later in more detail.

The most common type of buffer is that made by mixing a weak acid and a weak base; the conjugate-pair buffer is the simplest example of this type. To illustrate this type of buffer behavior, let us answer the following question. How would the pH change produced by adding 1.0 ml of 0.10 M HCl to 1,000 ml of pure water compare with the pH change if 1,000 ml of a conjugate-pair buffer which is 0.20 M in CH_3CO_2H and 0.30 M in $CH_3CO_2^-$ were used instead of the water?

The pure water would initially have a pH of 7.00. Addition of 1.0 ml of 0.10 M HCl to 1,000 ml of the water would result in a solution which was 10^{-4} M in H^+ (10^{-1} M HCl diluted a thousandfold). Thus, the pH would change from 7.00 to 4.00, a decrease of three full pH units.

The acetic acid-acetate buffer will be governed by the conjugate-pair proton competition expressed by

$$H_2O(l) + CH_3CO_2H(aq) \rightleftharpoons CH_3CO_2^-(aq) + H_3O^+(aq).$$

For this equilibrium

$$C_{H^+} = \frac{K_a \times C_{CH_3CO_2H}}{C_{CH_3CO_2^-}} \qquad \text{(from Eq. 4-7)} \qquad (4\text{-}20)$$

so that, initially,

$$C_{H^+} = \frac{1.8 \times 10^{-5} \times 0.20}{0.30} = 1.2 \times 10^{-5} \text{ mole liter}^{-1}$$

and the pH = 4.92.

Since the HCl is being added to a solution containing a weak base (acetate ion) most of the added protons will be captured and, when equilibrium is reestablished,

$$\begin{aligned} C_{CH_3CO_2H} &= (\sim 2{,}000 \times 10^{-4}) + (1 \times 10^{-4}) \\ &= \sim 0.2001 \\ &= \sim 0.20 \text{ mole liter}^{-1} \end{aligned}$$

and

$$C_{CH_3CO_2^-} = (\sim 3{,}000 \times 10^{-4}) - (1 \times 10^{-4})$$
$$= \sim 0.2999$$
$$= \sim 0.30 \text{ mole liter}^{-1}$$

Therefore,

$$C_{H^+} = \frac{1.8 \times 10^{-5} \times 0.20}{0.30} = 1.2 \times 10^{-5} \text{ mole liter}^{-1}$$

and pH = 4.92.

Thus, an amount of added acid that caused a pH change of three full pH units in pure water does not measurably affect the pH of the acetic acid-acetate buffer solution.

Buffer pH From Eq. 4-20 it is apparent that the pH of a conjugate-pair buffer is determined by the K_a involved and by the *ratio* of the concentration of the acid to the concentration of the conjugate base. For example, all the following buffers have a pH of 4.92: 0.20 M CH_3CO_2H/0.3 M $CH_3CO_2^-$, 0.10 M CH_3CO_2H/0.15 M $CH_3CO_2^-$, 0.020 M CH_3CO_2H/0.030 M $CH_3CO_2^-$.

Buffer Capacity Unlike pH, which is essentially independent of actual concentrations of buffer components (i.e., depends only on *relative* concentrations), the *quantity* of added acid or base that can be absorbed by the buffer within a specified pH change does depend on the *amount* of buffering substance present. If, for example, we had added 10 ml of 10 M HCl to a liter of the 0.20 M CH_3CO_2H/0.30 M $CH_3CO_2^-$ buffer considered above, then when equilibrium was reestablished,

$$C_{CH_3CO_2H} = \sim 0.20 + 0.10 = \sim 0.30 \text{ mole liter}^{-1}$$

and

$$C_{CH_3CO_2^-} = \sim 0.30 - 0.10 = \sim 0.20 \text{ mole liter}^{-1}$$

so that

$$C_{H^+} = \frac{1.8 \times 10^{-5} \times 0.30}{0.20} = 2.7 \times 10^{-5} \text{ mole liter}^{-1}$$

and

$$\text{pH} = 4.57.$$

This change of 0.35 pH units (that is, 4.92 − 4.57) might be quite acceptable for most purposes, but for some critical biochemical experiments it might be far too great. Thus, the desired buffer capacity is determined by the use for which the buffer is designed.

pH INDICATORS

Although the pH of a particular solution can often be calculated, provided sufficient data are available and provided the solution does not contain too many components, the process is commonly a laborious one and is not suited to routine laboratory practice. Electronic instruments can be used to measure the pH of any aqueous solution accurately, but it is often sufficient to employ an indicator compound for this purpose. Such a compound can be used in solution or impregnated on paper.

Indicators are naturally occurring or synthetically prepared organic compounds that undergo definite color changes in well-defined pH ranges. Certain of these are already familiar materials. Thus, litmus is a substance that is said qualitatively to be red in acidic solutions and blue in alkaline solutions, but which actually undergoes this color change in the pH range 4.5 to 8.3. Phenolphthalein is a substance that changes from colorless to red in the pH range 8.3 to 10.0, whereas methyl orange changes from red to yellow in the pH range 3.1 to 4.4. These and other less-well-known indicators are summarized in Table 4-5. The error in saying that an indicator shows whether a solution is acidic or alkaline is apparent from Table 4-5. Actually, there is no single indicator that embraces the entire pH scale, and an indicator shows a pH range rather than acidity or alkalinity.

Table 4-5 pH ranges for color changes of indicators

Indicator	*Acidic Color*	pH *range*	*Alkaline color*
Methyl violet	Yellow	0–2	Violet
Malachite green (acidic)	Yellow	0–1.8	Blue-green
Thymol blue (acidic)	Red	1.2–2.8	Yellow
Bromphenol blue	Yellow	3.0–4.6	Purple
Methyl orange	Red	3.1–4.4	Yellow-orange
Bromcresol green	Yellow	3.8–5.4	Blue
Methyl red	Red	4.4–6.2	Yellow
Litmus	Red	4.5–8.3	Blue
Bromcresol purple	Yellow	5.2–6.8	Purple
Bromthymol blue	Yellow	6.0–7.6	Blue
Phenol red	Yellow	6.4–8.2	Red
m-Cresol purple	Yellow	7.6–9.2	Purple
Thymol blue (alkaline)	Yellow	8.0–9.6	Blue
Phenolphthalein	Colorless	8.3–10.0	Red
Thymolphthalein	Colorless	9.3–10.5	Blue
Alizarin yellow	Yellow	10.1–11.1	Lilac
Malachite green (alkaline)	Green	11.4–13.0	Colorless
Trinitrobenzene	Colorless	12.0–14.0	Orange

Compounds that behave as acid-base indicators have structures which are sufficiently complicated for them to be omitted from this discussion.* There is in each molecular structure, however, some arrangement of electrons which changes when hydrogen ion is added or removed. It is this change in structure that produces the color change observed for the substance in question. In general, indicator molecules are either weak acids or bases, the behavior of which may be described by the equilibrium

$$\mathrm{HInd} \rightleftharpoons \mathrm{H^+} + \mathrm{Ind^-} \tag{4-21}$$

for which

$$K_{\mathrm{Ind}} = \frac{C_{\mathrm{H^+}} \times C_{\mathrm{Ind^-}}}{C_{\mathrm{HInd}}}$$

The colors of the species HInd and $\mathrm{Ind^-}$ are different. Thus, with methyl orange the acid, HInd, is red; whereas the ion, $\mathrm{Ind^-}$, is yellow. Each indicator is characterized by a definite value of K_{Ind}.

ILLUSTRATIVE EXAMPLES

Example 4-1 Determine the K_a for a monoprotic acid, the degree of dissociation of which, α, is 0.341 in 2.0×10^{-2} *M* solution.

Solution Consider that the equilibrium may be expressed as

$$\mathrm{HA} \rightleftharpoons \mathrm{H^+} + \mathrm{A^-}$$

and

$$K_a = \frac{C_{\mathrm{H^+}} \times C_{\mathrm{A^-}}}{C_{\mathrm{HA}}}$$

Then

$$C_{\mathrm{H^+}} = C_{\mathrm{A^-}} = 0.341 \times 2.0 \times 10^{-2} = 6.8 \times 10^{-3} \text{ mole liter}^{-1}$$

* For example, the pH indicator phenol red undergoes the reaction outlined below:

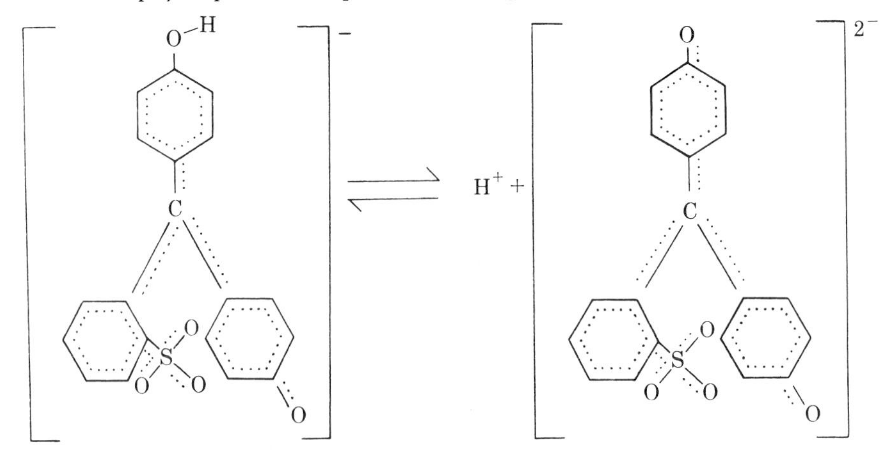

and

$$C_{HA} = (2.0 \times 10^{-2}) - (0.68 \times 10^{-2}) = 1.32 \times 10^{-2} \text{ mole liter}^{-1}$$

Thus

$$K_a = \frac{6.8 \times 10^{-3} \times 6.8 \times 10^{-3}}{1.32 \times 10^{-2}} = 3.5 \times 10^{-3} \text{ mole liter}^{-1}$$

Example 4-2 Determine the pH of a 0.10 M aqueous ammonia solution. The K_b for aqueous ammonia is 1.8×10^{-5} mole liter^{-1}.

Solution Aqueous ammonia solution is described by the equilibrium equation

$$NH_3(aq) + H_2O(l) \rightleftharpoons NH_4^+(aq) + OH^-(aq)$$

for which

$$K_b = \frac{C_{NH_4^+} \times C_{OH^-}}{C_{NH_3}} = 1.8 \times 10^{-5} \text{ mole liter}^{-1}$$

For this solution

$$C_{NH_4^+} = C_{OH^-}$$

and

$$C_{NH_3} = 0.10 \text{ mole liter}^{-1*}$$

Substitution then gives

$$\frac{C^2_{OH^-}}{0.10 \text{ mole liter}^{-1}} = 1.8 \times 10^{-5} \text{ mole liter}^{-1}$$

and

$$C^2_{OH^-} = 1.8 \times 10^{-6} \text{ mole}^2 \text{ liter}^{-2}$$

Thus

$$C_{OH^-} = 1.34 \times 10^{-3} \text{ mole liter}^{-1}$$

Then,

$$\text{pOH} = -\log(1.34 \times 10^{-3}) = -0.13 + 3 = 2.87$$

and

$$\text{pH} = 14.00 - \text{pOH} = 14.00 - 2.87 = 11.13$$

Example 4-3 Calculate the approximate concentration of an acetic acid solution the pH of which is 3.7.

Solution The equilibrium equation involved may be written as

$$CH_3CO_2H \rightleftharpoons H^+ + CH_3CO_2^-$$

for which

$$K_a = \frac{C_{H^+} \times C_{CH_3CO_2^-}}{C_{CH_3CO_2H}} = 1.8 \times 10^{-5} \text{ mole liter}^{-1}$$

If the pH is 3.7, then $\log C_{H^+} = -3.7 = 0.3 - 4$

* Since K_b is so small, most of the NH_3 must remain unreacted.

From a table of logarithms

$$C_{H^+} = 2.0 \times 10^{-4} \text{ mole liter}^{-1}$$

and, since $C_{H^+} = C_{CH_3CO_2^-}$,

$$\frac{2.0 \times 10^{-4} \times 2.0 \times 10^{-4}}{C_{CH_3CO_2H}} = 1.8 \times 10^{-5}$$

Hence,

$$C_{CH_3CO_2H} = \frac{4.0 \times 10^{-8}}{1.8 \times 10^{-5}} = 2.2 \times 10^{-3} \text{ mole liter}^{-1}$$

Example 4-4 Calculate the pH of a 0.10 M solution of sodium nitrite.

Solution Since nitrite ion is the conjugate base of a weak acid, nitrous acid, the equilibrium equation may be written as

$$NO_2^- + H_2O \rightleftharpoons HNO_2 + OH^-$$

for which

$$K_b = \frac{C_{HNO_2} \times C_{OH^-}}{C_{NO_2^-}}$$

and

$$K_b = \frac{K_w}{K_{a(HNO_2)}} = \frac{1.0 \times 10^{-14}}{4.5 \times 10^{-4}} = 2.2 \times 10^{-11}$$

Since $C_{HNO_2} = C_{OH^-}$ and since K_b is very small,

$$\frac{C^2_{OH^-}}{0.10 \text{ mole liter}^{-1}} = 2.2 \times 10^{-11} \text{ mole liter}^{-1}$$

$$C^2_{OH^-} = 2.2 \times 10^{-12} \text{ mole}^2 \text{ liter}^{-2}$$

and

$$C_{OH^-} = 1.5 \times 10^{-6} \text{ mole liter}^{-1}$$

Then

$$pOH = -\log (1.5 \times 10^{-6}) = -0.18 + 6 = 5.82$$

Thus

$$pH = 14.00 - pOH = 14.00 - 5.82 = 8.18$$

Example 4-5 A buffer solution is prepared by mixing 10 ml of 6 M acetic acid solution with 20 ml of 6 M potassium acetate solution. Calculate the hydrogen-ion concentration of a solution obtained by (*a*) diluting 10 ml of this buffer solution to 100 ml and (*b*) treating 10 ml of this buffer solution with 1 ml of 6 M hydrochloric acid solution and then diluting to 100 ml.

Solution In the original buffer solution of 30-ml volume,

$$\text{Quantity of acetic acid} = \frac{6 \text{ moles}}{1 \cancel{\text{liter}}} \times \frac{1 \cancel{\text{liter}}}{1{,}000 \cancel{\text{ml}}} \times 10 \cancel{\text{ml}}$$

$$= 0.06 \text{ mole}$$

and

$$\text{Quantity of sodium acetate} = \frac{6\ \text{moles}}{1\ \text{liter}} \times \frac{1\ \text{liter}}{1{,}000\ \text{ml}} \times 20\ \text{ml}$$
$$= 0.12\ \text{mole}$$

or, in each 10 ml of this solution there is 0.02 mole of acetic acid and 0.04 mole of acetate ion.

(*a*) Dilution of 10 ml of this buffer solution to 100 ml gives a solution for which

$$C_{CH_3CO_2H} = \frac{0.02\ \text{mole}}{100\ \text{ml}} \times \frac{1{,}000\ \text{ml}}{1\ \text{liter}}$$
$$= 0.2\ \text{mole liter}^{-1}$$

and

$$C_{CH_3CO_2^-} = \frac{0.04\ \text{mole}}{100\ \text{ml}} \times \frac{1{,}000\ \text{ml}}{1\ \text{liter}}$$
$$= 0.4\ \text{mole liter}^{-1}$$

Substitution of these quantities into the expression

$$\frac{C_{H^+} \times C_{CH_3CO_2^-}}{C_{CH_3CO_2H}} = 1.8 \times 10^{-5}\ \text{mole liter}^{-1}$$

gives

$$C_{H^+} = 9.0 \times 10^{-6}\ \text{mole liter}^{-1}$$

(*b*) Adding 1 ml of 6 *M* hydrochloric acid solution adds an

$$\text{Amount of hydrochloric acid} = \frac{6\ \text{moles}}{1\ \text{liter}} \times \frac{1\ \text{liter}}{1{,}000\ \text{ml}} \times 1\ \text{ml}$$
$$= 0.006\ \text{mole}$$

Since this material reacts with acetate ion giving acetic acid, in the 11-ml volume of solution produced

$$\text{Quantity of acetic acid} = 0.02 + 0.006 = 0.026\ \text{mole}$$

$$\text{Quantity of acetate ion} = 0.04 - 0.006 = 0.034\ \text{mole}$$

Dilution to 100 ml then gives a solution for which

$$C_{CH_3CO_2H} = \frac{0.026\ \text{mole}}{100\ \text{ml}} \times \frac{1{,}000\ \text{ml}}{1\ \text{liter}}$$
$$= 0.26\ \text{mole liter}^{-1}$$

and

$$C_{CH_3CO_2^-} = \frac{0.034\ \text{mole}}{100\ \text{ml}} \times \frac{1{,}000\ \text{ml}}{1\ \text{liter}}$$
$$= 0.34\ \text{mole liter}^{-1}$$

Substitution as before then gives

$$C_{H^+} = 1.4 \times 10^{-5}\ \text{mole liter}^{-1}$$

Example 4-6 Calculate the approximate K_{Ind} for phenol red, assuming that at the midpoint of the pH range of the indicator $C_{HInd^-} = C_{Ind^{2-}}$.

Solution The equilibrium equation which describes this reaction may be written as

$$HInd^- \rightleftharpoons H^+ + Ind^{2-}$$

for which

$$K_{Ind} = \frac{C_{H^+} \times C_{Ind^{2-}}}{C_{HInd^-}}$$

At the midpoint of the pH range of the indicator (Table 4-5), pH = 7.3. Then

$$\log C_{H^+} = -7.3 = 0.7 - 8$$

and from a logarithm table,

$$C_{H^+} = 5.0 \times 10^{-8} \text{ mole liter}^{-1}$$

Since

$$C_{HInd^-} = C_{Ind^{2-}}$$

$$K_{Ind} = C_{H^+} = 5.0 \times 10^{-8}$$

EXERCISES

4-1. Why is the equilibrium-constant concept applicable to aqueous solutions of weak acids but not to aqueous solutions of strong acids?

4-2. In the light of your answer to Exercise 4-1, what is the significance of saying that the degree of dissociation of hydrochloric acid in 0.1 *M* solution is about 0.9?

4-3. Discuss the advantages and disadvantages of formulating cations as hydrated species in chemical equations and equilibrium expressions. Is the proton a unique species in this respect?

4-4. Why might you expect the successive ionization constants of polyprotic acids to decrease in numerical magnitudes?

4-5. How does the fraction of molecules of a weak acid which are dissociated in aqueous solution vary with concentration? Suggest a possible reason or explanation for this variation.

4-6. Convert each of the following hydrogen-ion concentrations (in moles per liter) to pH. Specify whether for each pH the solution is acidic, alkaline, or neutral.

(*a*) 10^{-7}
(*b*) 3.0×10^{-5}
(*c*) 7.0
(*d*) 1.3×10^{-11}
(*e*) 5.5×10^{-8}

4-7. Convert each of the following notations to concentration of hydrogen ion in moles per liter.

(*a*) pH = 6.80
(*b*) pH = 11.4
(*c*) pH = 7.0
(*d*) pOH = 11.9
(*e*) pOH = 3.25

4-8. Calculate the K_a for a monoprotic acid whose degree of dissociation is 0.185 in 0.0100 *M* solution.

4-9. What is the pH of each of the following solutions?

(*a*) 0.30 *M* HCl
(*b*) 0.30 *M* NaOH
(*c*) 0.30 *M* acetic acid
(*d*) 0.30 *M* nitrous acid
(*e*) 0.30 *M* ammonia

4-10. Calculate the approximate concentration of undissociated acid, or free base for (*c*) and (*d*), in each of the following aqueous solutions.

(*a*) Hydrofluoric acid of pH = 3.0
(*b*) Acetic acid of pH = 4.28
(*c*) Ammonia of pH = 8.5
(*d*) Methyl amine of pH = 9.11

4-11. Calculate the pH of each of the following solutions.

(*a*) 0.050 *M* sodium fluoride
(*b*) 0.033 *M* barium acetate
(*c*) 0.0085 *M* ammonium nitrate

4-12. State whether each of the following aqueous solutions is expected to be acidic, alkaline, or neutral. Explain in each case.

(*a*) 0.10 *M* HCl
(*b*) 0.10 *M* NaCl
(*c*) 0.10 *M* $CaCl_2$
(*d*) 0.10 *M* $AlCl_3$
(*e*) 0.10 *M* $(CH_3NH_3)Cl$
(*f*) 0.10 *M* HCN
(*g*) 0.10 *M* NH_4CN
(*h*) 0.10 *M* NaCN
(*i*) 0.10 *M* $NH_4(CH_3CO_2)$
(*j*) A solution 0.10 *M* in both CH_3CO_2H and $CH_3CO_2^-$

4-13. What is the pH of a buffer solution that is 0.30 *M* in hydrofluoric acid and 0.40 *M* in sodium fluoride?

4-14. What would be the pH of the buffer solution in Exercise 4-13 if:

(*a*) 1.0 ml of 3 *M* HCl were added to 100 ml of the buffer?
(*b*) 1.0 ml of 2 *M* NaOH were added to 100 ml of the buffer?

4-15. What would be the pH of a solution obtained from mixing 50 ml of 0.30 *M* sodium hydroxide with 50 ml of 0.80 *M* acetic acid?

4-16. What volume of 0.30 *M* NaOH must be mixed with 50 ml of 0.60 *M* acetic acid to produce a solution of pH 7.0?

4-17. Describe in detail how you could prepare 500 ml of a pH 8.5 buffer using 15 *M* aqueous ammonia and solid ammonium chloride.

4-18. Describe in detail how you would prepare the buffer in Exercise 4-17 using a concentration of buffer components so that addition of 1.0 ml of 3.0 *M* HCl to the buffer would produce a pH change of 0.1 pH unit.

4-19. The amino acid glycine has the formula shown below at its *isoelectric point*.

$$H_3\overset{+}{N}CH_2CO_2^-$$

Assuming that the acidic group of this amino acid is similar to the methyl ammonium ion and the basic group is similar to acetate ion

(*a*) Write a net ionic equation for the reaction of glycine with hydrochloric acid.

(*b*) Write a net ionic equation for the reaction of glycine with sodium hydroxide.

(*c*) Explain how amino acids may act as buffers.

(*d*) Estimate the pH of the isoelectric point for glycine.

(*e*) Calculate the approximate pH of a solution made by dissolving 100 mg of glycine in 50 ml of 0.10 *M* acetic acid.

SUGGESTED SUPPLEMENTARY READING

Bard, A. J.: "Chemical Equilibrium," Harper and Row, New York, 1966 (paperback).

Clapp, L. B., "The Chemistry of the OH Group," Prentice-Hall, Englewood Cliffs, N.J., 1967 (paperback).

Morris, K. B.: "Principles of Chemical Equilibrium," Reinhold, New York, 1965 (paperback).

Nyman, C. J. and Hamm, R. E.: "Chemical Equilibrium," Raytheon Education Company, Boston, 1968 (paperback).

Sienko, M. J.: "Equilibrium," Benjamin, New York, 1964 (paperback).

Sisler, H. H.: "Chemistry in Non-Aqueous Solvents," Reinhold, New York, 1961 (paperback).

Vanderwerf, C. A.: "Acids, Bases, and the Chemistry of the Covalent Bond," Reinhold, New York, 1961 (paperback).

5
Heterogeneous Equilibria in Aqueous Systems

Most of the equilibrium processes considered thus far have been homogeneous in character. However, many important laboratory operations involve solid substances as well as solutions. Chemists are concerned, therefore, with the formation of solid substances by reactions which take place in solution, and with the dissolving of solids once they have been formed. In reactions involving precipitation and reactions involving dissolution of precipitates, equilibria of the heterogeneous type are encountered. The formation of stable colloidal dispersions is also a problem frequently encountered in attempts to obtain separable precipitates, so that some understanding of these systems is of importance in any study of heterogeneous equilibria.

THE PRECIPITATION PROCESS

When the ions of a sparingly soluble substance, e.g., barium sulfate, are brought together in solution, it is observed that, if the concentration of these ions is sufficiently large, a solid is formed that ultimately settles out.

Such a process is described by the equation

$$Ba^{2+}(aq) + SO_4{}^{2-}(aq) \rightleftharpoons \mathbf{BaSO_4}(s)$$

If the solutions are sufficiently dilute, or if the reactants are mixed carefully, it is often observed that the initially clear mixture becomes faintly opalescent and then increasingly opalescent before any visible solid particles can be detected. This observation suggests that the precipitation process may involve more than one stage.

Two such stages are generally recognized:

1. Condensation of ions to invisible clusters
2. Growth of these clusters as a result of diffusion from the surrounding solution

Although both these processes are reasonable, they are not completely understood and are rather difficult to demonstrate. If particles of sufficient size to settle out are to be produced, it is logical to consider that they must grow by the fitting of component ions into the crystal lattices of smaller clusters of ions. However, it is more difficult to visualize what causes such clusters to form initially and how large they can be. In a rough way, an aqueous solution of ions can be pictured as groups of water molecules aggregated around charged particles. Water molecules closest to the ions are probably strongly held in rather ordered arrangements, whereas molecules farther from the charge centers are more loosely held and in less regular arrays. These aggregates are engaged in a constant jostling motion, and the number and arrangement of water molecules are continuously changing. The water molecules shield the ions from each other both by "bulk" shielding and by reducing the charge densities of the ions by interaction with the appropriate polar region of the water molecule. In order for ion pairs to form as fundamental units for crystal growth, it is necessary for ions of opposite charge to attract each other strongly, even through their water-molecule shields, and to collide with sufficient energy to break loose solvent molecules in the region between the ions.* Both kinetic energy and degree of hydration vary considerably for individual ions in solution, and both factors are complex functions of temperature and the character of the ions. It has not proved feasible to predict which substances will precipitate and which will not except on the basis of experimental observations.

Although oppositely charged ions may have inherently a sufficiently large attraction for each other to start precipitation, this attraction can lead to precipitation only if the ions present are in solutions whose concentrations are sufficiently high that the chance of contact is good. Even if the materials are brought together in the exact quantities dictated by the stoichiometry of

* It should be noted that many crystals form in which lattice positions are occupied by hydrated ions, so it is not always necessary for *all* solvent molecules to be displaced.

the reaction and solubility of the expected product, there may be insufficient attraction to cause the solid to form. In fact, it appears that in every instance a certain degree of *supersaturation** with respect to the precipitated compound must be achieved before the solid is formed. This circumstance has been stated by P. P. von Weimarn as

$$w = k \times \frac{\text{precipitation pressure}}{\text{precipitation resistance}} = kU \tag{5-1}$$

where w is the velocity of precipitation
U is the percent supersaturation
and k is a constant.

Equation 5-1 says, in effect, that the greater the degree of supersaturation, the more rapid the precipitation process. Since a smaller particle size results as the solid is formed more rapidly, extremely finely divided products result if the degree of supersaturation is too large. The degree of supersaturation is determined by the ratio of the extent to which the solubility of the compound in question is exceeded to the solubility of that compound. Thus, a high degree of supersaturation is favored by the mixing of relatively high concentrations of reacting substances and by relatively small solubilities. The first of these can be controlled mechanically to give precipitates of desirable properties, but the second is a characteristic property of the compound and cannot be altered. Since small particle size often gives precipitates that are gelatinous in character, it is not surprising that very sparingly soluble compounds, such as metal sulfides and hydroxides, precipitate in this form regardless of the concentrations of the solutions used. Because small particles are, on the average, more soluble than large ones, it is often possible to transform gelatinous precipitates to more nearly crystalline ones by heating their suspensions. During *digestion* processes, material crystallizes on the larger particles as it is dissolved from the smaller ones.

It must be pointed out that there is a practical concentration limit below which precipitated particles cannot be detected visually without the aid of magnification. Although this limit is, of course, variable as the color, etc., of the precipitated material changes, it amounts on the average to about 10^{-4} mole of substance per liter. If, therefore, the total quantity of solid present lies below this value, no precipitate can be seen.

EQUILIBRIA IN SATURATED SOLUTIONS OF SPARINGLY SOLUBLE ELECTROLYTES

No compound, regardless of how slightly soluble it is, is completely insoluble. Every compound, therefore, can be made to form a saturated

* Supersaturation to a metastable condition in which the concentration of a substance in solution exceeds its equilibrium concentration when in contact with the pure substance as a separate phase.

solution. A saturated solution is, by definition, one in which undissolved solute is in equilibrium with solution. If the solute in question is a salt, the saturated solution will contain the ions of the salt, and equilibria will result, such as

$$\mathbf{BaSO_4}(s) \rightleftharpoons Ba^{2+}(aq) + SO_4{}^{2-}(aq)$$

for the specific case of barium sulfate, or

$$\mathbf{M_xA_y}(s) \rightleftharpoons xM^{y+}(aq) + yA^{x-}(aq)$$

for a general case.

It is confusing to treat these equilibria by direct application of the law of mass action because of difficulty in expressing meaningfully the concentration of a solid substance. Although such a concentration is commonly regarded as constant (= unity), an alternative approach is more nearly exact. The surface of a crystalline solid such as barium sulfate amounts to an orderly arrangement of ions. In a saturated solution the rate at which these ions leave the surface is equal to the rate at which they are deposited from the solution. Each of these rates can then be formulated in terms of the ions involved.

For the specific case of barium sulfate, let the fraction of the exposed surface covered by barium ions be z. The fraction covered by sulfate ions is then $1 - z$. The rate r_1 at which barium ions leave the surface is proportional to z and can be expressed as

$$r_1 = k_1 z \tag{5-2}$$

where k_1 is a rate constant. The rate r_2 at which barium ions deposit from the solution is proportional both to the concentration of barium ions in the solution and the remaining available spaces, $1 - z$, for barium ions on the surface. This is expressed mathematically as

$$r_2 = k_2(1 - z)C_{Ba^{2+}} \tag{5-3}$$

Because of the equilibrium condition, then,

$$k_1 z = k_2(1 - z)C_{Ba^{2+}} \tag{5-4}$$

This equation can be rearranged to

$$C_{Ba^{2+}} = \frac{k_1 z}{k_2(1 - z)} \tag{5-5}$$

The same reasoning applied to the sulfate ion leads to the expression

$$C_{SO_4{}^{2-}} = \frac{k_3(1 - z)}{k_4 z} \tag{5-6}$$

Multiplication of Eq. 5-5 by Eq. 5-6 gives

$$\frac{k_1 z}{k_2(1 - z)} \times \frac{k_3(1 - z)}{k_4 z} = \frac{k_1 k_3}{k_2 k_4} = C_{Ba^{2+}} \times C_{SO_4{}^{2-}} \tag{5-7}$$

or, combining k_1, k_2, k_3, and k_4 as K_{sp},

$$K_{sp} = C_{Ba^{2+}} \times C_{SO_4{}^{2-}} \tag{5-8}$$

The constant K_{sp} for barium sulfate at room temperature equals 1.5×10^{-9} mole^2 liter^{-2} and is called the "solubility-product constant."

Similarly, for the general case of M_xA_y, the same approach yields the expression

$$K_{sp} = C^x_{M^{y+}} \times C^y_{A^{x-}} \tag{5-9}$$

Significance of the solubility-product constant Each sparingly soluble electrolyte is thus characterized by a solubility-product constant of definite numerical magnitude at a given temperature. Inasmuch as the solubilities of substances are dependent upon temperature, the magnitudes of solubility-product constants also change as temperature is changed. The typical constants summarized in Appendix E are, in general, for temperatures around 20 to 25°C. It may be assumed in all subsequent discussions and problems that room temperature is meant and that no significant temperature change occurs. It is apparent that the solubility-product constant can have almost any numerical magnitude, although in practice, values larger than about 10^{-4} have little significance because the salts they describe are quite soluble. For compounds of the same formula type; for example, $AgCl$ and $BaSO_4$, or CaF_2 and Ag_2CrO_4; the numerical values of the solubility-product constants indicate the orders of solubilities, but as Eq. 5-9 indicates, such comparisons among salts of different formula types are not valid.

It is important to realize that for any sparingly soluble electrolyte, the product of the concentrations of the individual ions in a saturated solution raised to appropriate powers is constant,* regardless of the relative magnitudes of the individual ion concentrations. Thus, the equilibrium existing in saturated silver chloride solution amounts to

$$\mathbf{AgCl}(s) \rightleftharpoons Ag^+ + Cl^-$$

and is described by the expression

$$K_{sp} = C_{Ag^+} \times C_{Cl^-} = 2.8 \times 10^{-10} \text{ mole}^2 \text{ liter}^{-2} \tag{5-10}†$$

* The K_{sp} value actually varies somewhat with particle size of the solid. In addition, approximations are inherent in the use of concentration units such as moles liter^{-1}. See: L. Meites, J. S. F. Pode, and H. C. Thomas, *J. Chem. Ed.*, **43**:667 (1966).

† Like previously described equilibrium constants, the K_{sp} values used involve concentration units of mole liter^{-1} and will, therefore, have the dimensions of mole^n liters^{-n}. For the sake of convenience these units are omitted in the following discussions. The student can determine the appropriate dimensions for specific cases from consideration of the form of the solubility-product expressions.

This means that the concentration of the chloride ion determines the concentration of the silver ion, and vice versa. In a saturated solution of silver chloride in pure water the two concentrations are the same, but they need not be so in other solutions. Any increase in the concentration of one of these species must be accompanied by a sufficient decrease in the concentration of the other to render the product equal to 2.8×10^{-10}.

The solubility-product-constant expression, Eq. 5-9, applies to an equilibrium condition between ions and solid. It follows that, if the concentrations of ions of a particular electrolyte in a given solution are such that their product does not exceed the characteristic solubility-product constant, no precipitation can take place. On the other hand, if the concentrations of the ions are such that their product does exceed the solubility-product constant, precipitation takes place and continues until the concentrations of the ions are sufficiently reduced for their product to equal the solubility-product constant. Thus, using Eq. 5-10, no precipitate of silver chloride can be noted if the two ion concentrations are, say, 10^{-6} M since

$$10^{-6} \times 10^{-6} = 10^{-12} \qquad (<2.8 \times 10^{-10})$$

However, if the silver-ion concentration is 10^{-6} M and the chloride-ion concentration is 10^{-3} M, precipitation takes place because

$$10^{-6} \times 10^{-3} = 10^{-9} \qquad (>2.8 \times 10^{-10})$$

Evaluation of solubility-product constants Solubility-product constants can be evaluated from various types of physicochemical data, but for the purpose of this discussion it is sufficient to consider only calculations from data on solubilities in water. Since the solubilities of many compounds are known with considerable accuracy, this method is a generally useful one.

The procedure may be illustrated for the case of silver orthophosphate, Ag_3PO_4, the solubility of which is given as 6.50×10^{-3}g liter^{-1} at 20°C. This value is conveniently converted first to a mole base as

$$\text{Solubility} = \frac{6.50 \times 10^{-3}\text{g liter}^{-1}}{419\text{ g mole}^{-1}} = 1.55 \times 10^{-5} \text{ mole liter}^{-1}$$

In a saturated solution the equilibrium involved is

$$\mathbf{Ag_3PO_4}(s) \rightleftharpoons 3Ag^+ + PO_4^{3-}$$

for which

$$K_{sp} = C^3_{Ag^+} \times C_{PO_4{}^{3-}} \tag{5-11}$$

The solubility-product constant can then be calculated if the equilibrium concentrations of silver and orthophosphate ions can be expressed in terms of the known water solubility. The equilibrium expression indicates that *each* mole of silver orthophosphate that dissolves gives 3 moles of silver ion and 1 mole of orthophosphate ion. Therefore, since 1.55×10^{-5} mole of the salt dissolves,

$$C_{Ag^+} = 3 \times 1.55 \times 10^{-5} = 4.65 \times 10^{-5} \text{ mole liter}^{-1}$$

$$C_{PO_4^{3-}} = 1.55 \times 10^{-5} \text{ mole liter}^{-1}$$

Substitution in Eq. 5-11 then gives

$$K_{sp} = (4.65 \times 10^{-5})^3 \times (1.55 \times 10^{-5})$$
$$= 1.56 \times 10^{-18}$$

Other examples are treated analogously.

Calculations involving solubility-product constants Among the general applications of solubility-product constants are the following:

1. Evaluation of concentrations of ions essential for precipitation
2. Determination of equilibrium concentrations of ions in saturated solutions
3. Determination of solubilities in water
4. Determination of solubilities in solutions containing a common ion

Each of these can be illustrated appropriately by a numerical calculation.

Concentrations of ions essential for precipitation If for a compound containing two ions the concentration of one is known, the concentration of the other can be determined by a simple calculation from the solubility-product expression. Thus, suppose the concentration of hydroxide ion required to precipitate magnesium hydroxide ($K_{sp} = 8.9 \times 10^{-12}$ mole^3 liter^{-3}) from a solution containing 0.01 M magnesium ion is desired. The equilibrium involved is

$$\mathbf{Mg(OH)_2}(s) \rightleftharpoons Mg^{2+} + 2OH^-$$

for which

$$K_{sp} = C_{Mg^{2+}} \times C^2_{OH^-} = 8.9 \times 10^{-12} \qquad (5\text{-}12)$$

The maximum concentration of hydroxide ion that can exist in contact with 0.01 mole liter^{-1} of magnesium ion is then

$$C_{OH^-} = \sqrt{\frac{8.9 \times 10^{-12}}{1 \times 10^{-2}}} = 3.0 \times 10^{-5} \text{ mole liter}^{-1}$$

It follows, then, that precipitation will occur with this magnesium-salt solution whenever

$$C_{OH^-} > 3.0 \times 10^{-5} \text{ mole liter}^{-1}$$

Equilibrium concentrations of ions in saturated solutions It is sometimes important to know the equilibrium concentrations of the ions of a particular salt in its saturated aqueous solution. The method involved may be illustrated by determining the concentrations of magnesium and hydroxide ions in saturated magnesium hydroxide solution. Reference to the preceding example shows that under this condition

$$C_{\mathrm{OH}^-} = 2C_{\mathrm{Mg}^{2+}} \qquad (5\text{-}13)$$

Equation 5-12 may then be put in terms of $C_{\mathrm{Mg}^{2+}}$ by substituting as

$$C_{\mathrm{Mg}^{2+}} \times (2C_{\mathrm{Mg}^{2+}})^2 = 8.9 \times 10^{-12}$$

Therefore,

$$C_{\mathrm{Mg}^{2+}} = 1.3 \times 10^{-4} \text{ mole liter}^{-1}$$

and, employing Eq. 5-13,

$$C_{\mathrm{OH}^-} = 2 \times 1.3 \times 10^{-4} = 2.6 \times 10^{-4} \text{ mole liter}^{-1}$$

Solubilities in water Solubility-product constants can be used for the evaluation of water solubilities in calculations the reverse of those previously employed for arriving at these constants. Thus, to take a specific example, the equilibrium in a saturated aqueous solution of silver chromate

$$\mathbf{Ag_2CrO_4}(s) \rightleftharpoons 2\mathrm{Ag}^+ + \mathrm{CrO_4}^{2-}$$

is described by the relationship

$$K_{sp} = C^2_{\mathrm{Ag}^+} \times C_{\mathrm{CrO_4}^{2-}} = 1.9 \times 10^{-12} \qquad (5\text{-}14)$$

Then, if s equals the solubility of silver chromate in water in mole per liter, it follows that

$$C_{\mathrm{Ag}^+} = 2s$$

and

$$C_{\mathrm{CrO_4}^{2-}} = s$$

since *each* mole of silver chromate that dissolves gives 2 moles of silver ion and 1 mole of chromate ion. Hence,

$$(2s)^2(s) = 1.9 \times 10^{-12*}$$

or

$$s = 7.8 \times 10^{-5} \text{ mole liter}^{-1}$$

* Many students believe that in a situation of this type the concentration of an ion is being both doubled and squared. Such is not the case. It is the *total* concentration of the ion that is being squared. The inclusion of the 2 results from expressing this total ion concentration *in terms of another quantity*, in this case the solubility of the salt.

Solubilities in solutions containing a common ion The principles of equilibrium predict that any salt is less soluble in a solution containing a common ion than in water alone.* By use of the solubility-product constant for the salt in question, the exact reduction in solubility can be calculated. Thus for silver chromate, the substance considered in the previous example, the solubility (in mole per liter) in 0.1 M potassium chromate solution may be represented by y. It follows that

$$C_{Ag^+} = 2y$$

as before, but

$$C_{CrO_4{}^{2-}} = (C_{CrO_4{}^{2-}})_{\text{from } Ag_2CrO_4} + (C_{CrO_4{}^{2-}})_{\text{from } K_2CrO_4} \tag{5-15}$$

or

$$C_{CrO_4{}^{2-}} = y + 0.1$$

since both the dissolved silver chromate and the potassium chromate are responsible for chromate ion in the solution. Substituting in Eq. 5-14 gives

$$(2y)^2(y + 0.1) = 1.9 \times 10^{-12}$$

and solution of the resulting cubic equation gives

$$y = 2.23 \times 10^{-6} \text{ mole liter}^{-1}$$

A simplification may be effected by considering that in Eq. 5-15 the quantity y is sufficiently small in comparison with 0.1 for it to be neglected in this sum. The validity of this assumption is indicated by the fact that even in water alone the solubility of silver chromate is only 0.000078 mole liter^{-1}, a quantity about a thousandth of the 0.1 in this equation. In the presence of the common chromate ion the solubility is reduced, and the difference between the two quantities is even greater. The simplification is thus even easier to justify. Neglecting y, then, in $(y + 0.1)$ we have

$$(2y)^2(0.1) = 1.9 \times 10^{-12}$$

or

$$y = 2.23 \times 10^{-6} \text{ mole liter}^{-1}$$

the same solubility as calculated by the more nearly exact procedure. Comparison of this value with the quantity 7.8×10^{-5} mole liter^{-1} as calculated for water alone indicates to what a large degree the solubility of the compound has been decreased by the common chromate ion.

Since each formula weight of silver chromate contains 2 moles of silver to 1 mole of chromate, it is expected that solubility of this compound

* The formation of complex ions in the presence of a common ion is not taken into account in these considerations. This subject is treated in Chap. 6.

can be reduced even more strikingly by addition of silver ion. If the solubility of the substance in 0.1 *M* silver nitrate solution is represented by y', it follows that

$$C_{Ag^+} = 2y' + 0.1 \tag{5-16}$$

$$C_{CrO_4^{2-}} = y'$$

and that

$$(2y' + 0.1)^2(y') = 1.9 \times 10^{-12}$$

or, neglecting $2y'$ in relation to 0.1 in Eq. 5-16,

$$(0.1)^2(y') = 1.9 \times 10^{-12}$$

and

$$y' = 1.9 \times 10^{-10} \text{ mole liter}^{-1}$$

In general, the greater the number of ions of a particular type present in a mole of a compound, the more the solubility of that compound is decreased in a solution containing the common ion.

PRINCIPLES OF CONTROLLED PRECIPITATION

From the preceding discussion it is apparent that if an ion is present in solution in some definite concentration, it can be precipitated by a second ion only if that second ion is present in sufficient concentration for the solubility-product constant of the compound in question to be exceeded. This is the same as saying that precipitation of this compound can be controlled by controlling the concentration of one of its component ions. Furthermore, if two ions, both of which can give sparingly soluble compounds with a third ion, are present together in solution, the addition of a large quantity of the third ion can precipitate both species. However, if the two sparingly soluble compounds differ sufficiently from each other in solubility, and the concentration of the third ion is correctly controlled, one of the two ions can be precipitated and the other left in solution. Processes of this type may be called "controlled-precipitation" processes.

A controlled precipitation can be carried out in some cases by limiting the quantity of precipitant added, but this is not a generally useful approach because there is nothing to prevent the development of localized high concentrations where the reagent enters the solution. It is much better to utilize another reaction that limits the concentration of the precipitant in solution and maintains this concentration at some predetermined level. Such a reaction may involve the formation of a weak acid or base in solution. Suitable systems of this type are invaluable in chemical analysis because they permit separations to be effected through precipitation of certain components and prevention of precipitation of others. Systems involving hydroxides, carbonates, or sulfides are particularly useful.

Hydroxide precipitations Hydroxide precipitations can be controlled by controlling the hydroxide-ion concentration with ammonium ion in terms of the equilibrium

$$NH_4^+ + OH^- \rightleftharpoons NH_3 + H_2O$$

This equilibrium will be recognized as the reverse of that describing the electrolytic dissociation of aqueous ammonia solutions which is governed by

$$K_b = \frac{C_{NH_4^+} \times C_{OH^-}}{C_{NH_3}} = 1.8 \times 10^{-5} \tag{5-17}$$

Equation 5-17 can be used to calculate the quantity of ammonium ion necessary to prevent precipitation of a given hydroxide if the quantity of hydroxide ion needed has been determined.

The principles involved may be illustrated by an example involving magnesium hydroxide. Suppose it is desired to know the quantity of ammonium ion necessary to prevent the precipitation of magnesium hydroxide when a 0.01 M solution* of a magnesium salt is made 0.1 M in ammonia. As shown previously, precipitation of magnesium hydroxide takes place if

$$C_{OH^-} > 3.0 \times 10^{-5} \text{ mole liter}^{-1}$$

From Eq. 5-17 the concentration of ammonium ion needed to limit the concentration of hydroxide ion to this value is then

$$\begin{aligned} C_{NH_4^+} &= 1.8 \times 10^{-5} \times \frac{C_{NH_3}}{C_{OH^-}} \\ &= 1.8 \times 10^{-5} \times \frac{10^{-1}}{3.0 \times 10^{-5}} \\ &= 0.06 \text{ mole liter}^{-1} \end{aligned}$$

Thus, if sufficient ammonium salt to give an ammonium-ion concentration greater than 0.06 mole liter^{-1} is present in the magnesium salt solution before the ammonia is added, no precipitate can form.

The results of similar calculations applied to a number of metal hydroxides are given in Table 5-1. It is obvious that the precipitation of the more soluble hydroxides can be prevented by the presence of ammonium ion in concentrations easily attained in the laboratory. However, even as soluble

* Many solutions used in the laboratory operations of qualitative analysis are about 0.01 M.

a salt as ammonium nitrate (192g/100g water, or about 24 *M* at 20°C) cannot give a sufficiently large concentration of ammonium ion to prevent precipitation of the less-soluble hydroxides.*

Differences such as appear in Table 5-1 allow separations of metal ions by hydroxide precipitation. The best-known instance of this probably involves separation of iron(III) and chromium(III) ions from magnesium ion by addition of aqueous ammonia to a solution containing ammonium salts. Other possible separations are suggested by the data in Table 5-1.

Carbonate precipitations Carbonate precipitations can be controlled by controlling carbonate-ion concentration with ammonium ion in terms of the equilibrium

$$NH_4^+ + CO_3^{2-} \rightleftharpoons NH_3 + HCO_3^-$$

This equilibrium is exactly comparable with that just discussed in the section on hydroxide precipitations. Both result because of the strongly basic nature of the hydroxide and carbonate ions and because of the acidic nature of the ammonium ion and are, thus, examples of acid-base reactions.

An equilibrium constant describing this system can be obtained as shown below. The carbonate ion reacts with water as

$$CO_3^{2-} + H_2O \rightleftharpoons HCO_3^- + OH^-$$

Table 5-1 Effect of ammonium ion upon hydroxide precipitations†

Hydroxide	*Solubility-product constant*	C_{OH^-} *for precipitation of* $M(OH)_n$ *if* $C_{M^{n+}} = 0.01\ M$, *mole liter*$^{-1}$	$C_{NH_4^+}$ *to prevent precipitation, mole liter*$^{-1}$
$Mg(OH)_2$	8.9×10^{-12}	3.0×10^{-5}	>0.06
$Mn(OH)_2$	2.0×10^{-13}	4.5×10^{-6}	0.4
$Cd(OH)_2$	2.0×10^{-14}	1.4×10^{-6}	1.3
$Pb(OH)_2$	4.2×10^{-15}	6.5×10^{-7}	2.8
$Fe(OH)_2$	1.8×10^{-15}	4.2×10^{-7}	4.3
$Ni(OH)_2$	1.6×10^{-16}	1.3×10^{-7}	14
$Zn(OH)_2$	4.5×10^{-17}	6.7×10^{-8}	25
$Cu(OH)_2$	1.6×10^{-19}	4.0×10^{-9}	450
$Cr(OH)_3$	6.7×10^{-31}	4.1×10^{-10}	4,400
$Al(OH)_3$	5.0×10^{-33}	7.9×10^{-11}	23,000
$Fe(OH)_3$	6.0×10^{-38}	1.8×10^{-12}	1,000,000

† Assuming C_{NH_3} to be 0.1 *M*.

* In actual laboratory practice, the amounts of ammonium ion necessary to prevent precipitation of the various hydroxides may differ significantly from these values because of uncertainties in solubility-product-constant values and lack of rigorous application of equilibrium constants. However, the data in Table 5-1 are qualitatively correct.

for which

$$K_b(CO_3^{2-}) = \frac{K_w}{K_a(HCO_3^-)} = \frac{1 \times 10^{-14}}{4.8 \times 10^{-11}} = 2.1 \times 10^{-4}$$

Also, it is true that

$$NH_4^+ + OH^- \rightleftharpoons NH_3 + H_2O$$

for which

$$K_{\text{equil}} = \frac{1}{K_b(NH_3)}$$

Addition of these two equilibria gives

$$NH_4^+ + CO_3^{2-} \rightleftharpoons NH_3 + HCO_3^-$$

for which

$$K_{\text{equil}} = \frac{C_{NH_3} \times C_{HCO_3^-}}{C_{NH_4^+} \times C_{CO_3^{2-}}} = K_b(CO_3^{2-}) \times \frac{1}{K_b(NH_3)}$$

$$= \frac{2.1 \times 10^{-4}}{1.8 \times 10^{-5}} = 11.7 \qquad (5\text{-}18)$$

Equation 5-18, or its reciprocal if the equilibrium

$$NH_3 + HCO_3^- \rightleftharpoons NH_4^+ + CO_3^{2-}$$

is formulated, then indicates how carbonate-ion concentration is determined by ammonium-ion concentration.

It is convenient to consider in this light the effects of ammonium ion upon the precipitation of magnesium carbonate. This reaction involves the equilibrium

$$\mathbf{MgCO_3}(s) \rightleftharpoons Mg^{2+} + CO_3^{2-}$$

for which

$$K_{sp} = C_{Mg^{2+}} \times C_{CO_3^{2-}} = 4.0 \times 10^{-5} \qquad (5\text{-}19)$$

If a 0.01 M magnesium salt solution is involved, it is apparent that for precipitation

$$C_{CO_3^{2-}} > 4.0 \times 10^{-3} \text{ mole liter}^{-1}$$

The quantity of ammonium ion necessary to reduce the carbonate-ion concentration to this value if the solution is 0.1 M in ammonia and bicarbonate ion is then obtained from Eq. 5-18 as

$$C_{NH_4^+} = \frac{C_{NH_3} \times C_{HCO_3^-}}{C_{CO_3^{2-}}} \times \frac{1}{K_{\text{equil}}}$$

$$= \frac{0.1 \times 0.1}{4.0 \times 10^{-3}} \times \frac{1}{11.7} = 0.21 \text{ mole liter}^{-1} \qquad (5\text{-}20)$$

Table 5-2 Effect of ammonium ion upon carbonate precipitations*

Carbonate	*Solubility-product constant*	$C_{CO_3^{2-}}$ *for precipitation of* MCO_3 *if* $C_{M^{2+}} = 0.01\ M$, *mole liter*$^{-1}$	$C_{NH_4^+}$ *to prevent precipitation, mole liter*$^{-1}$
$MgCO_3$	4.0×10^{-5}	4.0×10^{-3}	0.21
$CaCO_3$	4.7×10^{-9}	4.7×10^{-7}	1,800
$BaCO_3$	1.6×10^{-9}	1.6×10^{-7}	5,300
$SrCO_3$	7.0×10^{-10}	7.0×10^{-8}	12,200
$PbCO_3$	1.5×10^{-13}	1.5×10^{-11}	570,000,000

* Assuming $C_{NH_3} = C_{HCO_3^-} = 0.1\ M$.

Thus, an ammonium-ion concentration in excess of about 0.21 M will inhibit precipitation of magnesium carbonate under these conditions.

Similar calculations for other carbonates yield results such as those given in Table 5-2.† Although precipitation of magnesium carbonate can be prevented quite easily, the quantities of ammonium ion required for complete prevention of precipitation of the others are too large to be of any practical significance.

Sulfide precipitations Sulfide precipitations can be controlled by controlling sulfide-ion concentration with hydrogen ion in terms of the equilibria

$$S^{2-} + H^+ \rightleftharpoons HS^-$$
$$HS^- + H^+ \rightleftharpoons H_2S$$

These equilibria will be recognized as essentially the reverse of the equilibria involved in the ionization of hydrogen sulfide in aqueous solution as tabulated on page 69. It will be apparent that for the summarized process

$$H_2S + 2H_2O \rightleftharpoons 2H_3O^+ + S^{2-} \qquad (H_2S \rightleftharpoons 2H^+ + S^{2-})$$

$$K_a = \frac{C_{H^+}^2 \times C_{S^{2-}}}{C_{H_2S}} = 1.3 \times 10^{-20} \tag{5-21}$$

† None of these calculations is completely accurate because of the complexities of the equilibria involved, hydrolysis reactions, and arbitrary assumptions as to ammonia and bicarbonate-ion concentrations; but the results are useful when compared with each other. Considering magnesium carbonate and calcium carbonate, it is apparent that, if the initial metal-ion concentrations are equal, the quantities of carbonate ion necessary for precipitation are in the same ratio as the solubility-product constants, that is,

$$\frac{(CO_3^{2-})Mg^{2+}}{(CO_3^{2-})Ca^{2+}} = \frac{(K_{sp})Mg^{2+}}{(K_{sp})Ca^{2+}} = \frac{4.0 \times 10^{-5}}{4.7 \times 10^{-9}} = 8{,}500$$

It follows that the quantity of ammonium ion necessary to prevent precipitation of calcium carbonate is some 8,500 times that necessary to prevent precipitation of magnesium carbonate.

Since in a saturated aqueous solution at atmospheric pressure and room temperature the solubility of hydrogen sulfide is roughly 0.1 mole liter^{-1},

$$C_{H_2S} = 0.1 \text{ mole liter}^{-1}$$

Including this value in Eq. 5-21 yields

$$K_{H_2S} = C_{H^+}^2 \times C_{S^{2-}} = 1.3 \times 10^{-21} \tag{5-22}$$

where K_{H_2S} is the net ion-product constant for hydrogen sulfide in its saturated aqueous solution.

It follows from Eq. 5-22 that fixing the hydrogen-ion concentration determines the sulfide-ion concentration and, therefore, controls any precipitation involving the equilibrium

$$2M^{y+} + yS^{2-} \rightleftharpoons \mathbf{M_2S_y}(s)$$

Quantitatively, the effects of hydrogen-ion concentration upon sulfide-ion concentration are shown by the data in Table 5-3 and Fig. 5-1. It follows from these data that when the hydrogen-ion concentration is large, only the most sparingly soluble sulfides can be precipitated; whereas in alkaline solution the sulfide-ion concentration is sufficiently large to precipitate even the more soluble sulfides.

The concentration of hydrogen ion necessary to prevent precipitation of a given metal sulfide can be calculated from Eq. 5-22 and the appropriate solubility-product constant. Thus, for copper(II) sulfide, where

$$\mathbf{CuS}(s) \rightleftharpoons Cu^{2+} + S^{2-}$$

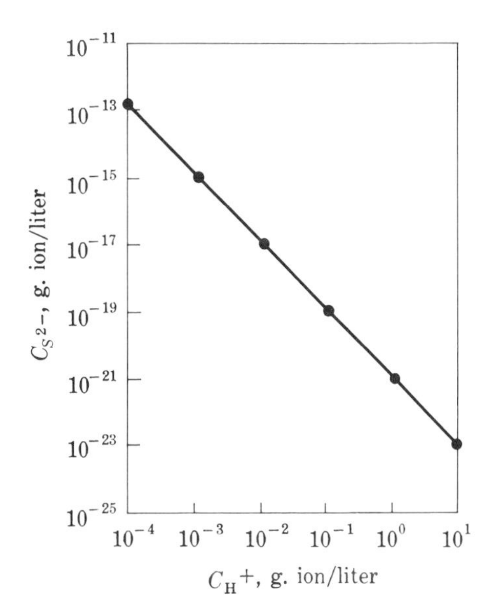

Fig. 5-1 Influence of hydrogen-ion concentration on sulfide-ion concentration.

Table 5-3 Effect of hydrogen-ion concentration upon sulfide-ion concentration

Solution	C_{H^+}, *mole liter*$^{-1}$	$C_{S^{2-}}$, *mole liter*$^{-1}$
0.1 M H_2S, 0.0001 M H^+	10^{-4}	1.3×10^{-13}
0.1 M H_2S, 0.001 M H^+	10^{-3}	1.3×10^{-15}
0.1 M H_2S, 0.01 M H^+	10^{-2}	1.3×10^{-17}
0.1 M H_2S, 0.1 M H^+	10^{-1}	1.3×10^{-19}
0.1 M H_2S, 1.0 M H^+	10^{0}	1.3×10^{-21}
0.1 M H_2S, 10 M H^+	10^{1}	1.3×10^{-23}
0.1 M Na_2S*		5.0×10^{-2}

* Alkaline solution, $C_{S^{2-}}$ reduced by hydrolysis.

and

$$K_{sp} = C_{Cu^{2+}} \times C_{S^{2-}} = 8 \times 10^{-37}$$

the concentration of sulfide ion necessary for precipitation when using a 0.01 M copper(II) salt solution is obviously

$$C_{S^{2-}} = 8 \times 10^{-35} \text{ mole liter}^{-1}$$

From Eq. 5-22 it then follows that

$$C_{H^+} = \sqrt{\frac{1.3 \times 10^{-21}}{8 \times 10^{-35}}}$$

$$= 4 \times 10^{6} \text{ moles liter}^{-1}$$

meaning that some concentration of hydrogen ion in excess of this quantity is necessary to inhibit precipitation completely. Obviously, this concentration is so fantastically large that it has no real significance, and the practical conclusion is that copper(II) sulfide precipitates under any pH condition obtainable in the laboratory. However, as the solubilities of sulfides increase, more practical values for hydrogen-ion concentrations result, and a useful range is approached. This circumstance is shown by the data in Table 5-4 for solutions 0.001 M in various metal ions. As solubility continues to increase, the quantity of hydrogen ion that prevents precipitation becomes so small that precipitation under even moderately acidic conditions is almost impossible.

Table 5-4 shows a significant break in solubilities in a useful range of hydrogen-ion concentration between tin(II) and zinc sulfides. To make a practical separation at this point, it would be necessary to maintain a concentration of sulfide ion larger than the 10^{-23} mole liter^{-1} necessary to precipitate tin(II) sulfide but smaller than the 10^{-20} mole liter^{-1} necessary to precipitate zinc sulfide. A sulfide-ion concentration of 10^{-21} mole liter^{-1} would be in the useful range. The same method of calculation then shows that if

$$C_{H^+} > 1.1 \text{ mole liter}^{-1}$$

a separation of sulfides at this point can be effected. This is the principle underlying the precipitation of a group of cations (cation group II) as sulfides from a solution of controlled acidity, and its separation from another group of cations (cation group IV), the sulfides of which are too soluble to precipitate under these conditions. This operation is one of the most important applications of principle that can be made to the laboratory operations of qualitative analysis. In practice, because of other factors not considered in these calculations (e.g., hydrolysis), a hydrogen-ion concentration of 0.25 to 0.30 *M* is adequate for this procedure.*

Although the above discussion is based upon use of hydrogen sulfide as a precipitant in either acidic or alkaline solution, the principles outlined are equally applicable to any system containing sulfide ion. It has been

Table 5-4 Effect of hydrogen ion upon sulfide precipitations

Sulfide	*Solubility-product constant*	$C_{S^{2-}}$ *for precipitation of* M_xS_y *if* $C_{M^{y+}} = 0.001$ *M, mole liter*$^{-1}$	C_{H^+} *to prevent precipitation, mole liter*$^{-1}$
HgS	1.6×10^{-54}	1.6×10^{-51}	$>9.0 \times 10^{14}$
Ag_2S	5.5×10^{-51}	5.5×10^{-45}	4.9×10^{11}
Cu_2S	1.2×10^{-49}	1.2×10^{-43}	1.0×10^{11}
CuS	8.0×10^{-37}	8.0×10^{-34}	1.3×10^{6}
Fe_2S_3	1.0×10^{-88}	4.6×10^{-28}	1.7×10^{3}
PbS	7.0×10^{-29}	7.0×10^{-26}	1.4×10^{2}
CdS	1.0×10^{-28}	1.0×10^{-25}	1.1×10^{2}
SnS	1.0×10^{-26}	1.0×10^{-23}	1.1×10^{1}
ZnS	1.6×10^{-23}	1.6×10^{-20}	2.9×10^{-1}
$CoS(\alpha)$†	5.0×10^{-22}	5.0×10^{-19}	5.0×10^{-2}
$NiS(\alpha)$†	3.0×10^{-21}	3.0×10^{-18}	2.1×10^{-2}
FeS	4.0×10^{-19}	4.0×10^{-16}	1.8×10^{-3}
MnS	7.0×10^{-16}	7.0×10^{-13}	4.3×10^{-5}

† Most soluble form.

* Solubility-product constants for sulfides are seldom known with accuracy, and, as a consequence, calculations involving them must not be regarded as absolutely correct. One factor contributing to uncertainty in these constants is change in solubility of the sulfide upon standing in contact with the mother liquor. These aging effects are often profound and, thus, cause sizable alterations in solubility-product constants. The values given in Table 5-4 apply, in general, to freshly precipitated sulfides. In all laboratory operations involving sulfides, aging effects should be avoided by treating the materials as soon as possible after precipitation. For discussions in greater depth, see J. R. Goates, M. B. Gordon, and N. D. Faux, *J. Am. Chem. Soc.*, **74**:835 (1952); L. Meites, J. S. F. Pode, and H. C. Thomas, *J. Chem. Educ.*, **43**:667 (1966).

shown* that sulfide precipitations can be effected by the hydrolysis of thioacetamide

$$CH_3\overset{\overset{\displaystyle S}{\|}}{C}-NH_2$$

in aqueous solution without the dangers and unpleasant conditions induced by the poisonous, foul-smelling hydrogen sulfide. Thiocetamide is a white, crystalline solid that dissolves readily in pure water to give stable solutions that hydrolyze very slowly. However, in acidic or alkaline solutions or at elevated temperatures, the rate of hydrolysis of thioacetamide is increased markedly, the rate being greater in alkaline solutions than in acidic solutions. These hydrolysis reactions may be formulated, for simplicity, as

$$CH_3\overset{\overset{\displaystyle S}{\|}}{C}-NH_2 + 2H_2O + H^+ \rightarrow$$

$$CH_3\overset{\overset{\displaystyle O}{\|}}{C}-OH + NH_4^+ + H_2S \qquad \text{(acidic solution)}$$

and

$$CH_3\overset{\overset{\displaystyle S}{\|}}{C}-NH_2 + 3OH^- \rightarrow$$

$$CH_3\overset{\overset{\displaystyle O}{\|}}{C}-O^- + NH_3 + H_2O + S^{2-} \qquad \text{(alkaline solution)}$$

It is apparent that control of hydrogen-ion concentration effects the same control of sulfide-ion concentration in thioacetamide solutions as it does in solutions treated with hydrogen sulfide. Thioacetamide has the additional advantages of altering the sulfide-ion concentration slowly, uniformly, and homogeneously throughout the solution, thus providing conditions most favorable for the formation of granular precipitates that settle more rapidly and more nearly completely than the gelatinous precipitates obtained with hydrogen sulfide gas.†

* H. H. Barber and E. Grzeskowiak, *Anal. Chem.*, **21**:192 (1949).

† Thioacetamide has the disadvantages of being destroyed by the oxidizing effects of ions such as arsenate and iron(III), of yielding sulfate ion when oxidized and thus precipitating sulfates, and of yielding buffering acetate ion upon hydrolysis. The presence of the latter necessitates the addition of extra acid for accurate sulfide-ion control, and precludes substitution of thioacetamide for hydrogen sulfide for precipitations under acidic conditions without some modification in procedure. See L. Lehman and P. Schneider, *J. Chem. Educ.*, **32**:474 (1955).

LIMITATIONS OF THE SOLUBILITY-PRODUCT CONCEPT

The solubility-product concept as outlined above is an idealized concept which, for a given compound, is accurate only when that compound is present alone in saturated aqueous solution at a fixed temperature and when its ions do not undergo hydrolysis.* Even then, approximations are inherent in neglecting the variation of solubility with particle size. Applicability of the concept is thus limited by temperature changes, the presence in solution of other compounds, the average particle size of the solid, and the hydrolysis of ions in equilibrium with the solid. Of these, only temperature has been discussed thus far.

Effect of other salts In the presence of a common ion, sparingly soluble salts become less soluble, as predicted. In certain cases beyond a limiting point, however, solubility may increase appreciably, and in fact, the precipitate may even dissolve. This circumstance is usually the result of complex-ion formation (Chap. 6), which tends to remove equilibrium quantities of metal ion from solution. Thus, the addition of hydrochloric acid solution to a lead(II) salt solution precipitates the white crystalline chloride

$$Pb^{2+} + 2Cl^- \rightleftharpoons \mathbf{PbCl_2}(s)$$

but as more hydrochloric acid solution is added, the precipitate dissolves (Fig. 5-2).

$$\mathbf{PbCl_2}(s) + Cl^- \rightleftharpoons [PbCl_3]^-$$

and

$$[PbCl_3]^- + Cl^- \rightleftharpoons [PbCl_4]^{2-}$$

Of course, many similar cases can be distinguished, and there are obviously instances where the formation of a complex ion from a precipitate does not necessitate the presence of an ion common to that precipitate.

It might be inferred from this discussion that the solubility of an electrolyte in a salt solution containing neither a common ion nor a complexing ion should be the same as in water since no equilibrium or chemical effect would be expected. Actually, the solubilities of electrolytes are found to increase with increasing salt concentrations under these conditions. This is shown in Fig. 5-3 by data for thallium(I) chloride. As the concentration of an added salt increases, solubility may increase so much that deviations of 20 percent or more from the solubility product are observed. This "uncommon" ion effect is a result of interionic attraction. Ions in equilibrium with the sparingly soluble substance are attracted by oppositely

* Hydrolysis refers to reaction with water, in this case as an acid or base (page 75).

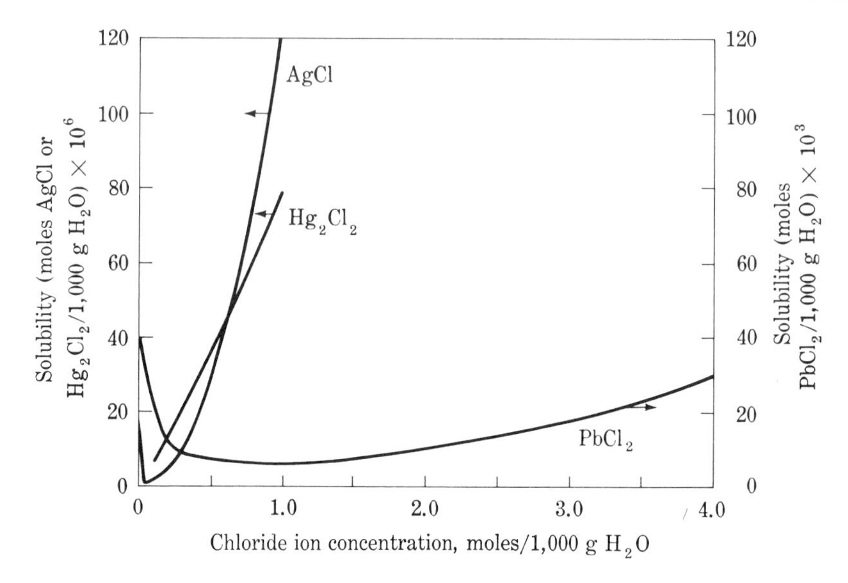

Fig. 5-2 Effect of chloride ion on solubilities of cation group I chlorides.

charged ions of the added salt, and the development of ion "atmospheres" then reduces the velocities of the ions in question and hinders their recombination to form the solid substance. This amounts, in effect, to a removal of these ions from solution, and to an increase in solubility. It is

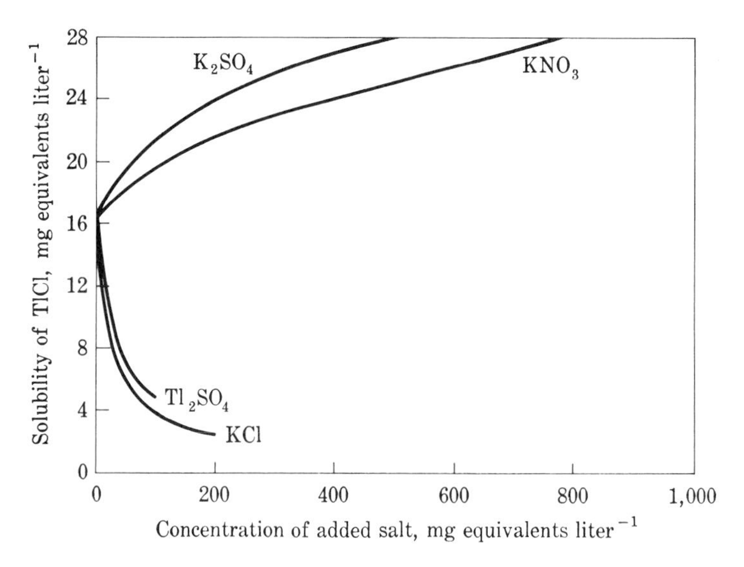

Fig. 5-3 Effects of various salts on the solubility of thallium(I) chloride.

most effective, of course, at high salt concentrations. Corrections for deviations in solubility so produced may be made by employing thermodynamic *activities* instead of concentrations, and using *activity-product constants* instead of solubility-product constants as previously defined. However, since these corrections become appreciable for salts of the common 1:1 valence type (for example, AgI) and 2:2 valence type (for example, CuS) only when the solubility-product constant is larger than about 10^{-10}, there is little to be gained in using them in this treatment.

Effect of hydrolysis If one or both of the ions of a sparingly soluble salt can hydrolyze (i.e., act as an acid or base by proton exchange with H_2O), the solubility of that salt in water is larger than that directly predicted by the solubility-product constant because of the removal of equilibrium ions. This difference is apparent from the scheme

$$\begin{array}{lccc} \mathbf{M}_x\mathbf{A}_y(s) \rightleftharpoons & xM^{y+} & + & yA^{x-} \\ & + & & + \\ & xH_2O & & yH_2O \\ & \downarrow\uparrow & & \downarrow\uparrow \\ & xM(OH)^{+y-1} + xH^+ & & yHA^{-x+1} + yOH^- \end{array}$$

for the general case of the salt $\mathbf{M}_x\mathbf{A}_y$. For salts derived from weakly acidic cations and weakly basic anions (for example, **AgCl** and **$BaSO_4$**) hydrolysis effects are negligible, and, for all practical purposes, hydrolysis of the cation is seldom sufficiently extensive to produce a major change in solubility. If, however, the salt contains a very strongly basic anion such as sulfide or cyanide, solubility is altered rather extensively. This situation may be illustrated for sulfides.

Consider the case of cadmium sulfide for which (Table 5-4)

$$\mathbf{CdS}(s) \rightleftharpoons Cd^{2+} + S^{2-}$$

and

$$K_{sp} = C_{Cd^{2+}} \times C_{S^{2-}} = 1.0 \times 10^{-28} \qquad (5\text{-}23)$$

Neglecting hydrolysis of the sulfide ion, the solubility of cadmium sulfide in water is then calculated as

$$\text{Solubility} = \sqrt{K_{sp}} = 1.0 \times 10^{-14} \text{ mole liter}^{-1}$$

However, it has been shown in Chap. 4 that hydrolysis of the sulfide ion

$$S^{2-} + H_2O \rightleftharpoons HS^- + OH^-$$

is described by the expression

$$K_b(S^{2-}) = \frac{C_{HS^-} \times C_{OH^-}}{C_{S^{2-}}} = 7.7 \times 10^{-2} \qquad (5\text{-}24)$$

The second hydrolysis step involving the HS^- ion is unimportant by comparison. Equation 5-24 says that conversion of S^{2-} ion to HS^- ion by hydrolysis is essentially complete, and that any sulfide ion formed by the dissolution of cadmium sulfide is converted quantitatively into HS^- ion. This means that in a saturated solution of cadmium sulfide

$$C_{Cd^{2+}} = C_{HS^-} \tag{5-25}$$

The exact relationship between the concentrations of sulfide and hydrogen sulfide ions depends, as shown by Eq. 5-24, upon the hydroxide-ion concentration. Although it is true that essentially all the sulfide ion is hydrolyzed, the quantity of hydroxide ion so produced must be small (e.g., about 10^{-14} mole liter^{-1} for CdS), except for very soluble sulfides in comparison with that produced by ionization of the water itself ($= 10^{-7}$ mole liter^{-1}). Substituting 10^{-7} mole liter^{-1} into Eq. 5-24 gives

$$C_{HS^-} = 7.7 \times 10^5 C_{S^{2-}}$$

or

$$C_{S^{2-}} = \frac{1}{7.7 \times 10^5} \times C_{HS^-} \tag{5-26}$$

Substitution of this equivalent into Eq. 5-23 gives

$$\frac{C_{Cd^{2+}} \times C_{HS^-}}{7.7 \times 10^5} = 1.0 \times 10^{-28}$$

which, upon substituting the equivalence shown by Eq. 5-25, reduces to

$$C_{Cd^{2+}} = 8.8 \times 10^{-12} \text{ mole liter}^{-1}$$

This quantity is equal to the solubility of cadmium sulfide in water, due allowance being made for hydrolysis. Comparison of this value with the 1.0×10^{-14}-mole liter^{-1} solubility obtained without regard to hydrolysis indicates clearly the importance of the hydrolysis effect.

It follows, therefore, that none of the values recorded in Table 5-4 is absolutely correct, although for purposes of general comparison the values are adequate.* No exact calculation of the solubility of a salt containing a strongly basic anion should neglect hydrolysis.

DISSOLUTION OF PRECIPITATES

The general equilibrium

$$\mathbf{M}_x\mathbf{A}_y(s) \rightleftharpoons x\mathrm{M}^{y+} + y\mathrm{A}^{x-}$$

* Further and more detailed information concerning the effects of hydrolysis upon the solubilities of sulfides may be found in P. Van Rysselberghe and A. H. Gropp, *J. Chem. Educ.*, **21**:96 (1944). Even the most exacting calculations of this type do not give results that agree completely with experimental observations for controlled precipitations of sulfides of intermediate solubilities (for example, CdS, ZnS, and SnS).

indicates that a precipitate M_xA_y can be dissolved if the equilibrium concentrations of the component ions are sufficiently reduced. These reductions in concentrations are effected by removing the ions via chemical reactions.

Removal of anions is more generally carried out than removal of cations. Anions can be removed by: (1) conversion into weak acids, H_xA; (2) oxidation; or (3) conversion into complex ions. These processes are considered in more detail in the following sections.

Conversion into weak acids A sparingly soluble salt dissolves in acidic solutions only when there is a favorable relationship between the solubility-product constant and the ionization constant of the resulting weak acid. Situations of this type are encountered with hydroxides and carbonates where, in general, the solubility-product constants are not extremely small and the resulting acids are comparatively weak. With sulfides, however, data recorded in Table 5-4 indicate clearly that only where the equilibrium quantity of sulfide ion is about 10^{-25} mole liter^{-1} or more can an acid of moderate concentration be used as a solvent. Salts derived from very weakly basic anions (for example, $BaSO_4$ and AgBr) show little tendency to dissolve in acids because of the lack of appreciable combination between the anion and the hydrogen ion.

Oxidation If the anion present is a reducing agent, it can be removed by oxidation to a component not common to the original equilibrium. The solubilities of sulfides in nitric acid solutions are cases in point. Here, because of the extent to which oxidation of the sulfide ion, as represented by the equation

$$3S^{2-} + 8H^+ + 2NO_3^- \rightarrow 2NO(g) + 4H_2O + 3S(s)$$

takes place, all the sulfides listed in Table 5-4 can be dissolved in, say, $>6\ M$ nitric acid. The extremely small quantity of sulfide ion in equilibrium with mercury(II) sulfide is such that 2 M nitric acid can be used to dissolve many other metal sulfides away from the mercury compound. Salts containing ions such as bromide, iodide, cyanide, arsenite, sulfite, and oxalate are also attacked by oxidizing agents.

Conversion into complex ions If a metal ion can be found that gives a complex ion yielding a lower concentration of anion than the solid compound in question, the precipitate will dissolve in a solution of that metal ion. These situations are not common and the technique is seldom employed.

Removal of cations almost invariably involves formation of complex ions derived from those cations. As such, the subject of complex ions is discussed in Chap. 6, and so an example or two of this behavior should be sufficient here. Mercury(II) sulfide, although only slightly soluble in nitric acid, dissolves readily in aqua regia. Since aqua regia is not a more powerful oxidizing agent than nitric acid, this enhanced solubility is probably due to the formation of stable complex species such as the $[HgCl_4]^{2-}$ ion, and the process may be formulated as

$$\begin{array}{lll} \mathbf{HgS}(s) \rightleftharpoons & Hg^{2+} & + \; S^{2-} \\ & \quad \downarrow\uparrow \; Cl^- & \quad \downarrow\uparrow \; H^+ + NO_3^- \\ & [HgCl_4]^{2-} & \quad S + NO + H_2O \end{array}$$

Other examples involve the dissolution of silver salts in aqueous ammonia solutions to give the species $[Ag(NH_3)_2]^+$, and the dissolution of a hydroxide such as that of tin(IV) in sodium hydroxide solutions to give the species $[Sn(OH)_6]^{2-}$. Situations of this type are described more nearly quantitatively in Chap. 6.

PROBLEMS FROM COLLOIDAL DISPERSIONS

In the discussion at the beginning of this chapter it was pointed out that in the precipitation process growth of primary particles must be rather extensive before the ultimate aggregates become sufficiently large to settle out. In such a process, then, there is a region of intermediate particle size where aggregates that are larger than ion pairs or molecules but smaller than precipitated particles must exist. Conversely, it might be considered that the same region could be entered by progressively breaking up precipitated particles. Particles of these intermediate sizes suspended in some medium in which they are insoluble are referred to as *colloidal particles* and the systems embracing particles and medium are referred to as *colloidal suspensions*, or *colloidal dispersions*.

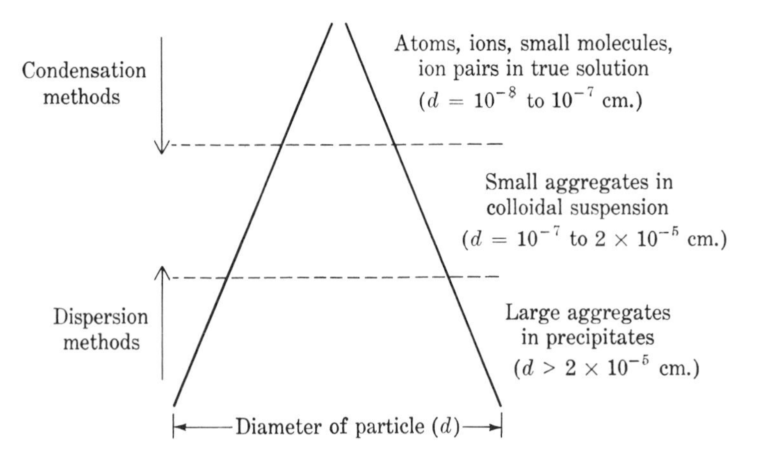

Fig. 5-4 Rough representation of size relationships for colloidal particles.

Relationships between the nature of particles and their size are summarized diagrammatically in Fig. 5-4, where the distance between the two diverging lines at any point represents particle diameter *d*. The ranges of size given are rather broad, and it must not be inferred that sharp breaks exist between the colloidal region and either of the other two. Quite to the contrary, each region shades off into the adjacent one. It is apparent that particles of colloidal dimensions can be obtained either by causing smaller particles to aggregate (*condensation* process) or by decreasing the size of precipitated particles (*dispersion* process). Condensation methods are exemplified by the formation of colloidal arsenic(III) sulfide by reaction of hydrogen sulfide with an aqueous solution of arsenic(III) oxide. Dispersion methods are exemplified by the formation of colloidal aluminum hydroxide by treatment of the precipitated hydroxide with either aluminum chloride solution or very dilute hydrochloric acid solution in insufficient quantity to bring about dissolution. Conversion of a precipitate into a colloidal suspension is called "peptization."

Types of colloidal suspensions Whereas all pure substances existing in a single phase (solid, liquid, gas) may be considered *continuous* and *homogeneous*, solutions or other mixtures may be considered as *discontinuous* and *heterogeneous*; i.e., properties vary from one portion to the other. In true solutions this heterogeneity is submicroscopic in that regions around dissolved particles are slightly different from nearby regions containing only pure solvent, but these regions are so close together in all but extremely dilute solutions that for all practical purposes true solutions are treated as homogeneous mixtures. In coarse mixtures, the heterogeneity may be so pronounced as to be directly observable, as in a mixture of sand and coarse salt. Somewhere between these extremes lie the systems referred to as colloidal dispersions. Since the particles in suspension are from 10 to 10,000 times as large as the molecules of the suspending medium, the degree of heterogeneity is more pronounced than in the case of the true solution.

Table 5-5 Some types of colloidal suspensions

Classification	*Suspended phase*	*Suspension medium*	*Examples*
Foam	Gas	Liquid	Air in water, whipped cream
Solid foam	Gas	Solid	Pumice, Ivory soap
Aerosol (fog)	Liquid	Gas	Fog, insecticide sprays
Aerosol (smoke)	Solid	Gas	NH_4Cl in air, smoke
Emulsion	Liquid	Liquid	Oil in soapy water
Sol	Solid	Liquid	Sulfur in water, Fe_2O_3 in water
Gel	Liquid	Solid	Both phases continuous (Jello)

The colloidal system may involve all combinations of the states of matter except gas in gas (since particle sizes in true gases are much smaller than those in colloidal dispersions), and are frequently classified in terms of the *suspended phase* (discontinuous) and the *suspension medium* (continuous). Some examples of these various combinations are given in Table 5-5. Only the colloidal *sols* are of particular importance in precipitation phenomena, and these are discussed in the following sections.

Properties of colloidal sols Colloidal sols encountered in the laboratory most commonly involve water as the suspending medium. To the eye they may appear perfectly clear (e.g., gold or dilute iron(III) oxide) or opalescent (e.g., sulfur or aluminum hydroxide). However, unlike true solutions, they are heterogeneous in character. This heterogeneity can be demonstrated by the clearly defined beam of light scattered from the particles (*Tyndall beam*) when light is passed through such a suspension. The particles in true solution are too small to scatter light in this fashion. If a Tyndall beam is examined with a microscope, tiny dots of light can be observed. These dots are not the particles themselves, but rather the reflections of light from these particles. These dots are in constant and erratic motion and appear to follow zigzag paths of no regularity. This so-called "Brownian movement" appears to be the result of bombardment of suspended particles by molecules of the suspending medium, and is a clear indication of the fundamental validity of the *kinetic molecular theory*.

Colloidal sols often remain as stable suspensions without detectable settling for indefinitely long periods of time. Such stability suggests that any tendency toward the coalescence of suspended particles to produce particles of precipitate size must be opposed in some way. It is probable that the very small sizes of the suspended particles and their constant motion help to prevent aggregation, but a much more important factor is present when all particles in a given suspension bear like electrical charges and, thus, mutually repel each other. That particles in sols are charged is easily demonstrated by the mass migration of the suspended material (*electrophoretic migration*) under influence of a direct current. Sols involving suspended metals or sulfides commonly migrate to the positive pole and, thus, contain negatively charged colloidal particles, whereas oxide and hydroxide sols commonly contain positively charged colloidal particles.

Charges on colloidal particles appear to be due to adsorbed ions. It is a well-known fact that the amount of surface which a solid presents, and, therefore, the extent to which that solid can adsorb other materials, increases tremendously as particle size is decreased. Particles of colloidal dimensions thus present very large surfaces for adsorption. Considering that a colloidal particle amounts to an aggregate of molecules probably containing water as well, and that if any ions are closely attached to the surface of the particle

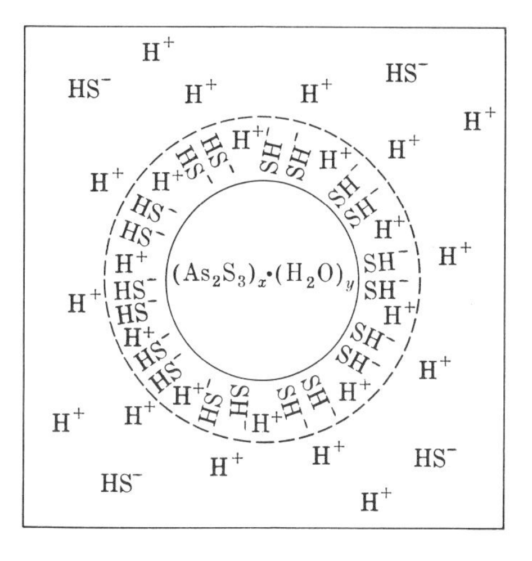

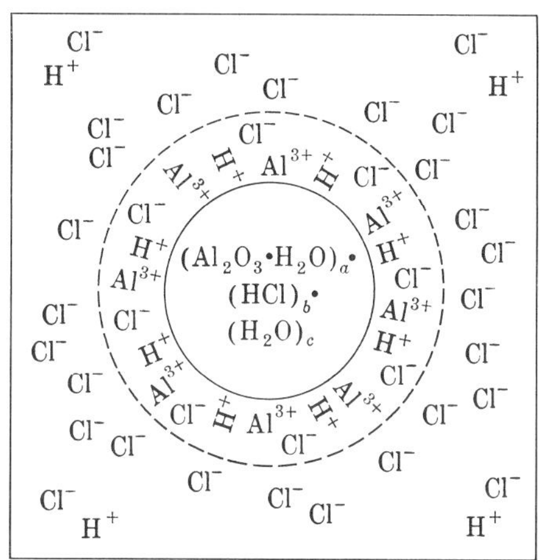

Fig. 5-5 Idealized composition of sol particles. (*a*) Arsenic(III) sulfide; (*b*) aluminum oxide.

they must be balanced by other ions in the surrounding medium to preserve overall electrical neutrality, typical sol particles may be represented diagrammatically as in Fig. 5-5. It is believed that the ions distribute themselves between a closely associated layer and a diffuse outer layer (the so-called "double layer"). The importance of small quantities of electrolyte in contributing to sol stability is shown by the flocculation of a sol upon prolonged dialysis where the electrolyte is completely removed. Peptization appears to result from adsorption of ions by precipitated particles.

Although small quantities of electrolyte are essential for stability, colloidal sols cannot usually tolerate large quantities and generally undergo flocculation under these conditions. The addition of electrolyte may be regarded as neutralizing the charges produced by adsorbed ions and, thus, removing the barrier to flocculation.* Highly charged ions are, of course, most effective in coagulating sols. In general, flocculation of *positive* sols increases in ease and rapidity with increasing charge of added *anion*, and of *negative* sols with increasing charge of added *cation*. Sols can be similarly flocculated by adding sols of opposite charge or by electrolyzing.

Colloidal sols and chemical analysis The formation of colloidal sols often complicates a laboratory procedure that depends upon the removal of a precipitate. It is seldom indeed that the actual precipitation process stops in the colloidal region, but peptization phenomena are particularly

* Actually, the process is somewhat more involved and is associated with the electrical potential (zeta potential) in the diffuse part of the double layer. A certain potential, resulting from the ions present, is essential for stability; but altering this potential by means of additional electrolyte destroys stability.

common. Thus, in the precipitation of hydroxides the precipitate initially formed may disperse, either under the influence of excess hydroxide ion or because of the presence of excess unprecipitated metal ion. In the precipitation of sulfides, excess hydrogen sulfide may cause similar peptization. More commonly, when precipitates are washed, the quantity of excess inert electrolyte may be reduced to such a level that the traces of precipitating ion remaining can effect peptization. Dispersion of precipitates during washing on filter paper is often observed. Less commonly, it occurs also during centrifuging. Freshly precipitated and washed precipitates can be peptized in many cases by adding water and a little of the original precipitating reagent.

To minimize losses of material, prevention of sol formation is generally desirable. Peptization can be avoided in the majority of instances by maintaining a reasonable concentration of some otherwise inert electrolyte, e.g., an ammonium salt, in the suspension or in the wash water. Sols that have formed can be flocculated by adding an electrolyte such as an ammonium salt, by boiling, or by combining the two processes. Unfortunately, the conditions under which sols form are so dependent upon the surface characteristics of the precipitates that it is almost impossible to devise conditions that will always avoid the difficulty.

ILLUSTRATIVE EXAMPLES

Example 5-1 The solubility of barium chromate in water at room temperature is 0.000233 g/100 ml. Evaluate the solubility-product constant for barium chromate.

Solution Since each mole of barium chromate represents 253 g,

$$\text{Water solubility} = \frac{0.000233\ \cancel{\text{g BaCrO}_4}}{100\ \cancel{\text{ml}}} \times \frac{1{,}000\ \cancel{\text{ml}}}{1\ \text{liter}} \times \frac{1\ \text{mole}}{253\ \cancel{\text{g BaCrO}_4}}$$

$$= 9.2 \times 10^{-6}\ \text{mole liter}^{-1}$$

The equilibrium involved

$$\mathbf{BaCrO_4}(s) \rightleftharpoons Ba^{2+} + CrO_4^{2-}$$

is described by the expression

$$K_{sp} = C_{Ba^{2+}} \times C_{CrO_4^{2-}}$$

Since each mole of barium chromate that dissolves gives 1 mole of barium ion and 1 mole of chromate ion, it follows that

$$C_{Ba^{2+}} = C_{CrO_4^{2-}} = \text{water solubility of } BaCrO_4 = 9.2 \times 10^{-6}\ \text{mole liter}^{-1}$$

Hence,

$$K_{sp} = (9.2 \times 10^{-6}) \times (9.2 \times 10^{-6})$$

$$= 8.5 \times 10^{-11}$$

Example 5-2 The solubility-product constant for calcium orthophosphate is 1.3×10^{-32}. Determine the solubility of this compound in water.

Solution For the equilibrium

$$Ca_3(PO_4)_2(s) \rightleftharpoons 3Ca^{2+} + 2PO_4^{3-}$$

$$K_{sp} = C^3_{Ca^{2+}} \times C^2_{PO_4^{3-}} = 1.3 \times 10^{-32}$$

If, then, s = water solubility of $Ca_3(PO_4)_2$ in mole per liter, it follows in terms of the above equilibrium equation that

$$C_{Ca^{2+}} = 3s$$

and

$$C_{PO_4^{3-}} = 2s$$

Substitution then gives

$$(3s)^3(2s)^2 = 1.3 \times 10^{-32}$$

which solves to

$$s = 1.6 \times 10^{-7} \text{ mole liter}^{-1}$$

Or, since each mole of calcium orthophosphate represents 310 g,

$$\text{Water solubility} = \frac{1.6 \times 10^{-7} \cancel{\text{mole}}}{1 \text{ liter}} \times \frac{310 \text{ g}}{1 \cancel{\text{mole}}}$$

$$= 5.0 \times 10^{-5} \text{ g liter}^{-1}$$

Example 5-3 Determine the solubility of lead iodide ($K_{sp} = 8.3 \times 10^{-9}$) in 0.02 M potassium iodide solution, assuming negligible complex formation.

Solution For the equilibrium

$$PbI_2(s) \rightleftharpoons Pb^{2+} + 2I^-$$

the expression

$$K_{sp} = C_{Pb^{2+}} \times C^2_{I^-} = 8.3 \times 10^{-9}$$

holds. If, then, s = solubility of PbI_2 in mole per liter, at equilibrium

$$C_{Pb^{2+}} = s$$

$$C_{I^-} = 2s + 0.02$$

Neglecting the quantity $2s$ in the second expression, substituting, and solving gives

$$s = 2.1 \times 10^{-5} \text{ mole liter}^{-1}$$

Example 5-4 To a solution that is 0.01 M in chloride ion and 0.001 M in bromide ion, 0.01 M silver nitrate solution is added drop by drop. Determine which silver salt will precipitate first.

Solution For silver chloride

$$AgCl(s) \rightleftharpoons Ag^+ + Cl^-$$

and

$$K_{sp} = C_{Ag^+} \times C_{Cl^-} = 2.8 \times 10^{-10}$$

Hence, the equilibrium quantity of silver ion which could exist with 0.01 mole liter^{-1} of chloride ion is, by direct substitution,

$$C_{Ag^+} = 2.8 \times 10^{-8} \text{ mole liter}^{-1}$$

and silver chloride will precipitate whenever the silver-ion concentration exceeds this value. Similarly, for silver bromide

$$\mathbf{AgBr}(s) \rightleftharpoons Ag^+ + Br^-$$

and

$$K_{sp} = C_{Ag^+} \times C_{Br^-} = 5.0 \times 10^{-13}$$

In the same manner, then, precipitation of silver bromide from the 0.001 M bromide solution will occur if

$$C_{Ag^+} > 5.0 \times 10^{-10} \text{ mole liter}^{-1}$$

Since a smaller concentration of silver ion is required to precipitate the bromide, silver bromide will precipitate first.

Example 5-5 For the system described in Example 5-4, determine the concentration of bromide ion remaining in the solution when precipitation of silver chloride begins.

Solution As shown in Example 5-4, precipitation of silver chloride requires that

$$C_{Ag^+} > 2.8 \times 10^{-8} \text{ mole liter}^{-1}$$

Substitution of this quantity into the solubility-product expression for silver bromide gives

$$C_{Br^-} = \frac{5.0 \times 10^{-13}}{2.8 \times 10^{-8}}$$

$$= 1.7 \times 10^{-5} \text{ mole liter}^{-1}$$

as the quantity of bromide ion which could be present when silver chloride begins to precipitate. This represents 1.7 percent of the bromide ion originally present.

Example 5-6 Determine how many miligrams of silver ion can remain per 100 ml of solution if sodium chromate is added to a silver nitrate solution until $C_{CrO_4{}^{2-}} = 0.02$ mole liter^{-1}.

Solution For the precipitation of silver chromate

$$\mathbf{Ag_2CrO_4}(s) \rightleftharpoons 2Ag^+ + CrO_4{}^{2-}$$

and

$$C^2_{Ag^+} \times C_{CrO_4{}^{2-}} = 1.9 \times 10^{-12}$$

Substituting the given chromate-ion concentration and solving gives

$$C_{Ag^+} = 9.7 \times 10^{-6} \text{ mole liter}^{-1}$$

This equilibrium concentration of silver ion is then converted to the desired units as

$$\frac{9.7 \times 10^{-6} \text{ mole}}{1 \text{ liter}} \times \frac{108 \text{ g}}{1 \text{ mole}} \times \frac{1{,}000 \text{ mg}}{1 \text{ g}} \times \frac{10^{-1} \text{ liter}}{100 \text{ ml}} = 0.105 \text{ mg}/100 \text{ ml}$$

Example 5-7 A solution which is 0.01 *M* in Fe^{2+} ion and 0.1 *M* in acetic acid is saturated with hydrogen sulfide. Determine the concentration of acetate ion that must be present for precipitation to take place.

Solution For iron(II) sulfide, the equilibrium

$$FeS(s) \rightleftharpoons Fe^{2+} + S^{2-}$$

is described by the relationship

$$C_{Fe^{2+}} \times C_{S^{2-}} = 4 \times 10^{-19}$$

Hence, it follows that for precipitation from 0.01 *M* solution

$$C_{S^{2-}} > 4 \times 10^{-17} \text{ mole liter}^{-1}$$

Using the expression

$$C_{H^+}^2 \times C_{S^{2-}} = 1.3 \times 10^{-21}$$

substitution of this sulfide-ion concentration then gives

$$C_{H^+} = 5.7 \times 10^{-3} \text{ mole liter}^{-1}$$

Substitution of this quantity and the acetic acid concentration of 0.1 *M* in the ionization relationship for acetic acid,

$$\frac{C_{H^+} \times C_{CH_3CO_2^-}}{C_{CH_3CO_2H}} = 1.8 \times 10^{-5}$$

and simplification gives

$$C_{CH_3CO_2^-} = 3.1 \times 10^{-4} \text{ mole liter}^{-1}$$

EXERCISES

5-1. What practical use is made of von Weimarn's precipitation law in laboratory operations?

5-2. Why is it difficult to predict the solubility of ionic substances in water solely from considerations of ionic charge and ionic radii?

5-3. Suggest an experiment that could be used to determine whether solubility equilibria are static or dynamic.

5-4. Suggest a reasonable explanation for variation of solubility with particle size of the solid. (*Hint*: Consider energy required to break large particles into smaller ones.)

5-5. In what respects is the solubility-product constant, as commonly expressed, only approximately constant?

5-6. Formulate solubility-product-constant expressions for the following:
$Ca_3(PO_4)_2$, SrF_2, $Ag_4[Fe(CN)_6]$, Cu_2S, Hg_2Br_2, HgI_2, Ag_3AsO_4, BaC_2O_4, $Al(OH)_3$, $CrPO_4$.

5-7. Give examples of application of Le Châtelier's principle to solubility-product relationships in terms of temperature variation of solubility and common-ion effects.

5-8. Give examples of two different situations in which acid-base or complex-ion equilibria compete with solubility equilibria.

5-9. In calculating the solubility of a salt in a solution containing a common ion, the quantity of this ion produced by the dissolving salt is often neglected. Why is this justifiable? What limitations exist upon this practice?

5-10. Cite three general approaches to the dissolution of precipitates, and illustrate each by means of a balanced ionic equation.

5-11. Ordinary filter paper has a pore diameter of about 10^{-14} cm. What would happen to a typical colloidal sol poured through such paper?

5-12. Give specific examples of two different ways in which colloidal sols can be converted to mixtures separable by simple centrifugation.

5-13. What effect would adding a crystal of ZnS have on a colloidal ZnS sol?

5-14. Outline clearly conditions under which peptization of precipitates may occur in laboratory operations. Indicate steps that may be taken to avoid this difficulty.

5-15. From each of the following values of solubility in water, evaluate the corresponding solubility-product constant: $\mathbf{PbCrO_4}$, 4.3×10^{-5} g liter^{-1}; $\mathbf{Ag_3PO_4}$, 0.0065 mg ml^{-1}; $\mathbf{MgNH_4PO_4}$, 6.3×10^{-5} mole liter^{-1}; $\mathbf{SrCrO_4}$, 1.22 g liter^{-1}.

5-16. Determine the concentration of metal ion in a saturated solution of each of the following: lead(II) chloride, silver(I) orthophosphate, cadmium carbonate, calcium fluoride, silver(I) carbonate, lead(II) orthophosphate.

5-17. Compare the relative concentrations of M^+ ions in saturated solutions of compounds of the types $\mathbf{MCl}$, $\mathbf{M_2SO_4}$, $\mathbf{M_3PO_4}$, and $\mathbf{M_4[Fe(CN)_6]}$, if the solubility-product constant is 10^{-20} in each case.

5-18. Compare the calculated solubilities of magnesium fluoride in (*a*) water, (*b*) 0.01 *M* magnesium chloride solution, and (*c*) 0.01 *M* potassium fluoride solution. Account for differences among the values obtained.

5-19. Determine the minimum concentration of ammonium nitrate solution needed to prevent precipitation when a 0.01 *M* iron(II) salt solution is made 0.05 *M* in ammonia.

5-20. Will a precipitate form if a solution 0.02 *M* in chloride ion is made 0.01 *M* in lead(II) ion? Give necessary calculations.

5-21. One-liter volumes of 0.04 *M* silver(I) nitrate solution and 0.08 *M* magnesium sulfate solution are mixed. Will a precipitate of silver(I) sulfate form? Explain.

5-22. To 1.0 ml of a solution 0.001 *M* in each of the ions chloride, bromide, and iodide, is added 0.10 ml of 0.001 *M* silver(I) nitrate. What is the probable composition of the precipitate? Explain.

5-23. Determine the pH of a saturated aqueous solution of magnesium hydroxide.

5-24. Solid sodium sulfate is added to 1 liter of a solution 0.01 *M* in lead(II) ion and 0.01 *M* in barium ion. Calculate:

(*a*) The sulfate-ion concentrations at which lead(II) and barium sulfates begin to precipitate.

(*b*) The concentration of barium ion remaining in solution when lead(II) sulfate just starts to precipitate.

5-25. Calculate the quantity of ammonium nitrate that must be added to 25 ml of 0.4 *M* aqueous ammonia solution to prevent precipitation of magnesium hydroxide when this solution is added to 50 ml of 0.02 *M* magnesium chloride solution.

5-26. On the basis of solubility-product-constant data, could one separate lead(II) ion from cadmium ion in 0.01 *M* solution of each ion (buffered at pH 7) by saturation of the solution with H_2S? Explain.

5-27. The solubility of barium chromate in water is approximately 0.0037 g $liter^{-1}$ at 30°C. If 3 ml of 0.01 *M* barium nitrate solution is mixed with 1 ml of 0.1 *M* potassium chromate solution, what concentration of barium ion will remain at equilibrium at 30°C?

5-28 A 0.010 *M* solution of copper(II) sulfate is buffered at pH 2 and saturated with H_2S. What percentage of the copper(II) ion should be removed as copper(II) sulfide?

5-29. When 1.0 ml of 0.1 *M* sodium carbonate solution is added to 3.0 ml of a solution 0.01 *M* in each of the ions magnesium, calcium, and strontium (buffered at pH 9), what will be the composition of the precipitate?

5-30. A 3.0-ml sample of 0.01 *M* aluminum nitrate solution is mixed with 2.0 ml of 0.50 *M* ammonia, and a colloidal sol is formed.

(*a*) Calculate the concentration of aluminum ion remaining.

(*b*) Would you expect the actual concentration of aluminum ion to be greater than, less than, or equal to the calculated value? Explain.

5-31. A 1.0-ml sample of a solution that is 0.10 *M* in each of the ions Fe^{2+}, Zn^{2+}, Cd^{2+}, and Cu^{2+} is mixed with 9.0 ml of a 0.11 *M* thioacetamide solution buffered at pH 2.0. Estimate the final equilibrium concentrations of Fe^{2+}, Zn^{2+}, Cd^{2+}, Cu^{2+}, and S^{2-}.

SUGGESTED SUPPLEMENTARY READINGS

Bard, A. J.: "Chemical Equilibrium," Harper & Row, New York, 1966 (paperback).

Morris, K. B.: "Principles of Chemical Equilibrium," Reinhold, New York, 1965 (paperback).

Nyman, C. J. and Hamm, R. E.: "Chemical Equilibrium," chap. 5, Raytheon Education Company, Boston, 1968 (paperback).

Sienko, M. J.: "Equilibrium," Benjamin, New York, 1964 (paperback).

Vold, M. J. and Vold, R. D.: "Colloid Chemistry," Reinhold, New York, 1964 (paperback).

6
Complex Ions and Coordination Compounds

In the majority of the preceding discussions, the cations present in aqueous solutions have been written as simple species, for example, H^+, Na^+, Cu^{2+}, and Fe^{3+}. Yet it has been pointed out that each such ion does associate with water molecules, and that the species actually present might better be formulated as $H(H_2O)^+$ (that is, H_3O^+),* $Na(H_2O)_x{}^+$, $Cu(H_2O)_y{}^{2+}$, and $Fe(H_2O)_z{}^{3+}$. The number of water molecules present in a given aggregate of this type is difficult to determine and is probably somewhat variable, depending upon conditions such as concentration and temperature. Rather sophisticated experiments have been devised to estimate the compositions of principal species in aqueous solution in terms of the number of "nearest neighbors" of water molecules associated with specific ions. Thus, many aqueous ions are now well-characterized. In crystalline hydrated salts the number of water molecules associated with a particular cation is fixed and is easily determined. Species such as $Cu(H_2O)_4{}^{2+}$ and $Fe(H_2O)_6{}^{3+}$ are well-defined in crystalline compounds. It is found that certain other neutral

* See footnote, page 16.

molecules, notably ammonia and the amines, also associate with cations in definite proportions in crystalline compounds. Furthermore, it is not uncommon for these associations to persist in solution, and species such as $H(NH_3)^+$ (that is, NH_4^+), $Cu(NH_3)_4^{2+}$, and $Ni(NH_3)_6^{2+}$ are familiar in aqueous solutions.

In addition, anions often associate with cations in definite proportions, both in the solid state and in solution, to give new species with new properties; for example, such ions as $Fe(CN)_6^{3-}$, $Fe(CN)_6^{4-}$, $Co(NCS)_4^{2-}$, $AuCl_4^-$, and $Al(OH)_4^-$. Indeed, even the common oxygen-containing acid radicals such as the sulfate or nitrate may be regarded, at least pictorially, as dehydration products of similar associations involving hydroxide ions. A situation of this type might be formulated for the specific case of the metaphosphate ion, $(PO_3)_n^{n-}$, as follows:

$$nP(V) + 6nOH^- \longrightarrow nP(OH)_6^- \xrightarrow{-2nH_2O} nH_2PO_4^- \xrightarrow{-nH_2O} (PO_3)_n^{n-}$$

It is apparent that there exist numerous ionic species that amount to intimate combinations, in definite proportions, of other independently existing species, and possess properties distinct from those of their components. These ions are termed *complex ions.* Although both cations and anions of this type are very common, neutral molecules can also result from proper combinations of the simple components. Such a situation exists, for example, with the compound $Pt(NH_3)_2Cl_4$, a substance that dissolves in water to give nonconducting solutions and which, as a consequence, must be nonionic. Molecular compounds of this sort, or compounds containing complex ions are called "complex compounds," or "coordination compounds."

FORMATION OF COMPLEXES

Cations as Lewis acids The formation of complex ions can be approached in terms of the observed tendency of positive ions to attract regions of high electron density. These regions occur in anions or neutral molecules having unshared pairs of valence (outer-orbital) electrons. This tendency of cations to "seek" electron pairs is the basis for G. N. Lewis' description of these ions as acids. If the electron-pair donor can approach the cation closely enough, a *coordinate covalent bond* results in which the electron pair is influenced by the positive fields of the original nucleus in the donor and the nucleus of the cation in the same general way as any electron pair constituting a polar covalent bond. The strength of the bond depends on the effective charges of the two positive centers [thus, on the nature of both the Lewis base and the Lewis acid, the latter determined to a large extent by the charge-to-radius ratio of the cation (Table 6-6)] and on the nature of the bond itself, as described by the type of orbital overlap involved.

Ions which most effectively combine high charge density with readily available orbitals for bonding are, then, strong Lewis acids. The most common examples are the proton, the cations of the *d* transition elements (for example, V^{3+}, Cr^{3+}, Fe^{2+}, Fe^{3+}, Co^{3+}, Ni^{2+}, Pd^{2+}, Pt^{4+}) and the cations of the copper and zinc families. Alkali and alkaline-earth-metal cations are weak Lewis acids and do not form highly stable complexes with most simple Lewis bases.

Ligands: Lewis bases Electron-pair donors may be simple anions (for example, Cl^-, CN^-), neutral molecules (for example, H_2O, NH_3, $H_2NCH_2CH_2OH$), or molecular-type substances containing ionic groups [for example, $EDTA^{4-}$ (Fig. 6-1) or macromolecules such as proteins]. The term "ligand" has now been generally accepted to describe a substance acting as an electron-pair donor, either before or after formation of the coordinate bond. Lewis bases are substances having available electron pairs, and this term does not, strictly speaking, apply to the substance once its electrons are involved in a coordinate bond.

Ligands are frequently classified in terms of the number of electron pairs the ligand donates to a specific cation (or atom). A *unidentate ligand*, then, donates a single electron pair; a *bidentate ligand* two electron pairs; a *terdentate ligand* three pairs; etc. *Multidentate* is a general term describing ligands that donate more than one electron pair.

The ability of multidentate ligands to "clasp" around a central ion or atom has led to use of the term "chelation" (from the Greek *chele*, "claw") to describe the formation of cyclic structures containing the ligand and the metal atom or ion (Fig. 6-2).

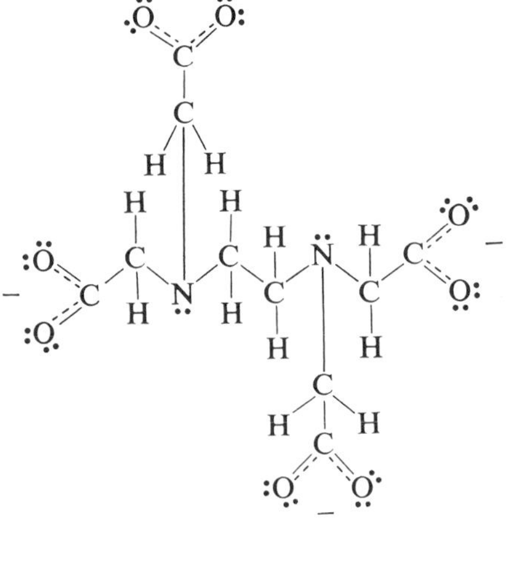

Fig. 6-1 $EDTA^{4-}$ (Ethylenediaminetetraacetate ion).

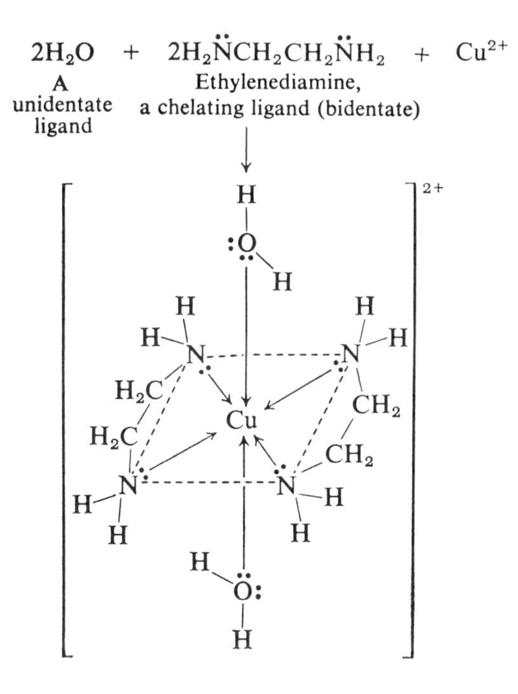

Fig. 6-2 Chelation.

The affinities of various ligands for any given metal ion depend on a number of factors, many of which are difficult to assess completely from simple considerations. These factors include:

1. *Base strength of the electron-pair centers in the ligand.* This term can be approximated from K_b values for the groups involved.
2. *Type of bond (i.e., sigma or pi) possible between the ligand group and the metal ion or atom.* Available orbitals on both the ligand group and the cation or atom must be considered.
3. *Steric factors.* Ligands must be able to approach the central ion or atom closely enough for effective coordination. If more than one ligand group is involved (as is usually the case), then the geometrical arrangement of other ligands, including solvent molecules, must be considered in determining whether the new group can fit properly into a suitable region for bonding.
4. *Thermodynamic factors.* Changes in order (related to entropy) associated with complex formation, as well as changes in other thermodynamic quantities, are useful considerations although often difficult to evaluate quantitatively. These factors are relegated to more-advanced texts.

Some generalizations are available that permit a qualitative discussion of complex formation, but there are numerous exceptions that require detailed study beyond the scope of this text. Simple electrostatic forces between ions and ligand electron-pair centers appear to account fairly well for complexes of the alkali, alkaline-earth, lanthanide-series, and actinide-series ions. For any one of these ions the general trend of ligand affinity closely parallels the K_b trend for the ligand group acting as a proton acceptor in water (for example, $F^- > Cl^-$, $NH_3 > CH_3COO^-$). For ions of the more-electronegative metals (for example, Pb, Au, Ag, Pt, Hg) and such *d* transition metal ions as Cu^{2+}, Ni^{2+}, and Co^{2+}, covalent-type bond interactions complicate the simple electrostatic considerations. These ions, unlike those of the former groups, form with such species as I^-, S^{2-}, CN^-, CO, and NO_2^- strong complexes in which rather complicated bond descriptions are applied. Values of stability (formation) or instability (dissociation) constants for complexes give the most reliable information on ligand-ion interactions in aqueous solution from an experimental point of view. Such constants are discussed in a later section on page 138.

NOMENCLATURE*

Nomenclature of complexes

Name endings The name of a positive complex ion or a neutral complex molecule ends with the common (English) name of the metal followed by the oxidation state as a Roman numeral in parentheses. For example,

$[PbCl]^+$ — Chlorolead(II) ion

$[Ni(CO)_4]$ — Tetracarbonylnickel(0)

The name of a complex anion ends in the suffix *ate*, preceded by the euphonious portion of the name of the metal as used to determine its symbol (Table 6-1). The oxidation state of the metal ion is indicated as before. For example

$[PbCl_3]^-$ — Trichloroplumbate(II) ion

Ligand names All ligand names are combined as a prefix to the metal ion name, using di-, tri-, tetra-, penta-, hexa-, etc., (or, for certain ligand names such as ethylenediamine, bis-, tris-, tetrakis-, etc.) to indicate the number of given ligand groups in the complex. Some common ligands are listed in

* For more details, see *J. Amer. Chem. Soc.*, **82**:5523 (1960).

Table 6-1 Names of some common metals as used in nomenclature of complexes

Symbol	*In complex cations or neutral molecules*	*In complex anions*
Ag	Silver	Argentate
Au	Gold	Aurate
Cu	Copper	Cuprate
Fe	Iron	Ferrate
Hg	Mercury	Mercurate*
Pb	Lead	Plumbate
Sb	Antimony	Antimonate*
Sn	Tin	Stannate

* Mercurate and antimonate are normally used rather than hydrargyrate and stibnate for reasons more musical than philosophical.

Table 6-2. In forming the prefix, negative ligands are listed before neutral ligands and, within each of these categories, ligands are named in order of increasing complexity. For example,

$[Co(NH_3)_3(NO_2)_3]$
Trinitrotriamminecobalt(III)

$[Pt(NH_3)Cl_5]^-$
Pentachloroammineplatinate(IV) ion

$[Cu(NH_3)_4(H_2O)_2]^{2+}$
Diaquotetraamminecopper(II) ion

Nomenclature of salts containing complex ions The name of the cation is given first; followed by, as a separate word, the name of the anion. No specification of relative numbers of cations and anions is made. For example,

$K_4[Fe(CN)_6]$
Potassium hexacyanoferrate(II)

$[Cr(NH_3)_6](NO_3)_3$
Hexaamminechromium(III) nitrate

GEOMETRIES OF COMPLEX IONS

As mentioned earlier, the geometries of complex ions can be described rather accurately when these ions are in crystals, but in solution they are more difficult to determine experimentally. In fact, any description of a complex substance must be considered as an idealized model, since any given species will vary somewhat from its most-stable structure as a consequence of normal internal vibrations and rotations and of collisions with other species. These motions are, of course, more pronounced for species in solution than for those in the solid state. The following geometries should, then, be considered as representations of isolated ions subject to varying degrees of distortion.

Table 6-2 Some common ligands

Formula	*Common name*	*Name as a ligand*
CH_3COO^-	Acetate ion	Acetato
Br^-	Bromide ion	Bromo
Cl^-	Chloride ion	Chloro
F^-	Fluoride ion	Fluoro
S^{2-}	Sulfide ion	Sulfo
CN^-	Cyanide ion	Cyano
OH^-	Hydroxide ion	Hydroxo
SCN^-	Thiocyanate ion	Thiocyanato—S (←SCN) Isothiocyanato or Thiocyanato—N (←NCS)
NO_2^-	Nitrite ion	Nitro ($\leftarrow NO_2$) Nitrito (← O—N—O)
$[O_2C{-}CO_2]^{2-}$	Oxalate ion	Oxalato
$[H_2\ddot{N}CH_2CO_2]^-$	Glycinate ion	Glycinato
SO_4^{2-}	Sulfate ion	Sulfato
$S_2O_3^{2-}$	Thiosulfate ion	Thiosulfato
$[CH_3C(=NOH){-}C(=NO)CH_3]^-$ (DMG)	Dimethylglyoximate ion	Dimethylglyoximato
CO	Carbon monoxide	Carbonyl
NO	Nitrogen(II) oxide	Nitrosyl
H_2O	Water	Aquo
$:NH_3$	Ammonia	Ammine
$CH_3CH_2\ddot{N}H_2$	Ethyl amine	Ethylamine
$H_2\ddot{N}CH_2CH_2\ddot{N}H_2$ (en)	Ethylenediamine	Ethylenediamine

Octahedral complexes The octahedral structure and the closely related tetragonal form (Fig. 6-3) are by far the most commonly encountered geometries proposed for complex ions. Since ligand-ion bond strengths and bond lengths vary, perfectly symmetrical octahedra would be expected only for

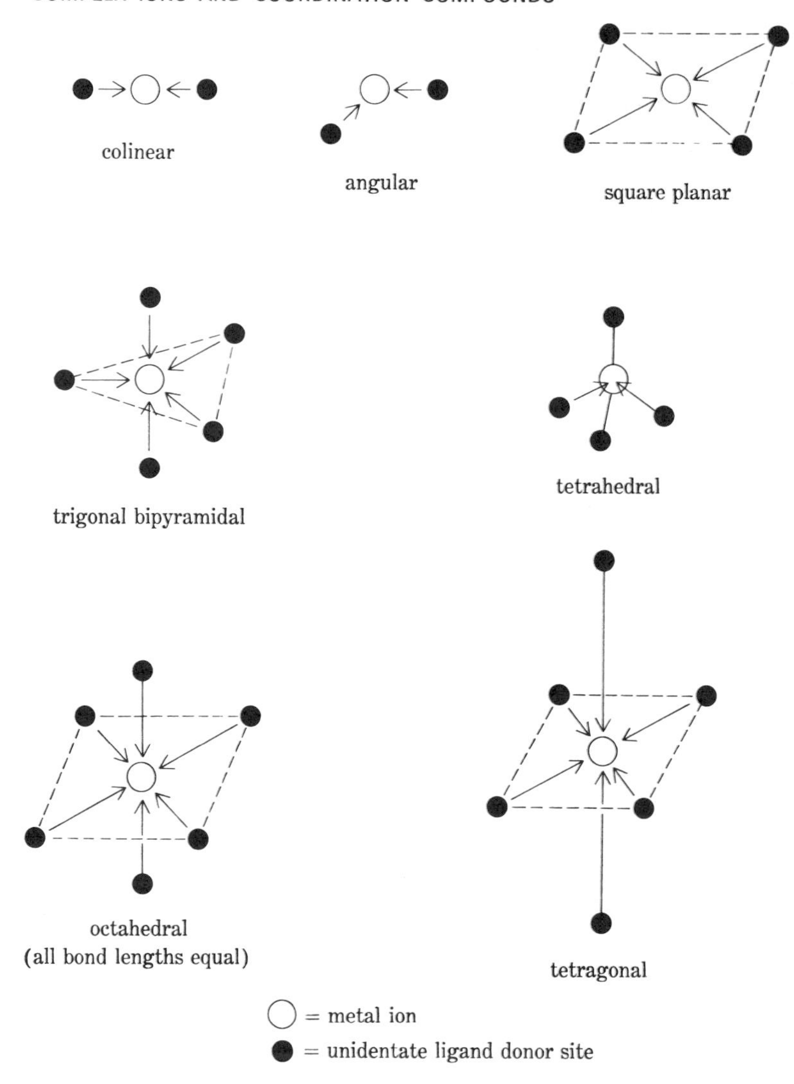

Fig. 6-3 Some common idealized geometries of complexes.

complexes in which all ligands are alike. Even in these cases distortions* occur that are attributed to partially filled electronic energy levels of the central ion. The general octahedral arrangement is, however, the most useful approximation for a large number of complex ions (Table 6-3).

Colinear complexes A few ions such as Ag^+, Au^+, and Cu^+ form linear complexes, although only the Ag^+ complexes are commonly encountered (Table 6-3).

* Jahn-Teller distortions.

Table 6-3 Some common complex ions in analytical chemistry

Formula	*Name of ion*	*Proposed geometry (idealized)*
$[Ag(CN)_2]^-$	Dicyanoargentate(I)	Colinear
$[Ag(NH_3)_2]^+$	Diamminesilver(I)	Colinear
$[Ag(S_2O_3)_2]^{3-}$	Dithiosulfatoargentate(I)	Colinear
$[AsS_4]^{3-}$	Tetrasulfoarsenate(V)	Tetrahedral
$[CdBr_4]^{2-}$	Tetrabromocadmate(II)	Tetrahedral
$[Co(EDTA)]^{2-}$	Ethylenediaminetetraacetatocobaltate(II)	Octahedral
$[Co(NH_3)_6]^{2+}$	Hexaamminecobalt(II)	Octahedral
$[Co(NO_2)_6]^{3-}$	Hexanitrocobaltate(III) (cobaltinitrite)	Octahedral
$[Cu(en)_2]^{2+}$	Bis(ethylenediamine)copper(II)	Square planar*
$[Cu(NH_3)_4]^{2+}$	Tetraamminecopper(II)	Square planar*
$[Fe(CN)_6]^{4-}$	Hexacyanoferrate(II) (ferrocyanide)	Octahedral
$[Fe(CN)_6]^{3-}$	Hexacyanoferrate(III) (ferricyanide)	Octahedral
$[Fe(H_2O)_5F]^{2+}$	Fluoropentaaquoiron(III)	Octahedral (distorted)
$[Fe(H_2O)_5NCS]^{2+}$	Thiocyanato-N-pentaaquoiron(III)	Octahedral (distorted)
$[Fe(H_2O)_5NO]^{2+}$	Nitrosylpentaaquoiron(II)	Octahedral (distorted)
$[Ni(en)_3]^{2+}$	Tris(ethylenediamine)nickel(II)	Octahedral
$[Ni(NH_3)_6]^{2+}$	Hexaamminenickel(II)	Octahedral
$[Pb(H_2O)Cl_3]^-$	Trichloroaquoplumbate(II)	Tetrahedral
$[Pb(CH_3COO)_4]^{2-}$	Tetraacetatoplumbate(II)	Tetrahedral
$[PtCl_4]^{2-}$	Tetrachloroplatinate(II)	Square planar
$[PtCl_6]^{2-}$	Hexachloroplatinate(IV)	Octahedral
$[SiF_6]^{2-}$	Hexafluorosilicate(IV)	Octahedral
$[SnCl_6]^{2-}$	Hexachlorostannate(IV)	Octahedral
$[Sn(OH)_6]^{2-}$	Hexahydroxostannate(IV)	Octahedral
$[Zn(NH_3)_4]^{2+}$	Tetraamminezinc(II)	Tetrahedral

* In aqueous solution as, respectively, tetragonal $[Cu(H_2O)_2(en)_2]^{2+}$ and tetragonal $[Cu(H_2O)_2(NH_3)_4]^{2+}$.

Tetrahedral complexes Ions forming stable complexes with only four ligands might be expected to produce tetrahedral complexes since this geometry provides maximum distance of separation among the four ligands (thus minimizing ligand-ligand repulsions). A number of ions do, in fact, have such structures (Table 6-3), but in others the electronic structure of the ion results in arrangements approximated by the square-planar geometry (Fig. 6-3).

Other structures The angular, trigonal bipyrimidal, and other more complicated structures (Fig. 6-3) are rarely encountered in aqueous solutions, and these geometries are of principal concern in advanced studies of inorganic chemistry. Square-planar ions (for example, $[Cu(NH_3)_4]^{2+}$) are not uncommon in crystalline coordination compounds, but these ions frequently associate with water molecules in solution to form tetragonal complexes.

BONDING IN COMPLEX IONS

Electron-dot structures can be used conveniently to indicate the number of electron pairs involved in complex formation, for example,

$$4\ \begin{array}{c} \mathrm{H} \\ \mathrm{H}:\ddot{\mathrm{N}}: \\ \mathrm{H} \end{array} + Cu^{2+} \rightarrow \left[\begin{array}{c} \mathrm{H} \\ \mathrm{H}:\ddot{\mathrm{N}}:\mathrm{H} \\ \mathrm{H}\quad\ \ \mathrm{H} \\ \mathrm{H}:\ddot{\mathrm{N}}:\mathrm{Cu}:\ddot{\mathrm{N}}:\mathrm{H} \\ \mathrm{H}\quad\ \ \mathrm{H} \\ \mathrm{H}:\ddot{\mathrm{N}}:\mathrm{H} \\ \mathrm{H} \end{array} \right]^{2+}$$

These designations are of limited utility since they indicate nothing about the geometry of the complex or the character of the bonds formed (other than that they are of the coordinate covalent type). In describing compounds of the nonmetals by means of electron-dot structures, attempts were commonly made to arrange electrons (dots) in pairs to give octets (or pairs for

Table 6-4 Usual coordination numbers of some of the common metal ions

Ion	*Coordination No.*	*Ion*	*Coordination No.*
Ag^+	2	Ni^{2+}	4, 6
Au^+	2, 4	Ca^{2+}	6
Cu^+	2, 4	Co^{3+}	6
		Cr^{3+}	6
Au^{3+}	4	Fe^{2+}	6
Cd^{2+}	4	Fe^{3+}	6
Pt^{2+}	4	Pt^{4+}	6
Zn^{3+}	4		
Al^{3+}	4, 6		
Co^{2+}	4, 6		

hydrogen) in the formulas, but many complexes involve six coordinate bonds (six pairs of electrons) around the central ion. A few involve two coordinate bonds; and complexes with three, five, seven, or eight coordinate bonds are known. The *coordination number* of an ion (or atom) is the number of electron pairs that it accepts in forming the complex species in question. This number must be known before the formula of the complex can be predicted or its electron-dot structure drawn. Usual coordination numbers of some of the common metal ions are given in Table 6-4.

An electron-dot formulation is useful only as outlined above. However, it gives no indication as to the way in which electrons are involved in the bond between the metal ion and the donor atom. A chemical bond is usually described in terms of a model that attempts to picture as a real system something that cannot be directly observed. Inasmuch as complex species are derived from metal ions of all types of electronic configuration, it is reasonable to assume that no single type of electronic model can be even approximately correct for every example. Thus, an aquo complex derived from a noble-gas-type ion such as Mg^{2+} differs markedly in properties from one derived from a *d* transition-metal-type ion such as Cr^{3+}. Since in the former the Mg^{2+} ion has no unfilled shells to accommodate electrons, the bond may well be largely electrostatic and result only from the coulombic attraction between the cation and a negative center in the dipolar water molecule (page 4). In the latter, however, the $3d^3$ electronic arrangement of the Cr^{3+} ion is a valence-shell arrangement and capable of interacting in some way with electrons from the oxygen atom of a water molecule. In this instance, the resulting bond could be either strongly covalent or partially covalent, and a different modular description would be required for each case.

The majority of the known complex species, and in particular those that are colored and persist through series of reactions, are derived from *d*-type transition metal ions. At least three bonding pictures have been offered to describe both the properties of these complexes and the geometrical arrangements of ligand atoms around central metal ions. There is space here only to point out the significant features of each theory.*

Valence-bond theory This theory is based upon the idea that electron pairs from donor atoms occupy unfilled orbitals of the *d*, *s*, and *p* types in the valence shell of the transition-metal ion, producing, thereby, hybrids that determine the geometry of the complex ion or molecule. For example, the Fe^{2+} ion ($3d^6 4s^0 4p^0$) reacts with six CN^- ions (each donating a pair of electrons) to form the diamagnetic octahedral ion $[Fe(CN)_6]^{4-}$. Using a small

* The interested student may wish to consult a more advanced text such as M. C. Day, Jr. and J. Selbin, "Theoretical Inorganic Chemistry," 2d ed., chap. 10, Reinhold, New York, 1969.

box to designate an orbital, and vertical arrows to represent electrons of the two spin types (↑ and ↓) the reaction can be formulated as

Fe^{2+}			+	$6CN^-$	→
3*d*	4*s*	4*p*			
[↑↓|↑|↑|↑|↑]	[ ]	[| |]	+	$12e^-$	→

$[Fe(CN)_6]^{4-}$		
3*d*	4*s*	4*p*
[↑↓|↑↓|↑↓|↑↓|↑↓]	[↑↓]	[↑↓|↑↓|↑↓]
d^2	s	p^3

The resulting d^2sp^3 hybrid accounts for the octahedral geometry and the absence of unpaired electron spins for the diamagnetism.*
Similarly we can formulate

	3*d*	4*s*	4*p*	
$[Fe(CN)_6]^{3-}$	[↑↓|↑↓|↑|↑↓|↑↓]	[↑↓]	[↑↓|↑↓|↑↓]	d^2sp^3

octahedral, one unpaired electron, paramagnetic

	3*d*	4*s*	4*p*	
$[CoCl_4]^{2-}$	[↑↓|↑↓|↑|↑|↑]	[↑↓]	[↑↓|↑↓|↑↓]	sp^3

tetrahedral, three unpaired electrons, paramagnetic

	4*d*	5*s*	5*p*	
$[Pd(NH_3)_2Cl_2]$	[↑↓|↑↓|↑↓|↑↓|↑↓]	[↑↓]	[↑↓|↑↓|]	dsp^2

square planar, no unpaired electrons, diamagnetic

However, the complex ion $[FeF_6]^{3-}$ is paramagnetic to the extent of five unpaired electrons rather than one as in the ion $[Fe(CN)_6]^{3-}$, and is also octahedral. The new (outer-orbital) formulation

	3*d*	4*s*	4*p*	4*d*
$[FeF_6]^{3-}$	[↑|↑|↑|↑|↑]	[↑↓]	[↑↓|↑↓|↑↓]	[↑↓|↑↓| | |]

is essential to account for these properties. Furthermore, acceptance of electron pairs by the cation imparts to that species an unlikely large negative

* A substance is said to be diamagnetic if, when placed in a magnetic field, it tends to move away from that field; paramagnetic when it tends to move into and align itself with that field. Diamagnetic behavior is associated with the complete pairing of electron spins; paramagnetic behavior with the presence of unpaired electron spins. For *d* transition-metal species, the magnitude of the observed paramagnetism is often directly proportional to the number of unpaired electrons present.

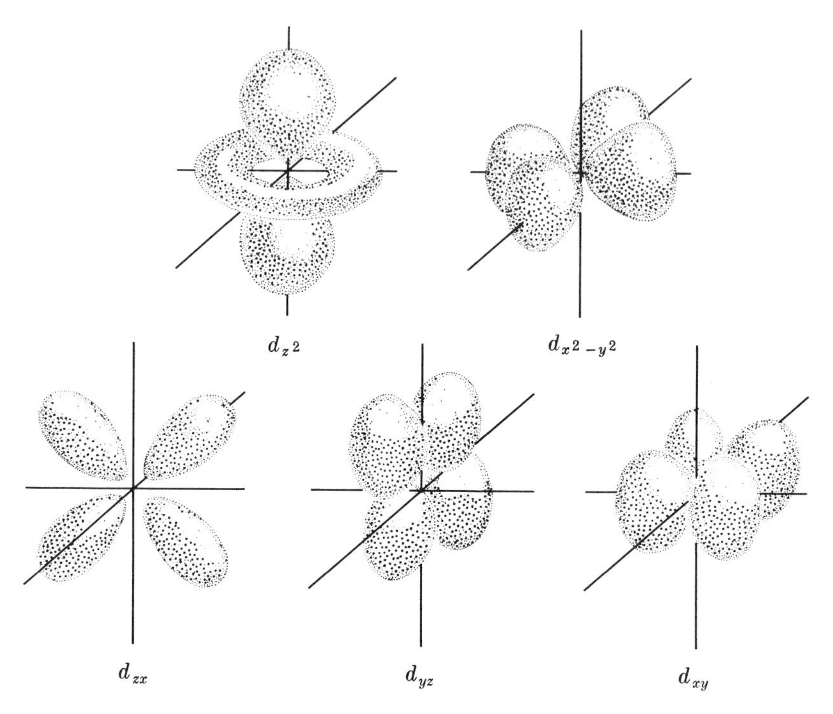

Fig. 6-4 Spatial orientation of *d* orbitals.

charge density, and there is no correlation between the color of an ion and this bonding picture.

Crystal field theory This theory is based upon the spatial orientation of the *d* orbitals, as shown in Fig. 6-4, and the relative energy changes that result when donor atoms, treated as point negative charges, approach occupied and/or unoccupied *d* orbitals. Fig. 6-4 indicates that the $d_{x^2-y^2}$ and d_{z^2} orbitals are directed along the x, y, and z axes of an imaginary cube circumscribed about the atomic nucleus as an origin; whereas the d_{xy}, d_{xz}, and d_{yz} orbitals are directed between these axes. If the *d* orbitals of the cation are occupied, electrons from six ligands approaching that cation along these axes (octahedral geometry) will be repelled by the d_{z^2} and $d_{z^2-y^2}$ orbitals but not by the other *d* orbitals. The result is a redistribution of the energies of the *d* orbitals, as shown for octahedral geometry in Fig. 6-5. Certain ligands (strong-field or spin-paired ligands) favor preferential occupancy of the lower-energy orbitals (t_{2g}) by *d* electrons from cations containing 4 to 7 *d* electrons and, thus, cause pairing of electron spins. Ligands of this type include CN^- and NO_2^- ions. Other ligands (weak-field or spin-free ligands) favor lack of pairing of *d*-electron spins and thus allow

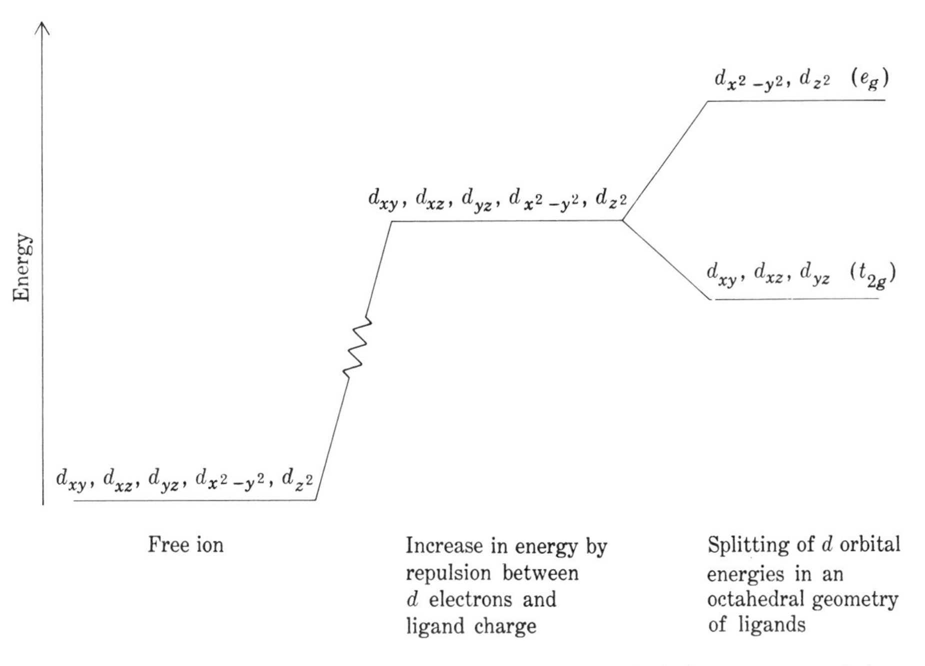

Fig. 6-5 Energy relationships among *d* orbitals in an octahedral arrangement of six ligands.

electrons from d^4 to d^7 species to occupy the higher energy (e_g) orbitals.* Ligands of this type include the halide and OH^- ions. The SCN^- ion and ammonia occupy an intermediate position. The energy gained by the preferential filling of lower-lying *d* levels over that of random occupancy is the *crystal field stabilization energy*. Its magnitude determines the overall stability of the complex.

The crystal field approach accounts for both the magnetic properties of complex species and their colors, as determined by the positions of bands in their absorption spectra. However, it treats all species in terms of aggregations of charged particles and does not allow for covalent bonding.

Ligand field theory Certain complex ions, notably those of cobalt(III), chromium(III), and platinum(IV), have many covalent properties and must contain bonds that are at least partially covalent. Covalent bonds involving carbon have been very successfully accounted for by a molecular orbital approach, that is, by an approach that says that in a covalent bond the atomic orbitals involved lose their identity and combine to form molecular

* There is no distinction in electron occupancy between strong- and weak-field ligands for d^1, d^2, d^3, d^8, d^9, and d^{10} species.

orbitals characteristic of the species as a whole. The same broad concept has been combined with the crystal field theory to yield the *ligand field theory* of bonding in complex ions. This approach accomplishes the same result as the crystal field theory but does allow for a covalent contribution to a metal ion-ligand bond.

COLORED COMPLEXES

A number of complex ions and coordination compounds are colored. Some of the more common colored species encountered in the analytical chemistry of solutions are listed in Table 6-5. Unfortunately, no simple model for bonding in complexes provides a satisfactory basis for explanation or prediction of color in these species. The best approach developed yet requires the more sophisticated models of *ligand field theory*, but the details are beyond the scope of this textbook.

COMPLEX IONS IN AQUEOUS SYSTEMS

The formation or destruction of complex ions is frequently utilized in analytical chemistry to effect separations of cations; to identify specific ions; or, in quantitative work, to determine the concentration or amount of a given substance in a mixture. Complex ions and coordination compounds are also of importance in biochemistry, e.g., in studies of ion transport through cell membranes or of the role of metal ions in certain enzyme systems.

Although there is much interesting chemistry concerned with non-ionic complexes such as pentacarbonyliron(0) and with topics such as isomerism in complexes, our discussion will be restricted to aspects of complex-ion chemistry commonly encountered in the qualitative analysis of aqueous solutions.

Table 6-5 Colors of some common complex ions in aqueous solution

Ion	*Color*	*Ion*	*Color*
$[Co(H_2O)_6]^{2+}$	Pink	$[Fe(H_2O)_5NO]^{2+}$	Brown
$[Co(NH_3)_6]^{2+}$	Tan	$[Fe(H_2O)_5NCS]^{2+}$	Deep red
$[Co(H_2O)_2(NCS)_4]^{2-}$	Blue	$[Fe(CN)_6]^{4-}$	Yellow
$[Co(NO_2)_6]^{3-}$	Yellow	$[Fe(CN)_6]^{3-}$	Red
$[Cr(H_2O)_6]^{3+}$	Violet	$[Mn(H_2O)_6]^{2+}$	Pink
$[Cu(H_2O)_6]^{2+}$	Light blue	$[Ni(H_2O)_6]^{2+}$	Green
$[Cu(H_2O)_2(NH_3)_4]^{2+}$	Dark blue	$[Ni(NH_3)_6]^{2+}$	Violet
$[Fe(H_2O)_6]^{2+}$	Pale green		

Table 6-6 Predicted relative strengths of water-ion bonds*

	Ion	*Ionic radius, Å*	*Relative charge*	*Charge: radius ratio*
↑ Increasing strength	Al^{3+}	0.51	3	5.9
	Fe^{3+}, Cr^{3+}	0.64	3	4.7
	Fe^{2+}	0.74	2	2.7
	Ca^{2+}	0.99	2	2.0
	Sr^{2+}	1.12	2	1.8
	Ba^{2+}	1.34	2	1.5
	Li^{+}	0.68	1	1.5
	S^{2-}	1.84	2	1.1
	Na^{+}	0.97	1	1.0
	K^{+}	1.33	1	0.75
	Cl^{-}	1.81	1	0.55
	Br^{-}	1.96	1	0.51
	I^{-}	2.20	1	0.45

* Although the predicted strengths are in general agreement with experimental values, deviations do occur that require explanations beyond the scope of this text. For more information see the suggested supplementary reading list at the end of this chapter.

METAL IONS IN AQUEOUS SOLUTIONS

X-ray and neutron-diffraction studies of crystalline hydrates have accurately determined the arrangement of water molecules around both positive and negative ions in many solid substances. From determinations of interparticle distances it has even been possible to estimate the relative strengths of forces of attraction between water molecules and several common ions. These investigations indicate that, in general, water-ion orientations are probably stable as

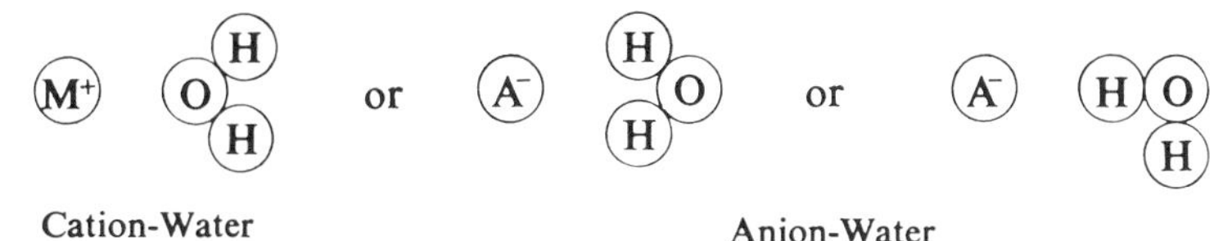

and that the strengths of the forces of attraction between the water molecules and the ions increase with increasing charge density of the ion, a factor that can be approximated by the charge-to-radius ratio of the ion (Table 6-6).

Unfortunately, accurate investigations of solutions are experimentally much more difficult than are studies of crystals, partly because of the continuously changing character of submicroscopic regions in solutions. It is also difficult to extrapolate from crystal structures to structures in solution. For example, the geometrical restrictions and *effective* charges of ions (in the close proximity of ions of opposite charge) in crystals differ widely from the corresponding characteristics in solution.

Current evidence supports the visualization of ions in aqueous solution as charge centers closely surrounded by more-or-less regular arrays of water molecules. The strengths of the water-ion bonds, the number of water molecules in closest proximity to the ions, and the preferred geometries of these clusters all depend on the effective charge of the ion, its size, and certain characteristics of its electronic configuration (e.g., types of available orbitals). One should not, however, visualize an aqueous solution of ions as a uniform and permanent set of hydrate clusters of specific geometry. Thermal motions in solutions result in collisions between hydrated species and solvent molecules (or between different hydrated clusters), so that any given hydrated ion must frequently change its shape. In many instances, the ion-to-water ratio is also changed as molecules are broken loose from or forced into the hydrate cluster. Discussions, then, of geometry of hydrates (or of any complexes in solution) can lead only to postulates of most stable structures and, thus, to models of most common species under given conditions.

EQUILIBRIA INVOLVING COMPLEX IONS

Although the discussion thus far has indicated that complex ions undergo changes in structure in solution, only relatively minor distortions of geometry have been considered. A much more profound effect concerns the dynamic equilibria between complexes of various structures and free ligands. In aqueous solutions complex ions are constantly being formed through replacement of water by other ligands (or by other water molecules) or broken down by the reverse process. An understanding of equilibria of this type reinforces the idea that solutions of complex ions are far more complicated than one would expect on the basis of a simple model of an isolated complex.

Stepwise formation The actual mechanisms for the formation of complex ions in aqueous solution are largely unknown. It is, however, unlikely that a concerted attack by surrounding ligands simultaneously converts a hydrated ion into a different complex. The more probable route involves a stepwise displacement of coordinated solvent molecules by new ligand groups. For a given system a series of steps could be postulated, for example, as

$$[Ni(H_2O_6]^{2+} + NH_3 \rightleftharpoons [Ni(H_2O)_5(NH_3)]^{2+} + H_2O$$
$$[Ni(H_2O)_5(NH_3)]^{2+} + NH_3 \rightleftharpoons [Ni(H_2O)_4(NH_3)_2]^{2+} + H_2O$$
$$[Ni(H_2O)_4(NH_3)_2]^{2+} + NH_3 \rightleftharpoons [Ni(H_2O)_3(NH_3)_3]^{2+} + H_2O$$
$$[Ni(H_2O)_3(NH_3)_3]^{2+} + NH_3 \rightleftharpoons [Ni(H_2O)_2(NH_3)_4]^{2+} + H_2O$$
$$[Ni(H_2O)_2(NH_3)_4]^{2+} + NH_3 \rightleftharpoons [Ni(H_2O)(NH_3)_5]^{2+} + H_2O$$
$$[Ni(H_2O)(NH_3)_5]^{2+} + NH_3 \rightleftharpoons [Ni(NH_3)_6]^{2+} + H_2O$$

For a solution prepared by dissolving a small amount of a nickel(II) salt (such as $NiSO_4$) in a very large excess of aqueous ammonia, it is possible to approximate experimentally the fraction of "bound" ammonia and, on this basis, to state with reasonable assurance that the principal nickel(II) species present is $[Ni(NH_3)_6]^{2+}$. It is considerably more difficult to determine the fractions of various complexes in solutions where neither the Ni^{2+} ion nor the NH_3 molecule is present in excess. In such cases the measurement of bound ammonia, for example, can tell the *average composition* (for example, $[Ni(H_2O)_2(NH_3)_4]^{2+}$), but such data might equally apply to a 50:50 mixture of $[Ni(H_2O)_4(NH_3)_2]^{2+}$ and $[Ni(NH_3)_6]^{2+}$, or to virtually endless combinations of the various formulas. Thus, any quantitative evaluation of stepwise equilibrium constants requires rather sophisticated procedures. For a large number of cases (although there are important exceptions) it appears that the stepwise equilibrium (formation) constants vary according to the trend $K_1 > K_2 > K_3 > K_4$, etc. (Table 6-7).

Table 6-7 Approximate stepwise equilibrium constants* for formation of some complex ions

"*Complete*" *complex* (*omitting* H_2O)	K_1	K_2	K_3	K_4	K_5	K_6
$[Ag(NH_3)_2]^+$	2.0×10^3	7.9×10^3†				
$[AlF_6]^{3-}$	1.4×10^6	1.0×10^5	6.3×10^3	5×10^2	40	3.2
$[CdCl_4]^{2-}$	1.0×10^2	5.0	4.0×10^{-1}	2.0×10^{-1}		
$[Cd(CN)_4]^{2-}$	3.0×10^5	1.3×10^5	4.3×10^4	3.5×10^3		
$[Cd(NH_3)_6]^{2+}$	4.5×10^2	1.3×10^2	28	8.5	4.8×10^{-1}	46†
$[Co(NCS)_4]^{2-}$	1.0×10^3	1.0	2.0×10^{-1}	1.0†		
$[Co(NH_3)_6]^{2+}$	1.0×10^2	30	8.0	4.0	1.3	2.0×10^{-1}
$[CuCl_4]^{2-}$	6.3×10^2	40	3.1	5.4†		
$[Cu(NH_3)_4]^{2+}$	1.9×10^4	3.9×10^3	1.0×10^3	1.5×10^2		
$[HgCl_4]^{2-}$	5.5×10^6	3.0×10^6	7.1	10†		
$[Ni(NCS)_3]^-$	1.6×10^2	3.0	1.6			
$[Ni(NH_3)_6]^{2+}$	6.3×10^2	1.7×10^2	50	16	6.0	1.1
$[Ni(en)_3]^{2+}$	2.0×10^7	1.0×10^6	2.0×10^4			
$[SnCl_4]^{2-}$	30	5.0	6.0×10^{-1}	2.0×10^{-1}		
$[Zn(NH_3)_4]^{2+}$	2.3×10^2	2.8×10^2†	3.2×10^2†	1.4×10^2		

* True equilibrium constants involve *activity* terms and are quantitatively correct with molar concentration values only for very dilute solutions. The more commonly employed *concentration* constants are quantitatively correct only for solutions of concentrations near those at which the constants were measured. In actual practice, however, approximations are generally satisfactory for semiquantitative work.

† Note deviations from general trend $K_1 > K_2 > K_3$, etc.

Equilibrium approximations The equilibria involving complex ions may be treated in terms of stability (formation) or instability (dissociation) constants. These terms are reciprocally related as

$$K_{\text{instab}} = \frac{1}{K_{\text{stab}}} \tag{6-1}$$

and either approach may be selected for a given situation as dictated by convenience.

Most frequently in the analytical chemistry of aqueous solutions one deals with a relatively large excess of the ligand in question. Under these conditions, the equilibria may be approximated by the overall equation and *net* equilibrium constant,* for example,

Stepwise

$$Ag^+ + NH_3 \rightleftharpoons [Ag(NH_3)]^+$$
$$[Ag(NH_3)]^+ + NH_3 \rightleftharpoons [Ag(NH_3)_2]^+$$

Overall equation

$$Ag^+ + 2NH_3 \rightleftharpoons [Ag(NH_3)_2]^+$$

for which†

$$K_{\text{stab}} = \frac{C_{[Ag(NH_3)_2]^+}}{C_{Ag^+} \times C^2_{NH_3}} \tag{6-2}$$

$$K_{\text{stab}} = K_1 \times K_2 = \frac{C_{[Ag(NH_3)]^+}}{C_{Ag^+} \times C_{NH_3}} \times \frac{C_{[Ag(NH_3)_2]^+}}{C_{[Ag(NH_3)]^+} \times C_{NH_3}} \tag{6-3}$$

Equilibria of these types may be handled in ways similar to simpler acid-base or solubility equilibria, but care must be exercised in assigning values to various concentration terms. The stepwise nature of the equilibria must be remembered. For example, for

$$Ag^+ + 2NH_3 \rightleftharpoons [Ag(NH_3)_2]^+$$

one cannot safely set $C_{NH_3} = 2 \times C_{Ag^+}$.‡ It is sometimes necessary to consult values of stepwise equilibrium constants in order to make reasonably accurate approximations. Simplified calculations are more readily justified in cases where the ligand is in reasonable excess so that the "complete" complex is the major species.

*$K_{\text{net}} = K_1 \times K_2 \times K_3 \times \cdots$ (= β_n, Appendix **F**)

† In a similar way one could write

$$[Ag(NH_3)_2]^+ \rightleftharpoons Ag^+ + 2NH_3$$

for which

$$K_{\text{instab}} = \frac{C_{Ag^+} \times C^2_{NH_3}}{C_{[Ag(NH_3)_2]^+}}$$

‡ See Example 6-1.

AMPHOTERIC SUBSTANCES

Chemical species that may act as either acids or bases are said to be amphoteric. Strictly speaking, water itself can be considered amphoteric.

$$\underset{\text{Acid}}{\underline{H_2O}} + \underset{\text{Base}}{NH_3} \rightleftharpoons \underset{\text{Acid}}{NH_4^+} + \underset{\text{Base}}{OH^-}$$

$$\underset{\text{Base}}{\underline{H_2O}} + \underset{\text{Acid}}{HF} \rightleftharpoons \underset{\text{Acid}}{H_3O^+} + \underset{\text{Base}}{F^-}$$

However, the term amphoteric is more commonly reserved for species other than water. These may be relative simple ions, for example,

$$\underset{\text{Base}}{HCO_3^-} + H^+ \rightleftharpoons H_2CO_3 \rightleftharpoons H_2O + CO_2$$

$$\underset{\text{Acid}}{HCO_3^-} + OH^- \rightleftharpoons H_2O + CO_3^{2-}$$

or very complicated macromolecular ions such as proteins or nucleic acids.

Some of the more common complex ions may be involved in amphoteric behavior. The most notable examples in routine analytical chemistry are the hydroxo complexes of zinc, aluminum, and chromium(III) (Fig. 6-6) and the thio complexes of arsenic, antimony, and tin(IV).

It is apparent from Fig. 6-6 that in discussing the amphoterism of hydroxo species another aspect of coordination chemistry is being dealt with. The same is true of amphoterism based upon any other chemical group. The correlation with acid-base behavior is apparent in the Lewis sense (page 40) that an electron-pair acceptor (e.g., a cation) is an acid, and an electron-pair donor (or ligand) is a base.

$$Zn^{2+} \underset{H^+}{\overset{OH^-}{\rightleftharpoons}} [Zn(OH)]^+ \underset{H^+}{\overset{OH^-}{\rightleftharpoons}} Zn(OH)_2 \underset{H^+}{\overset{OH^-}{\rightleftharpoons}} [Zn(OH)_3]^- \underset{H^+}{\overset{OH^-}{\rightleftharpoons}} [Zn(OH)_4]^{2-}$$

$$Al^{3+} \underset{H^+}{\overset{OH^-}{\rightleftharpoons}} [Al(OH)]^{2+} \underset{H^+}{\overset{OH^-}{\rightleftharpoons}} [Al(OH)_2]^+ \underset{H^+}{\overset{OH^-}{\rightleftharpoons}} Al(OH)_3 \underset{H^+}{\overset{OH^-}{\rightleftharpoons}} [Al(OH)_4]^-$$

(Cr^{3+} behaves similarly to Al^{3+}.)

Fig. 6-6 Common amphoteric hydroxides. (Coordinated water molecules are not indicated.)

Hydroxo compounds It is proper to inquire as to what renders a hydroxo compound an acid, a base, or an amphoteric substance. This problem may be approached in terms of the linkage

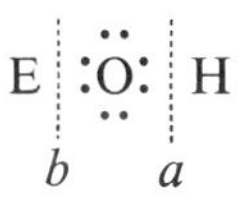

$$\underset{b}{E \,\vdots}\, :\ddot{\underset{..}{O}}: \underset{a}{\vdots\, H}$$

characteristic of all —OH-type compounds. If, when this material reacts, the bond breaks at *a*, the material is providing protons and acting as an acid. This type of behavior is favored by a high degree of polarity in the O—H bond, which in turn must be the result of a general displacement of electrons away from H and toward E. Any property of E that increases the attraction of that material for electrons then favors acidic behavior. Obviously, comparatively small size and comparatively high charge* are properties of E favoring acidity. Conversely, if the bond in this compound breaks at *b*, the material is providing hydroxide ions to neutralize protons and is behaving as a base. This type of behavior is favored by ionic character in the E—O bond, which in turn must be the result of lack of attraction of the element E for electrons. Obviously, comparatively large size and comparatively small charge are properties of E that promote base behavior. It follows, then, that when size and charge have intermediate values, amphoterism results.

Some examples should illustrate these generalizations. For example, for the second short series of the periodic system, the hydrogen-oxygen compounds and their properties may be summarized as

I	II	III	IV	V	VI	VII
$NaOH$	$Mg(OH)_2$	$Al(OH)_3$	$Si(OH)_4$	$PO(OH)_3$	$SO_2(OH)_2$	ClO_3OH
Strongly basic	Basic	Amphoteric	Weakly acidic	Acidic	Strongly acidic	Very strongly acidic

Change from basic to acidic behavior parallels decrease in size and increase in charge of the central element E very closely. The same line of reasoning is in agreement with the observed basic behavior of the compound KOH as opposed to the acidic character of the compound BrOH. Furthermore, it is helpful in predicting trends among the oxygen-hydrogen compounds of any individual element. Thus, for the element manganese

II	III	IV	VI	VII
$Mn(OH)_2$	$MnOOH$	MnO_2	$MnO_2(OH)_2$	MnO_3OH
Basic	Weakly basic	Amphoteric	Strongly acidic	Very strongly acidic

the acidic character increases with increasing positive oxidation number of the manganese. It is invariably true that if a given element forms more than one oxide or oxygen-hydrogen compound, acidity increases as positive oxidation number of that element increases.

* Charge is perhaps best thought of here as an effective charge. It may arise from high positive oxidation number of the element E in the compound in question, from a high nuclear charge in this element, or from both of these.

Table 6–8 Acidic Characteristics of Sulfides

	*Product of treatment with**					
Sulfide	*Aqueous* NH_3	$(NH_4)_2S$	$(NH_4)_2S_2$	NaOH	Na_2S (NaOH)	Na_2S_2 (NaOH)
As_2S_3	$AsOS^-$†	AsS_2^-	AsS_4^{3-}	$AsOS^-$	AsS_2^-	AsS_4^{3-}
As_2S_5	AsO_3S^{3-}†	AsS_4^{3-}	AsS_4^{3-}	AsO_3S^{3-}	AsS_4^{3-}	AsS_4^{3-}
Sb_2S_3	$SbOS^-$†	SbS_2^-	SbS_4^{3-}	$SbOS^-$	SbS_2^-	SbS_4^{3-}
Sb_2S_5	SbO_3S^{3-}†	SbS_4^{3-}	SbS_4^{3-}	SbO_3S^{3-}	SbS_4^{3-}	SbS_4^{3-}
SnS	SnS	SnS	SnS_3^{2-}	SnS	SnS	SnS_3^{2-}
SnS_2	SnS_2	SnS_3^{2-}	SnS_3^{2-}	SnO_2S^{2-}†	SnS_3^{2-}	SnS_3^{2-}
HgS	HgS	HgS	HgS	HgS	HgS_2^{2-}	HgS_2^{2-}

* In the sense that all these ions are complex ions, square brackets could be used to enclose their formulas.

† Products obtained are probably mixtures, for example, $SbO_2^- + SbS_2^-$, $AsO_4^{3-} + AsS_4^{3-}$, etc.

Basic properties among oxides and hydroxides are characteristic of compounds derived from large metal ions of relatively small charge (for example, 1+, 2+, 3+), whereas acidic properties are characteristic of compounds derived from nonmetals or metals in high states of oxidation (for example, 5+, 6+, 7+). Amphoteric oxides are derived from the metalloids or metals in intermediate states of oxidation (for example, 3+, 4+, 5+).

Sulfides From the analogies existing between the chemistry of oxygen and sulfur, one might expect amphoterism among sulfides as well as oxides. That this is true is indicated by the equations‡

$$\underset{\text{Base}}{SnS_2} + \underset{\text{Acid}}{4H^+} \rightarrow Sn^{4+} + 2H_2S$$

$$\underset{\text{Acid}}{SnS_2} + \underset{\text{Base}}{S^{2-}} \rightarrow SnS_3^{2-}$$

Metal sulfides are fundamentally basic, as shown by their tendency to react with hydrogen ion, but reaction with the basic sulfide ion shows acidic character. Experimentally, it is found that the compounds **HgS**, As_2S_3, As_2S_5, Sb_2S_3, Sb_2S_5, and SnS_2 all dissolve in solutions containing excess sulfide ion (i.e., solutions of sodium or potassium sulfide). On the other hand, the sulfides of lead, bismuth, copper, cadmium, manganese, etc., are sufficiently basic in property to be insoluble under comparable conditions. The behavior of these sulfides is summarized in Table 6-8. It is apparent that bases other than sulfide ion can cause these materials to behave as acids.

‡ These equations are written in simplified form for purposes of illustration. Thus, the ion Sn^{4+} has no existence as such; rather, in the hydrochloric acid solution used to dissolve the sulfide, the ion $[SnCl_6]^{2-}$ is more probable.

The generalizations regarding acid-base behavior cited above for oxides appear to apply equally well to sulfides. Thus, in the series **As_2S_3–Sb_2S_3–Bi_2S_3**, the arsenic compound dissolves very readily in solutions containing sulfide ion, whereas the antimony compound dissolves less readily and the bismuth compound is insoluble. This increase in basic character parallels increase in size in the series As(III)–Sb(III)–Bi(III). Futhermore, increase in positive oxidation state of the metal again increases acidity, as is shown by the enhanced solubility in sulfide ion of the compound **Sb_2S_5** and by the fact that although tin(II) sulfide does not dissolve in this reagent tin(IV) sulfide is readily soluble.

It is apparent that the acidic properties of certain sulfides can be utilized to separate these substances from other more basic sulfides. This is an important laboratory operation in qualitative analysis.

Miscellaneous substances Examples of amphoterism are found among many other types of compounds. Thus, for cyanides, the equations

$$\underset{\text{Base}}{Fe(CN)_2} + \underset{\text{Acid}}{2H^+} \rightarrow Fe^{2+} + 2HCN\uparrow$$

and

$$\underset{\text{Acid}}{Fe(CN)_2} + \underset{\text{Base}}{4CN^-} \rightarrow [Fe(CN)_6]^{4-}$$

show clearly the existence of both basic and acidic properties. Similarly, for fluorides, the equations

$$\underset{\text{Base}}{AlF_3} + \underset{\text{Acid}}{3H^+} \rightarrow Al^{3+} + 3HF$$

and

$$\underset{\text{Acid}}{AlF_3} + \underset{\text{Base}}{3F^-} \rightarrow [AlF_6]^{3-}$$

demonstrate the same characteristics. Commonly, however, among these and other similar examples, the term amphoteric is not applied. This is largely a matter of custom more than anything else.

APPLICATIONS OF COMPLEX SUBSTANCES TO QUALITATIVE ANALYSIS

Reactions involving the formation or destruction of complex species are frequently useful in laboratory work in qualitative analysis. In general, these reactions are employed either to effect separations or to permit identifications.

Separations Reactions of this type may be brought about if one of the ions in question forms a complex ion under a given set of conditions, whereas the other ion does not. Usually this involves either the selective precipitation of one ion or the selective dissolution of one solid. Thus, bismuth(III) ion does not form ammine complexes, whereas copper(II) and cadmium ions do. Hence, if a solution containing all three of these cations is treated with excess aqueous ammonia, bismuth is precipitated as the hydroxide $Bi(OH)_3$, and the other two ions give the ammine complex ions $[Cu(H_2O)_2(NH_3)_4]^{2+}$ and $[Cd(NH_3)_4]^{2+}$ which remain in solution. Similarly, cobalt(II), nickel, and zinc ions all give soluble ammine complex ions with aqueous ammonia; whereas iron(II or III), aluminum, and manganese(II) ions are precipitated as hydroxides under the identical conditions. Ions such as tin(II or IV), lead(II), zinc, or aluminum give soluble complex species with excess hydroxide ion; whereas ions such as cadmium, bismuth(III), iron(II or III), or cobalt(II) form insoluble hydroxides under comparable conditions. Silver chloride dissolves in aqueous ammonia, whereas mercury(I) chloride does not; so aqueous ammonia can be used to separate these two water-insoluble compounds. The sulfides of arsenic(III or V), antimony(III or V), and tin(IV) dissolve in solutions containing sulfide ion because of the formation of soluble sulfo-complex ions. Similar complexes are not formed by lead(II), bismuth(III), copper(II), and cadmium ions; so the sulfides of these cations are insoluble in sulfide ion. Other examples, which may be encountered in laboratory work, are equally useful.

It is sometimes possible to effect separations by taking advantage of differences in the stabilities of two or more complex ions. Thus, the cyano complex $[Cd(CN)_4]^{2-}$ provides a sufficiently high equilibrium concentration of cadmium ion, even in solutions containing excess cyanide ion, to permit precipitation of cadmium sulfide with hydrogen sulfide. On the other hand, the cyano complex $[Cu(H_2O)(CN)_3]^{2-}$ gives too small a concentration of copper(I) ion to precipitate copper(I) sulfide under comparable conditions. In the presence of a large excess of chloride ion the concentration of cadmium ion in equilibrium with the complex $[CdCl_4]^{2-}$ is too small to permit precipitation of cadmium sulfide, whereas the concentration of copper(II) ion in equilibrium with the complex $[CuCl_4]^{2-}$ is sufficient to permit precipitation of copper(II) sulfide.

Identifications Inasmuch as many complex species are highly colored, many identifications based upon this property are possible. Some of the more useful species are listed in Table 6-5. Often these colors are sufficiently selective for one ion to be identified in this way in the presence of many others. Indeed, there exists for nearly every ion some reagent (usually organic) that will give a colored complex useful for purposes of identification. An entire approach to qualitative analysis based upon the use of selective organic reagents has been developed, but the complexity and highly specific character of these reactions limit sharply the number of such reagents that should be employed in a systematic study of the properties of ions.

ILLUSTRATIVE EXAMPLES

Example 6-1 A solution is prepared by mixing equal volumes of 0.02 *M* $AgNO_3$ and 0.04 *M* NH_3. (*a*) Estimate C_{Ag^+}, C_{NH_3}, $C_{[Ag(NH_3)_2]^+}$, and pH, using the overall instability constant for $[Ag(NH_3)_2]^+$ of 6×10^{-8} and K_b for $NH_3 = 1.8 \times 10^{-5}$. (*b*) The actual pH of the solution is determined to be 10.5. Estimate the actual concentrations of Ag^+, NH_3, and $[Ag(NH_3)_2]^+$. (Hint: See Table 6-7.) (*c*) Account for any discrepancies between values calculated for parts *a* and *b*.

Solution (*a*) For $[Ag(NH_3)_2]^+ \rightleftharpoons Ag^+ + 2NH_3$

$$K_{\text{instab}} = \frac{C_{Ag^+} \times C^2_{NH_3}}{C_{[Ag(NH_3)_2]^+}} = 6 \times 10^{-8}$$

With only the data given, one must assume $C_{[Ag(NH_3)]^+}$ is negligible. Hence,

$$C_{NH_3} = 2 \times C_{Ag^+}$$

and, since K_{instab} is very small, nearly all the original Ag^+ and NH_3 are converted to $[Ag(NH_3)_2]^+$, so that

$$C_{[Ag(NH_3)_2]^+} \approx 0.01 \text{ mole liter}^{-1}$$

Then

$$6 \times 10^{-8} = \frac{(C_{Ag^+}) \times (2C_{Ag^+})^2}{10^{-2}}$$

from which

$$C_{Ag^+} = \sqrt[3]{1.5 \times 10^{-10}} = 5.3 \times 10^{-4} \text{ mole liter}^{-1}$$

and

$$C_{NH_3} = 2 \times C_{Ag^+} = 1.06 \times 10^{-3} \text{ mole liter}^{-1}$$

$$K_b = \frac{C_{NH_4^+} \times C_{OH^-}}{C_{NH_3}} = 1.8 \times 10^{-5}$$

$$C_{NH_4^+} = C_{OH^-}$$

so that

$$1.8 \times 10^{-5} = \frac{C^2_{OH^-}}{1.06 \times 10^{-3}}$$

$$C_{OH^-} = \sqrt[2]{1.9 \times 10^{-8}} = 1.38 \times 10^{-4} \text{ mole liter}^{-1}$$

$$\text{pOH} = -\log(1.38 \times 10^{-4}) = 3.86$$

$$\text{pH} = 14.00 - \text{pOH}= = 10.14 \ (\sim 10.1)$$

(*b*) if pH = 10.5, then pOH = 14.00 − pH = 3.5 and $C_{OH^-} = 3.2 \times 10^{-4}$ mole liter^{-1}

Then, from K_b for NH_3, since $C_{OH^-} = C_{NH_4^+}$,

$$1.8 \times 10^{-5} = \frac{(3.2 \times 10^{-4})^2}{C_{NH_3}}$$

$$C_{NH_3} = \frac{(3.2 \times 10^{-4})^2}{1.8 \times 10^{-5}} = 5.7 \times 10^{-3} \text{ mole liter}^{-1}$$

Since C_{NH_3} was originally 0.04 M and the two reactant solutions were mixed in equal volumes,

$$C_{\text{all NH}_3\text{-containing species}} = 0.02 = C_{NH_3} + C_{[Ag(NH_3)]^+} + 2 \times C_{[Ag(NH_3)_2]^+}$$

$$0.02 = 5.7 \times 10^{-3} + C_{[Ag(NH)_3]^+} + 2 \times C_{[Ag(NH_3)_2]^+}$$

from which,

$$C_{[Ag(NH_3)]^+} = 1.43 \times 10^{-2} - 2 \times C_{[Ag(NH_3)_2]^+}$$

From Table 6-7 for

$$[Ag(NH_3)_2]^+ \rightleftharpoons [Ag(NH_3)]^+ + NH_3$$

$$K_{\text{instab, 1}} = \frac{1}{K_{\text{stab, 2}}} = \frac{1}{8 \times 10^{-3}} = 1.25 \times 10^{-4}$$

Then,

$$1.25 \times 10^{-4} = \frac{C_{[Ag(NH_3)]^+} \times C_{NH_3}}{C_{[Ag(NH_3)_2]^+}}$$

$$1.25 \times 10^{-4} = \frac{(1.43 \times 10^{-2} - 2 \times C_{[Ag(NH_3)_2]^+})(5.7 \times 10^{-3})}{C_{[Ag(NH_3)_2]^+}}$$

$$(1.25 \times 10^{-4})(C_{[Ag(NH_3)_2]^+}) = 8.15 \times 10^{-5} - (1.14 \times 10^{-2})(C_{[Ag(NH_3)_2]^+})$$

$$(1.15 \times 10^{-2})(C_{[Ag(NH_3)_2]^+}) = 8.15 \times 10^{-5}$$

$$C_{[Ag(NH_3)_2]^+} = \frac{8.15 \times 10^{-5}}{1.15 \times 10^{-2}} = 7.1 \times 10^{-3} \text{ mole liter}^{-1}$$

Now, using the overall instability constant,

$$6 \times 10^{-8} = \frac{C_{Ag^+} \times C_{NH_3}^2}{C_{[Ag(NH_3)_2]^+}}$$

$$6 \times 10^{-8} = \frac{(C_{Ag^+})(5.7 \times 10^{-3})^2}{7.1 \times 10^{-3}}$$

$$C_{Ag^+} = \frac{7.1 \times 10^{-3} \times 6 \times 10^{-8}}{(5.7 \times 10^{-3})^2}$$

$$C_{Ag^+} = 1.3 \times 10^{-5} \text{ mole liter}^{-1}$$

(*c*) Comparison then shows

	[*Neglecting* $C_{[Ag(NH_3)]^+}$], *mole liter*$^{-1}$	[*Considering* $C_{[Ag(NH_3)]^+}$], *mole liter*$^{-1}$
C_{Ag^+}	5.3×10^{-4}	1.3×10^{-5}
$C_{[Ag(NH_3)_2]^+}$	10^{-2}	7.1×10^{-3}
C_{NH_3}	1.06×10^{-3}	5.7×10^{-3}

The discrepancies result from neglecting $C_{[Ag(NH_3)]^+}$. *Use of any overall equilibrium constant should be considered only a gross approximation, the accuracy of which depends on relative values of stepwise equilibrium constants and on relative concentrations of species involved.*

Example 6-2 Estimate the $C_{Cu^{2+}}$ in a solution prepared by mixing equal volumes of 0.010 M $CuSO_4$ and 3.0 M NH_3.

Solution Since K_{instab} for

$$[Cu(NH_3)_4]^{2+}* \left[= \frac{1}{K_1 \times K_2 \times K_3 \times K_4} \qquad \text{(Table 6-7)}\right]$$

is so small and NH_3 is present in large excess, one may safely assume

$$C_{[Cu(NH_3)_4]^{2+}} \approx 0.01\ M$$

and

$$C_{NH_3} \approx 3.0 - 0.04 \approx 3.0$$

Then, for the equilibrium

$$[Cu(NH_3)_4]^{2+} \rightleftharpoons Cu^{2+} + 4NH_3$$

$$K_{instab} = \frac{C_{Cu^{2+}} \times C^4_{NH_3}}{C_{[Cu(NH_3)_4]^{2+}}}$$

the overall

$$K_{instab} = \frac{1}{K_{stab,1} \times K_{stab,2} \times K_{stab,3} \times K_{stab,4}}$$

From Table 6-7

$$K_{instab} = \frac{1}{1.9 \times 10^4 \times 3.9 \times 10^3 \times 1.0 \times 10^3 \times 1.5 \times 10^2} = 1 \times 10^{-13}$$

Hence,

$$1 \times 10^{-13} = \frac{C_{Cu^{2+}} \times (3.0)^4}{10^{-2}}$$

$$C_{Cu^{2+}} = \frac{1 \times 10^{-15}}{81} = 1.2 \times 10^{-17} \text{ mole liter}^{-1}$$

Example 6-3 Will AgCl precipitate from a solution prepared by mixing equal volumes of 0.03 M $AgNO_3$, 6.0 M NH_3, and 0.09 M NaCl?

Solution The solubility-product constant for AgCl is 1.7×10^{-10}. After mixing, if no reaction occurs, $C_{Cl^-} = 0.03$ mole liter^{-1}.

Then, AgCl will precipitate if

$$C_{Ag^+} > \frac{K_{sp}}{C_{Cl^-}}$$

* Simplified expression for $[Cu(H_2O)_2(NH_3)_4]^{2+}$

That is, if

$$C_{Ag^+} > \left(\frac{1.7 \times 10^{-10}}{0.03} = 5.7 \times 10^{-9} \text{ mole liter}^{-1}\right)$$

Since C_{NH_3} is initially $\gg C_{Ag^+}$, one may safely consider that nearly all Ag^+ is converted to $[Ag(NH_3)_2]^+$, with $C_{[Ag(NH_3)]^+}$ negligible. Thus, at equilibrium for

$$[Ag(NH_3)_2]^+ \rightleftharpoons Ag^+ + 2NH_3$$

$$C_{[Ag(NH_3)_2]^+} \approx 0.01\ M$$

and

$$C_{NH_3} \approx 2.0 \text{ mole liter}^{-1}$$

Then,

$$K_{instab} = 6 \times 10^{-8} = \frac{C_{Ag^+} \times (2.0)^2}{10^{-2}}$$

from which

$$C_{Ag^+} = \frac{6 \times 10^{-8} \times 10^{-2}}{4} = 1.5 \times 10^{-10} \text{ mole liter}^{-1}$$

Since $1.5 \times 10^{-10} < 5.7 \times 10^{-9}$, AgCl will not precipitate.

EXERCISES

6-1. Discuss the relationship between charge density and hydration of ions. Predict the relative strengths of hydration for Na^+, Mg^{2+}, and Ca^{2+}.

6-2. A certain crystalline hydrate is found by x-ray and neutron-diffraction studies to contain a square-planar $[M(H_2O)_4]^{2+}$ complex ion. Suggest at least two reasons why the aqueous M^{2+} ion cannot definitely be described as a rigid square-planar complex.

6-3. Arrange the following in order of increasing strength as Lewis acids, considering H_2O as the Lewis base in each case. Al^{3+}, Ba^{2+}, Fe^{3+}, Na^+. Explain.

6-4. What is the distinction between the terms "Lewis base" and "ligand"? Give examples.

6-5. Arrange the following in order of increasing strength as Lewis bases, considering Cu^{2+} as the Lewis acid in each case.

H_2O, NH_3, Cl^-, CN^-.

Would the same order be used if the Lewis acid were Na^+? Explain.

6-6. Give an example of each of the following:

(*a*) A unidentate ligand
(*b*) A bidentate ligand
(*c*) A chelate ligand

6-7. Give an example of a structure involving chelation by $EDTA^{4-}$.

6-8. Give correct formulas for:

(*a*) Tetrachloroaurate(III) ion
(*b*) Dichlorotetraamminecobalt(III) ion
(*c*) Trihydroxotriaquochromium(III)
(*d*) Diaquobis(ethylenediamine)copper(II) ion
(*e*) Bromochlorotetraammineplatinum(IV) ion
(*f*) Potassium hexacyanoferrate(II)
(*g*) Ethylenediaminetetraacetatocobaltate(III) ion

6-9. Give correct names for:

(*a*) $[Ag(S_2O_3)_2]^{3-}$
(*b*) $[Co(NH_3)_5NCS]^{2+}$
(*c*) $[Co(NH_3)_2(NO_2)_4]^-$
(*d*) $[Fe(CO)_4I_2]$
(*e*) $[Ni(en)_3]Cl_2$
(*f*) $K[Pt(NH_3)Cl_5]$
(*g*) $[Pt(en)(NO_2)ClBrI]$

6-10. Suggest reasonable geometries for the complexes in Exercises 6-8 and 6-9. Draw structural formulas for 6-8*a*, *b*, *c* and 6-9*a*, *c*, *e*.

6-11. The bond angles in $[Co(NH_3)_2(NO_2)_4]^-$ in its crystalline potassium salt are those corresponding to the octahedral structure. Explain why it is probably an oversimplification to consider the aqueous complex ion as a simple symmetrical octahedron.

6-12. Draw electron-dot representations of:

(*a*) $[Ag(CN)_2]^-$
(*b*) $[Cu(CN)_4]^{2-}$
(*c*) $[Fe(CN)_6]^{3-}$
(*d*) $[Fe(CN)_6]^{4-}$

Which, if any, of these complexes contain unpaired electrons? Which, if any, of these complexes do not conform to the simple octet arrangement of electrons? What types of information are not indicated by electron-dot formulas?

6-13. What geometry would correspond to each of the following orbital descriptions?

(*a*) sp^2
(*b*) sp^3
(*c*) d^2sp^3
(*d*) dsp^3

6-14. Write the stepwise equilibrium equations and stepwise equilibrium-constant expressions for:

(*a*) Dissociation of $[Cu(NH_3)_4]^{2+}$
(*b*) Formation of $[Fe(CN)_6]^{3-}$

6-15. Describe conditions under which overall stability constants cannot be used to make reasonably accurate calculations of concentrations of solution components involving complex ions.

6-16. Estimate the $C_{Ni^{2+}}$ in a solution prepared by mixing equal volumes of 0.01 *M* $NiSO_4$ and 0.5 *M* NH_3.

6-17. Estimate the pH of a solution prepared by dissolving 5.0 g of $[Ni(NH_3)_6]Cl_2$ and 2.0 g of $Ni(NO_3)_2$ in 100 ml of water, assuming negligible concentrations of intermediate species.

6-18. When 6.0 *M* HCl is saturated with H_2S gas the C_{H_2S} is approximately 0.09 *M*. Will CuS precipitate from a solution prepared by mixing 1.0 ml of 0.02 *M* $CuSO_4$ with 1.0 ml of 12.0 *M* HCl and saturating the solution with H_2S?

6-19. Will silver chloride precipitate from a solution prepared by mixing equal volumes of 0.03 *M* silver nitrate, 3.0 *M* ammonia, and 0.03 *M* ammonium chloride?

6-20. If 0.30 g of $Cu(OH)_2$ is added to 100 ml of 1.0 *M* ammonia, will any solid remain at equilibrium?

SUGGESTED SUPPLEMENTARY READING

Basolo, F., and Johnson, R.: "Coordination Chemistry," Benjamin, New York, 1964 (paperback).
Hunt, J. P.: "Metal Ions in Aqueous Solution," Benjamin, New York, 1965 (paperback).
Murmann, R. K.: "Inorganic Complex Compounds," Reinhold, New York, 1964 (paperback).
Quagliano, J. V., and Vallarino, L. M.: "Coordination Chemistry," D. C. Heath, Boston, 1969 (paperback).

7
Electron-Transfer Reactions

OXIDATION AND REDUCTION

When any chemical change occurs, electrons are redistributed to some extent relative to the nuclei of participating particles. The extent to which this redistribution occurs varies widely.

Consider as an example a hypothetical case in which a single isolated lithium atom reacts with an isolated atom of fluorine (Fig. 7-1). If we define *oxidation* as *electron loss* and *reduction* as *electron gain*, then the lithium atom has been oxidized and the fluorine atom has been reduced. The realities of this reaction may be somewhat less simple, however, than this description indicates. It would be rare to study a reaction even closely approaching the situation of completely isolated atoms and ions. Much more likely would be the case in which gaseous fluorine reacts with solid metallic lithium to form a white crystalline substance, lithium fluoride (Fig. 7-2). Chemists still refer to such a reaction as an electron transfer or oxidation-reduction reaction, but it is no longer a simple loss of one electron by a lithium atom and gain of one electron by a fluorine atom.

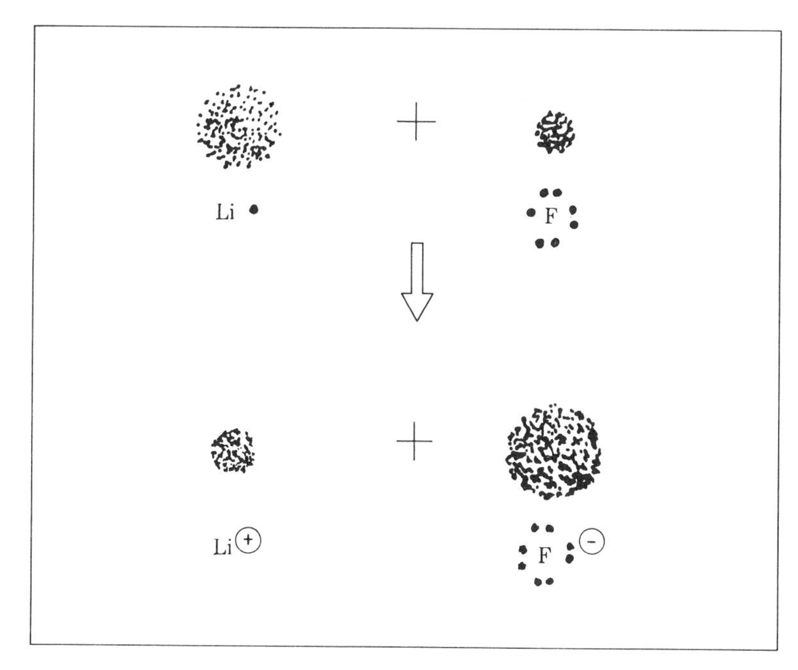

Fig. 7-1 Simplified electron-transfer reaction.

Charge on "Li^+" and "F^-" ions in the crystal lattice would be less than a full charge (1.6×10^{-19} coulomb) because of some electron sharing among adjacent ions. "Complete" electron transfer has not occurred.

Since the distinction between oxidation-reduction (electron transfer) and metathetical (no electron transfer) reactions is less precise on an atomic scale than one might generally realize, what arbitrary rules are followed to classify reactions? Experimentally, chemists consider a reaction to involve oxidation and reduction whenever that reaction can be made to produce electron flow through an external circuit. The science of electrochemistry, which is concerned with such systems, is discussed in a later section. It is obviously not feasible to perform every chemical reaction personally to see if current can be produced and, indeed, many reactions that involve oxidation and reduction would require very sensitive methods of current detection. Fortunately, there is a simple, but arbitrary, way of classifying reactions as oxidation-reduction, i.e., if a change in oxidation state (or oxidation number) occurs. The related material on the assignment of oxidation numbers and the balancing of oxidation-reduction equations, already presented in Chap. 2, should be reviewed at this point.

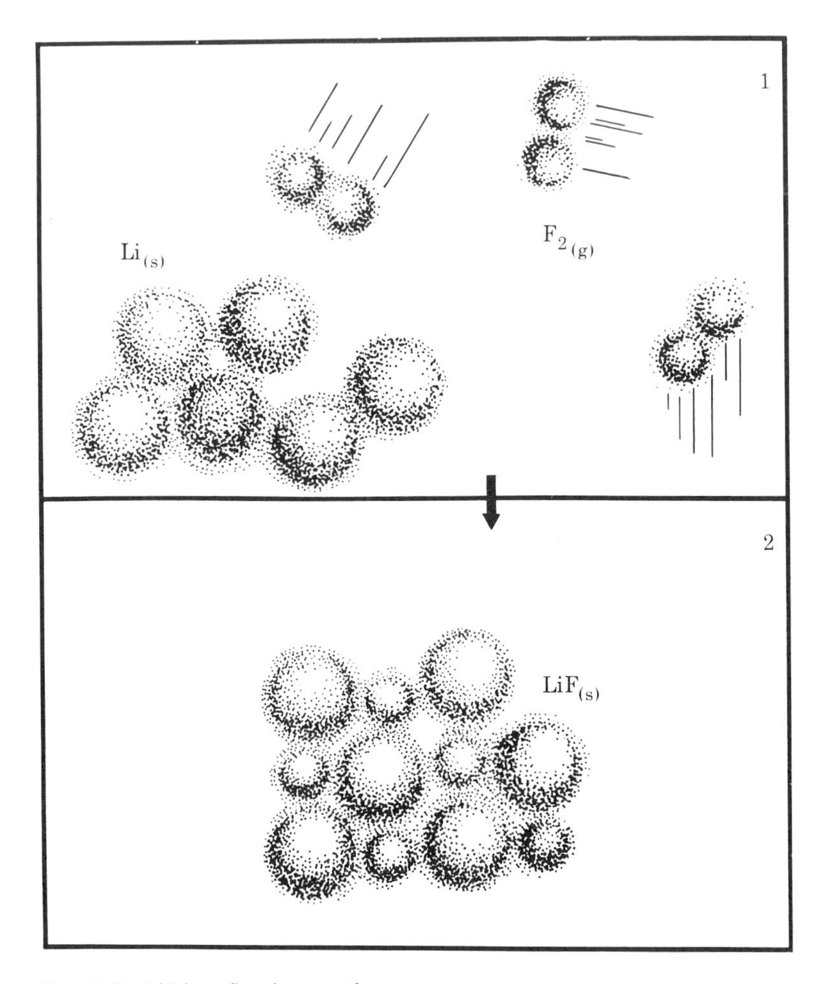

Fig. 7-2 Lithium-fluorine reaction.

Half-reactions Electron-transfer processes can be described by means of "half-reactions," one of which denotes electron loss (oxidation) and the other electron gain (reduction). Such a description is quite artificial for direct transfer processes involving collisions between reactants and products, since in these cases it is unlikely that a "free" electron actually "jumps" from one particle to another.

Consider as an example the reaction which occurs when a piece of zinc metal is placed in a 1.0 *M* solution of copper(II) sulfate (Fig. 7-3). The net process can be described by the equation:

$$\mathbf{Zn}(s) + Cu^{2+}(aq) \rightarrow Zn^{2+}(aq) + \mathbf{Cu}(s)$$

or in terms of two half-reactions as

$$\mathbf{Zn}(s) = \mathrm{Zn}^{2+}(aq) + 2e^-$$

$$2e^- + \mathrm{Cu}^{2+}(aq) = \mathbf{Cu}(s)$$

The first equation seems, perhaps, more appropriate since it shows simply that zinc and copper(II) ions are consumed while zinc(II) ions and metallic copper are produced. The half-reactions would appear more descriptive of an indirect electron-transfer (Fig. 7-4) in which there is direct evidence that electrons move through some significant distance rather than being transferred by direct collision of reactant atoms or ions.

To maintain current flow, electrons and cations move toward the right and anions toward the left for the system illustrated in Fig. 7-4. The "salt bridge" contains ions in solution (for example, K^+ and Cl^-) which migrate when the electrical circuit is closed. This type of arrangement minimizes direct reaction of Cu^{2+} with Zn.

Equations for half-reactions are useful for any type of oxidation-reduction process, whether electron transfer is direct or indirect. Again, this approach is a valuable way of "bookkeeping" and may not necessarily be descriptive of an actual reaction mechanism. The use of equations for half-reactions in balancing equations is discussed in Chap. 2, which discussion should be reviewed. Their value in electrochemical situations is discussed in a later section.

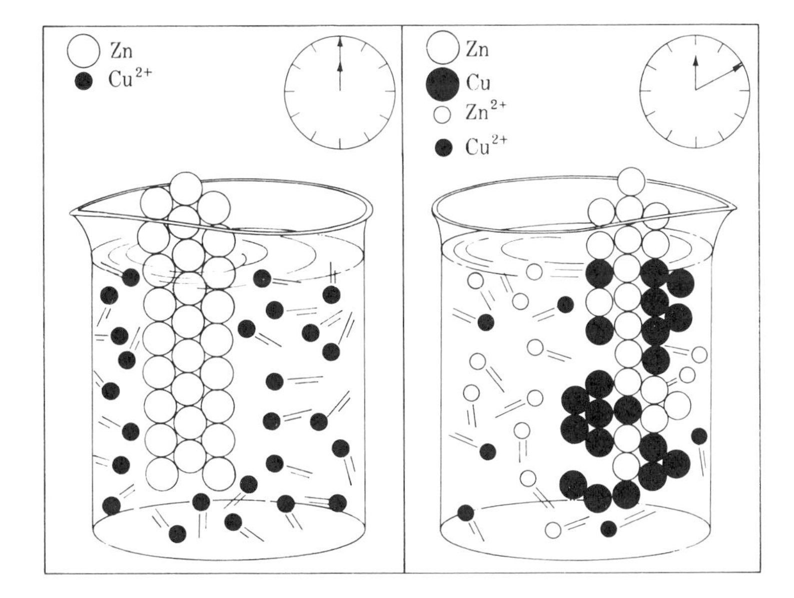

Fig. 7-3 Direct electron transfer.

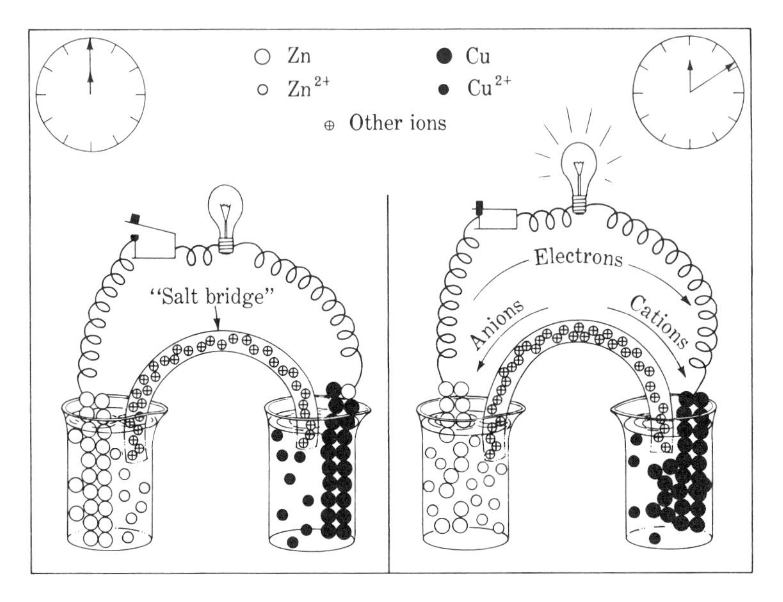

Fig. 7-4 Indirect electron transfer.

Oxidation-reduction equilibria Consider an electrochemical cell similar to that diagrammed in Fig. 7-4. Suppose in this case that each piece of metal consists of 1 mole of the metal (65.4 g of zinc and 63.5 g of copper) and that they are immersed, respectively, in 100 ml of 1.0 *M* Zn^{2+} and 100 ml of 1.0 *M* Cu^{2+} solutions. Stoichiometric calculations from the equation for this process indicate that all the Cu^{2+} (0.1 mole) would be used up before more than 10 percent of the zinc metal was consumed. At this time, electron flow would cease, and the cell would be described as completely discharged. In actual practice, current flow stops before all of the Cu^{2+} ions have been converted to copper atoms, i.e., a small, but detectable, concentration of Cu^{2+} remains when the cell ceases to function. One of the conditions of chemical equilibrium is observed—*constant macroscopic properties.* Removal of the copper strip and replacement by a different piece of copper does not alter the concentration of Cu^{2+} in the system. If a new piece of copper containing a radioactive isotope of copper is used to replace the original electrode, the concentration of Cu^{2+} still remains constant, but the solution gradually accumulates radioactive Cu^{2+} ions. Thus, evidence is available indicating that the second condition of chemical equilibrium exists—*a dynamic situation involving submicroscopic particles.*

Oxidation-reduction reactions, like other chemical changes studied, may attain a condition of chemical equilibrium. In this case, the condition will be characterized by no *net* electron transfer. At equilibrium one can write for the process in question:

$$\mathbf{Zn}(s) + Cu^{2+}(aq) \rightleftharpoons Zn^{2+}(aq) + \mathbf{Cu}(s)$$

for which,

$$K_{\text{equil}} = \frac{C_{Zn^{2+}}}{C_{Cu^{2+}}}$$

Note that solids are omitted as in solubility equilibria and that this concentration ratio is an approximation like the other equilibrium expressions discussed earlier.

Like other systems in the equilibrium condition, oxidation-reduction equilibria can be altered by appropriate changes (Le Chatelier's principle). In the copper-zinc cell (Fig. 7-4), for example, addition of solid copper(II) sulfate to the solution surrounding the copper electrode would "rejuvenate" the cell. As the $CuSO_4$ dissolved, the concentration of Cu^{2+} would increase, and electron flow would resume until a new equilibrium was established. Addition of some reagent capable of removing Cu^{2+} from the equilibrium solution would also disturb the equilibrium. This time electron flow would be the reverse of that for the original cell as copper metal was converted to Cu^{2+} ions until a new equilibrium condition was reached.

ELECTROCHEMISTRY

When a copper wire is placed in a solution which is 1.0 M in silver nitrate, the solution begins to turn blue, crystals of silver begin to grow on the wire (Fig. 7-5), and heat is evolved.

The energy released is difficult to harness for useful work. Direct electron transfer is usually an inefficient energy source. Combustion reactions involving hydrocarbon fuels are examples of direct electron transfer. They represent one of the major problems of modern society in that they are inefficient both in an energy sense and in a chemical sense. In a typical electrical generator powered by oxidation of hydrocarbon fuel only about 35 percent of the theoretical free energy of the reaction is converted to electrical energy. In addition, the combustion reaction is usually chemically inefficient, producing carbon monoxide and other components of industrial "smog." The same reactants can be employed for a "fuel cell" (discussed later) involving indirect electron transfer with about double the efficiency of energy conversion and a significant decrease in unwanted chemical products. Unfortunately, the technology of fuel cell development has not yet reached the stage where it is practical to supplant the less efficient processes now used. Indirect electron transfer does promise hope for more efficient utilization of "fossil fuels" and elimination of much of the air pollution problem associated with their oxidation.

Fig. 7-5 Displacement of silver by copper.

Stoichiometry of electrochemical processes Historically, electrochemical experiments were among the earliest quantitative studies of chemical reactions, being particularly associated with the work of Michael Faraday around 1830. For chemical changes produced by electrical current (electrolysis reactions), it was observed that the quantity of chemical product obtained was directly related to the magnitude of the electrical current flow (amperes) and the time during which this flow was maintained. A fixed quantity of charge, known as a Faraday (9.65×10^4 coulombs), was found to be associated with the same degree of chemical change (on a *molar* scale) for all electrochemical reactions:

$$\text{No. Faradays} = \frac{\text{No. of moles of product}}{\text{No. of electrons transferred per unit particle of product}} \qquad (7\text{-}1)$$

The modern interpretation of electron-transfer reactions and the knowledge of the charge on an electron, as determined by the Millikan oil-drop experiment, allow one to see the origin of this magic number, 9.65×10^4 coulombs.

When silver is electroplated from a solution of silver ions, the half-reaction occurring at the internal cathode of the cell can be expressed as:

$$Ag^+(aq) + e^- \rightarrow \mathbf{Ag}(s)$$

This equation indicates that each silver ion requires one electron for conversion to an atom of elemental silver. Each electron, according to experimental determinations, is associated with a charge of 1.60×10^{-19} coulomb. Remember that one mole (gram formula weight) of any substance contains 6.02×10^{23} unit particles of substance (Avogadro's number). Then:

$$\frac{1.60 \times 10^{-19}\text{ coulomb}}{1e^-} \times \frac{1e^-}{1\cancel{Ag^+}} \times \frac{6.02 \times 10^{23}\,\cancel{Ag^+}}{1\text{ mole }Ag^+}$$
$$= 9.63 \times 10^4 \text{ coulombs mole}^{-1}\ Ag^+$$

(If one uses more nearly precise values for the charge on the electron, 1.602×10^{-19} coulomb, and for Avogadro's number, 6.023×10^{23} particles per mole, then the value of a Faraday is 9.649×10^4 coulombs per mole to four significant figures).

A number of useful calculations can be made with respect to electrochemical processes, using the basic ideas suggested by the preceding discussion, Eq. 7-1, and the equivalence

$$1 \text{ coulomb} = 1 \text{ amp-sec} \qquad (7\text{-}2)$$

Example 7-1 What mass of copper could be electroplated from a solution of copper(II) sulfate per minute using a constant current flow of 6.00 amp? (Assuming sufficient copper(II) sulfate and 100 percent efficiency).

Solution

$$Cu^{2+}(aq) + 2e^- \rightarrow \mathbf{Cu}(s)$$

(Hence, 2 "moles" e^- per mole Cu)

$$\text{Weight of Cu deposited} = \frac{60\ \cancel{\text{sec}}}{1 \text{ min}} \times \frac{6.00\ \cancel{\text{amps}}}{1} \times \frac{1\ \cancel{\text{coulomb}}}{1\ \cancel{\text{amp-sec}}}$$
$$\times \frac{1\ \cancel{\text{mole } e^-}}{9.65 \times 10^4\ \cancel{\text{coulombs}}} \times \frac{1\ \cancel{\text{mole Cu}}}{2\ \cancel{\text{moles } e^-}} \times \frac{63.5 \text{ g Cu}}{1\ \cancel{\text{mole Cu}}}$$

Approximating:

$$\frac{6 \times 10^1 \times 6 \times 6 \times 10^1}{10^5 \times 2} \approx 108 \times 10^{-3} \approx 1 \times 10^{-1}\text{g Cu min}^{-1}$$

Answer:

$$1.18 \times 10^{-1} \text{ g Cu min}^{-1}$$

Example 7-2 An electrochemical cell is made by immersing a 100-g bar of zinc in 100 ml of 0.10 *M* $Zn(NO_3)_2$ and a 100-g bar of copper in 100 ml of 0.10 *M* $Cu(NO_3)_2$, connecting the solutions by a salt bridge, and connecting the metal electrodes by an external circuit. If it were possible to use the cell in such a way to maintain a constant current of 4.0 ma (milliamps), how many minutes would the cell run before it became completely discharged? (Assume 100 percent efficiency; also assume $C_{Cu^{2+}}$ becomes negligible exactly at the time when the cell is completely discharged. Note that the situation described only roughly approximates that of a real cell.)

Solution

Half-reactions:

$$\mathbf{Zn}(s) \rightarrow Zn^{2+}(aq) + 2e^-$$

$$Cu^{2+}(aq) + 2e^- \rightarrow \mathbf{Cu}(s)$$

Initial quantities:

$$\frac{100\ \text{g Zn}}{1} \times \frac{1\ \text{mole Zn}}{65.4\ \text{g Zn}} = 1.53\ \text{moles Zn}$$

$$\frac{100\ \text{g Cu}}{1} \times \frac{1\ \text{mole Cu}}{63.5\ \text{g Cu}} = 1.57\ \text{moles Cu}$$

$$\frac{0.10\ \text{mole Zn}^{2+}}{1\ \text{liter}} \times \frac{1\ \text{liter}}{1{,}000\ \text{ml}} \times \frac{100\ \text{ml}}{1} = 0.010\ \text{mole Zn}^{2+}$$

$$\frac{0.10\ \text{mole Cu}^{2+}}{1\ \text{liter}} \times \frac{1\ \text{liter}}{1{,}000\ \text{ml}} \times \frac{100\ \text{ml}}{1} = 0.010\ \text{mole Cu}^{2+}$$

Hence, Cu^{2+} is the limiting reagent and the cell will discharge until 0.010 mole Cu^{2+} is consumed (under the assumptions given).

Then,

$$\text{time} = \frac{0.010\ \text{mole Cu}^{2+}}{1} \times \frac{2\ \text{moles}\ e^-}{1\ \text{mole Cu}^{2+}} \times \frac{9.65 \times 10^4\ \text{coulombs}}{1\ \text{mole}\ e^-}$$

$$\times \frac{1\ \text{amp-sec}}{1\ \text{coulomb}} \times \frac{1\ \text{min}}{60\ \text{sec}} \times \frac{1{,}000\ \text{ma}}{1\ \text{amp}} \times \frac{1}{4.0\ \text{ma}}$$

Approximating:

$$\frac{10^{-2} \times 2 \times 10^5 \times 10^3}{6 \times 10^1 \times 4} \approx 0.08 \times 10^5 \approx 8 \times 10^3\ \text{min}$$

Answer:

8.0×10^3 min

Example 7-3 A commercial operation plans to produce aluminum by an electrolysis process at the rate of 70 tons per week. What average current must be maintained if the operation is to work on a 40-hour week and anticipates a 70 percent efficiency for the process?

Solution

$$Al^{3+}(aq) + 3e^- \rightarrow \mathbf{Al}(s)$$

$$\text{Moles Al per week} = \frac{70\ \text{tons}}{1\ \text{week}} \times \frac{2{,}000\ \text{lb}}{1\ \text{ton}} \times \frac{454\ \text{g}}{1\ \text{lb}} \times \frac{1\ \text{mole}}{27\ \text{g}}$$

$$= 2.4 \times 10^6\ \text{moles week}^{-1}$$

for 100 percent efficiency:

$$\text{Current required} = \frac{2.4 \times 10^6\ \cancel{\text{moles Al}}}{1\ \cancel{\text{week}}} \times \frac{3\ \cancel{\text{moles } e^-}}{1\ \cancel{\text{mole Al}}} \times \frac{9.65 \times 10^4\ \cancel{\text{coulombs}}}{1\ \cancel{\text{mole } e^-}} \times \frac{1\ \text{amp}\ \cancel{\text{sec}}}{1\ \cancel{\text{coulomb}}} \times \frac{1\ \cancel{\text{week}}}{40\ \cancel{\text{hrs}}} \times \frac{1\ \cancel{\text{hr}}}{3{,}600\ \cancel{\text{sec}}}$$

$$= 4.8 \times 10^6\ \text{amp}$$

But, this represents the current which is effective and the process is only 70 percent efficient. Thus,

$$4.8 \times 10^6\ \text{amp} = 0.7\,y$$

(where y is the actual current required)
from which,

$$y = \frac{4.8 \times 10^6}{0.7} = 6.9 \times 10^6\ \text{amp}$$

Standard electrode potentials The tendency for an electron transfer reaction to occur can be measured experimentally for an electrochemical cell, under conditions such that a negligible current flow occurs, in terms of an electrical potential difference (voltage). Figure 7-6 illustrates such a measurement for a particular cell.

Since a voltage measurement requires a *difference* of potential between two electrodes, it is not possible to determine an absolute potential for a single half-cell. It is, of course, possible to compare half-cell electrode potentials by selecting an arbitrary reference electrode and *assigning* it a value against which other comparisons can be made. The universally

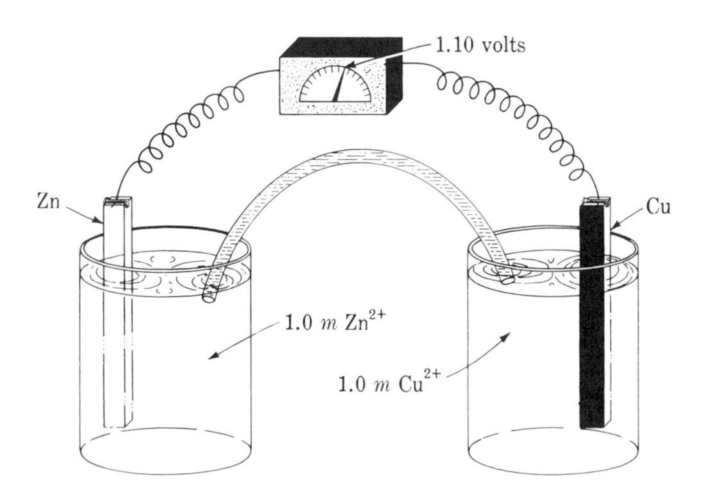

Fig. 7-6 Cell potential.

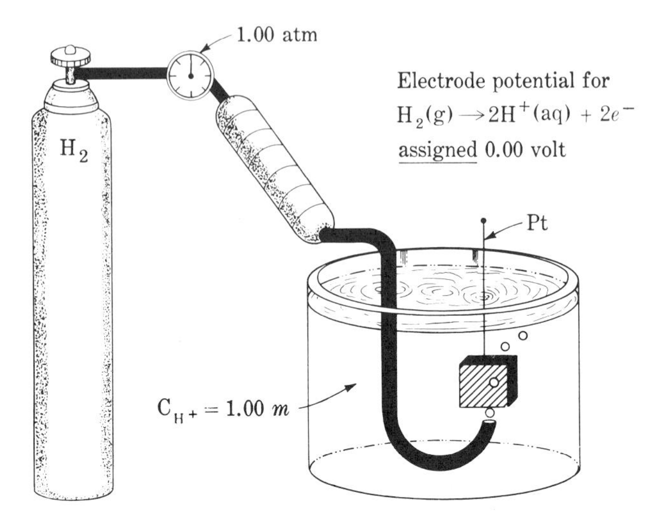

Fig. 7-7 Hydrogen electrode.

accepted reference standard is the hydrogen electrode (Fig. 7-7), which is assigned a potential of exactly 0 volts under conditions that require* a pressure of 1.00 atm of H_2 gas and an H^+ concentration of 1.00 *m* (approximately 1.00 *M*), at 25°C. Standard states for all substances are approximately: pure element for solids, 1.00 atm pressure for gases, and 1.00 *m* (~1.00 *M*) for species in solution. Now it is possible to measure cell voltages (Fig. 7-8)

* The exact requirement is expressed in terms of "thermodynamic" units which are only approximated by measured pressures and concentrations.

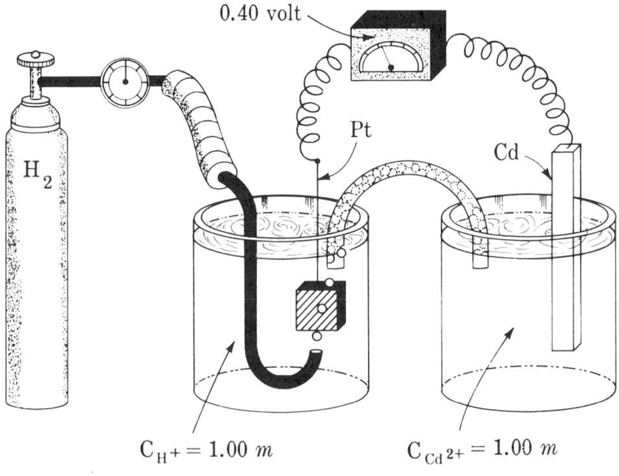

Fig. 7-8 Measurement of cell potential.

and, by comparisons, to assign a standard oxidation potential (Appendix G) for any half-reaction.

Standard oxidation potentials can be used to calculate a standard cell potential (E°) for a hypothetical cell utilizing a given overall reaction and thus to determine whether or not that reaction can proceed spontaneously. A prediction of this kind will, of course, apply only to the case in which the substances concerned are in their standard states. The prediction indicates *only* the tendency for a reaction to occur and says nothing about the rate to be expected. Within these limitations, predictions of this type are useful for qualitative studies.

Predicting spontaneity (Standard States)

Step 1. Describe the reaction by means of two simple half-reaction equations, one representing oxidation (electron loss) and one representing reduction (electron gain).

For example, for

$$6Br^-(aq) + 2Al^{3+}(aq) \rightarrow 3Br_2(l) + 2Al(s)$$
$$2Br^- \rightarrow Br_2 + 2e^-$$
$$Al^{3+} + 3e^- \rightarrow Al$$

Step 2. Calculate the theoretical cell potential involving the reactions given according to:

$$E^\circ_{cell} = E^\circ_{ox} + E^\circ_{red} \tag{7-3}$$

For example (from Appendix G)

$$2Br^- \rightarrow Br_2 + 2e^-, \; E^\circ_{ox} = -1.06 \quad \text{oxidation}$$
$$Al^{3+} + 3e^- \rightarrow Al, \; E^\circ_{red} = -1.66 \quad \text{reduction}$$

(from $Al \rightarrow Al^{3+} + 3e^-$, $E^\circ_{ox} = +1.66$)*

$$E^\circ_{cell} = -1.06 + (-1.66) = -2.72$$

Step 3. If the calculated cell potential is negative or zero, the reaction will not proceed spontaneously as described by the equation written (assuming standard states). It will proceed if E°_{cell} is positive, although the rate is not predicted.

For example, since the E°_{cell} calculated for $6Br^- + 2Al^{3+} \rightarrow 3Br_2 + 2Al$ is negative, this reaction will not proceed spontaneously under the conditions of standard states.

* Note that the electrode potential changes sign when the half-reaction equation is the reverse of that shown in Appendix G (for oxidation only), that is,

$$E^\circ_{red} = -E^\circ_{ox}\,(\text{tabulated})$$

Example 7-4 Should water be formed when hydrogen and oxygen gases are mixed in equal volumes at 25°C and a total pressure of 2 atm?

Solution

$$2H_2(g) + O_2(g) \rightarrow 2H_2O(l)$$

from Appendix G:

$$H_2 \rightarrow 2H^+ + 2e^-, \; E^\circ_{ox} = 0.00 \qquad \text{oxidation}$$

$$O_2 + 4H^+ + 4e^- \rightarrow 2H_2O, \; E^\circ_{red} = +1.23 \qquad \text{reduction}$$

Thus,

$$E^\circ_{cell} = 0 + (+1.23) = +1.23$$

and the reaction should proceed spontaneously. The actual reaction at 25°C is too slow to observe. Ignition of the mixture by spark or flame, however, produces a rather dramatic explosion.

Example 7-5 A chemistry student hopes to interest investors in the device for generating chlorine for swimming pools illustrated in Fig. 7-9 by a proposed reaction described by the equation

$$MnO_4^-(aq) + Cl^-(aq) \rightarrow Mn^{2+}(aq) + Cl_2(g) \qquad \text{(in acidic solution, not balanced as shown)}$$

Is the production of chlorine possible by the reaction indicated?

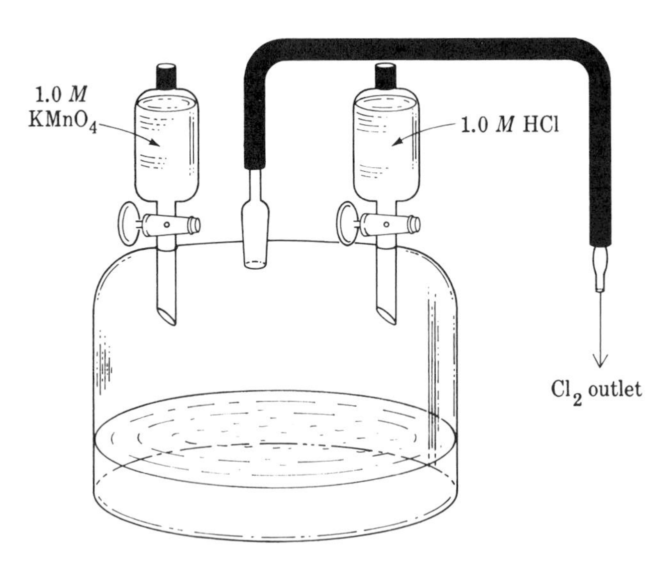

Fig. 7-9 Generation of elemental chlorine.

Solution

from Appendix G:

$$2Cl^- \rightarrow Cl_2 + 2e^-, E^\circ_{ox} = -1.36 \qquad \text{oxidation}$$

$$MnO_4^- + 8H^+ + 5e^- \rightarrow Mn + 4H_2O, E^\circ_{red} = 1.51 \qquad \text{reduction}$$

$$E^\circ_{cell} = -1.36 + (+1.51) = +0.25$$

The reaction should be spontaneous as indicated by the equation.

Some typical electrochemical cells Perhaps the most familiar electrochemical cell is that in the lead storage battery commonly used in automobiles. Unlike the simpler cells described in preceding sections, the lead storage cell typically consists of a number of cathode plates (packed with $\mathbf{PbO_2}$) alternately spaced with anode plates (packed with spongy **Pb**) (Fig. 7-10). This arrangement increases the possible current flow (number of electrons transferred per unit time) by increasing the surface area of the electrodes at which oxidation and reduction occur. A cell of this type produces electric current at a potential of about 2 volts. When three or six cells are arranged in series, a "battery" is formed which is rated at 6 or 12 volts, respectively. During discharge the principal half-reactions are formulated

$$\mathbf{Pb}(s) + SO_4^{2-}(aq) \rightarrow \mathbf{PbSO_4}(s) + 2e^-$$
[internal anode (external cathode)]

and

$$2e^- + \mathbf{PbO_2}(s) + SO_4^{2-}(aq) + 4H^+(aq) \rightarrow \mathbf{PbSO_4}(s) + 2H_2O$$
[internal cathode (external anode)]

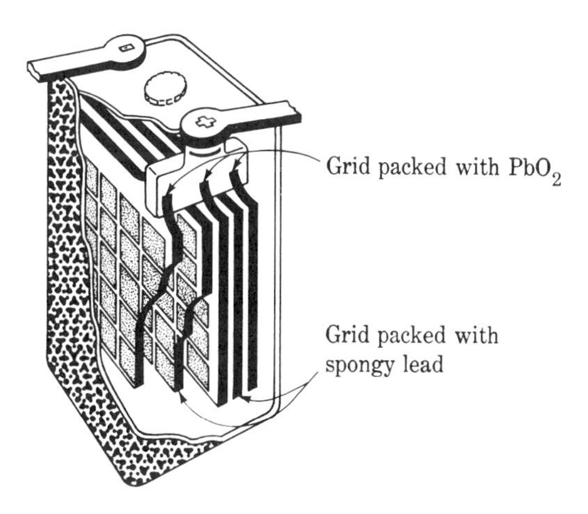

Fig. 7-10 A lead storage cell.

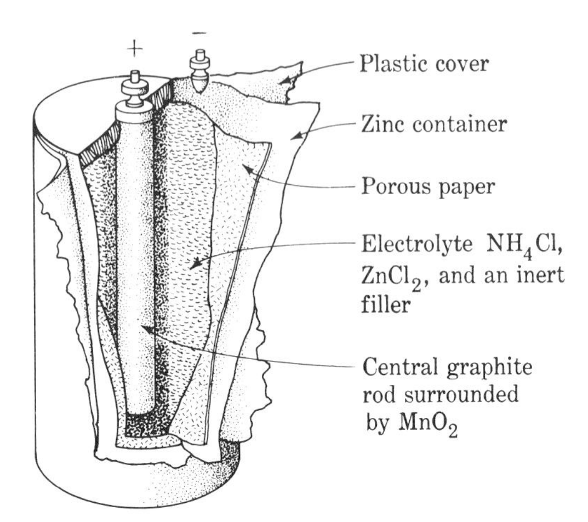

Fig. 7-11 A simple dry cell.

The above described reactions are reversed during "recharging," using an external source of current such as the automobile generator. The $PbSO_4$ which coats the electrodes during the discharge process is converted back to Pb and PbO_2, respectively. Recharging may fail to restore a battery completely, particularly when $PbSO_4$ deposits have partially dropped loose from electrodes. The recharging process also causes some electrolysis of the water in the battery, liberating hydrogen and oxygen gases, as is discovered occasionally by an unwary motorist checking his battery after dark by matchlight. The use of "tap" water, rather than distilled water, in the electrolyte for these batteries may reduce the sulfate concentration appreciably, e.g., by formation of insoluble $CaSO_4$.

The type of flashlight "battery" used almost exclusively until around 1960 was referred to as a "dry cell" (Fig. 7-11) because the electrolyte used was a moist paste rather than a liquid as in the lead cell. The cell reactions are complex, but are believed to be represented, for conditions of low current flow, by:

$$\mathbf{Zn}(s) \rightarrow \underset{\text{(in paste)}}{Zn^{2+}} + 2e^-$$

and

$$2e^- + 2\mathbf{MnO_2}(s) + 2NH_4^+ \text{ (in paste)} \rightarrow \mathbf{Mn_2O_3}(s) + H_2O \text{ (in paste)} + 2NH_3 \text{ (in paste)}$$

The Zn/MnO_2 cell has recently been supplemented by "mercury" cells and "nickel-cadmium" cells (Table 7-1). The latter can be conveniently recharged, a process which is difficult and inefficient for older type dry cells.

Table 7-1 Some discharge reactions

Mercury cell	$\mathbf{Zn}(s) + 2OH^-$ (in paste) $\rightarrow 2e^- + \mathbf{Zn(OH)_2}(s)$ (in paste) $2e^- + \mathbf{HgO}(s)$ (in paste) $+ H_2O$ (in paste) $\rightarrow$ $2OH^-$ (in paste) $+ Hg(l)$
Nickel-Cadmium cell	$\mathbf{Cd}(s) + 2OH^-(aq) \rightarrow \mathbf{Cd(OH)_2}(s) + 2e^-$ $2e^- + \mathbf{NiO_2}(s) + 2H_2O(l) \rightarrow 2OH^-(aq) + \mathbf{Ni(OH)_2}(s)$

One of the most interesting developments in recent years is the "fuel cell," which uses reactions normally associated with direct electron-transfer combustion processes for a more efficient indirect electron-transfer system (Fig. 7-12). The cell shown generates electrical energy in terms of half-reactions described as

$$CH_4(g) + 10OH^-(aq) \rightarrow CO_3^{2-}(aq) + 7H_2O(l) + 8e^-$$

and

$$4e^- + O_2(g) + 2H_2O(l) \rightarrow 4OH^-(aq)$$

The electrolyte must be replaced periodically as OH^- is depleted and CO_3^{2-} is accumulated. Fuel cells of this type are not reversible, i.e., they cannot be "recharged." Experimental automobiles and tractors have been constructed to use fuel cells in place of conventional internal combustion engines. Although there are still many problems—engineering, economic, and political—it is to be hoped that systems will soon be available to replace many of the current uses of direct combustion reactions in modern society.

The Nernst equation The use of standard oxidation potentials to calculate cell voltage or to predict spontaneity of reaction is limited by the

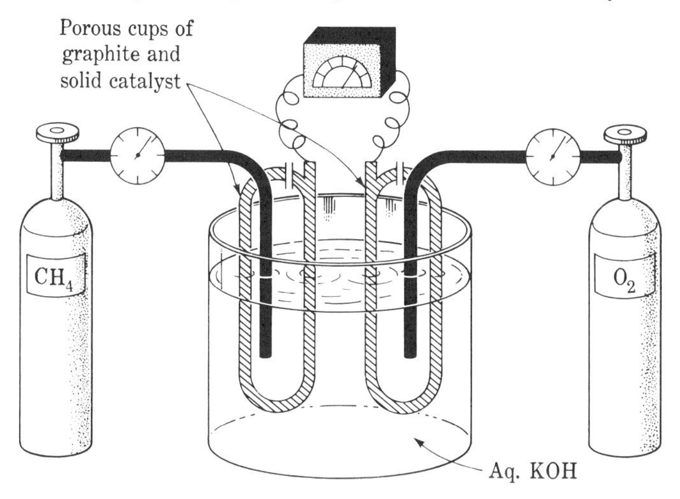

Fig. 7-12 A simple fuel cell.

requirements of standard states. It is obviously desirable to be able to make similar calculations for situations involving other concentrations. In 1899, Walther Nernst developed an equation which is exceedingly useful in relating electrode or cell potentials to concentrations of reacting substances. This *Nernst equation* may be expressed in the form which provides a satisfactory approximation* at 25°C:

$$E = E^\circ - \frac{0.05916}{n} \log Q \tag{7-4}$$

in which E is the cell or electrode potential under the conditions of the concentration ratio Q (product concentrations over reactant concentrations, *using the conventions developed for equilibrium constants*) at 25°C; E° is the standard potential involved; and n is the number of electrons transferred as required for a balanced equation.

Electrode potentials and concentrations The Nernst equation may be used to calculate an approximate potential for a half-cell not involving standard states.

Example 7-6 What is the approximate potential for the hydrogen electrode at 25°C when hydrogen gas at 2.0 atm is bubbled over a platinum electrode suspended in a solution of pH 4.0?

Solution

$$H_2 \rightleftharpoons 2H^+ + 2e^-, \quad E^\circ = 0.00 \text{ volt}$$

$$E = E^\circ - \frac{0.05916}{n} \log Q$$

$$E^\circ = \text{exactly } 0 \text{ volt (defined)}$$

from the equation for the half-reaction,

$$n = 2$$

and

$$Q = \frac{C^2_{H^+}}{p_{H_2}}$$

at pH 4.0, $C_{H^+} = 10^{-4}\ M$. Hence,

$$E = 0 - \frac{0.05916}{2} \log \frac{(10^{-4})^2}{2.0}$$

$$E = -0.0296 \log 10^{-8} - (-0.0296 \log 2)$$

$$E = 0.2368 + 0.0089$$

$$E = +0.25 \text{ volt (to two significant figures)}$$

* The exact form of the equation is

$$E = E^\circ - \frac{RT}{nF} \ln Q_a$$

in which R is the gas constant, T is the absolute temperature, F is the value of the Faraday, and Q_a is the "activity" ratio. At 25°C (= 298°K), 2.303 $RT/nF = 0.05916/n$.

Example 7-7 What is the approximate electrode potential for a half-cell involving liquid mercury and Hg_2^{2+} at a concentration of 6.5×10^{-7} M?

Solution

$$2Hg(l) \rightleftharpoons Hg_2^{2+}(aq) + 2e^-$$

from Appendix G, $E° = -0.79$ volt. Then,

$$E = -0.79 - \frac{0.05916}{2} \log (6.5 \times 10^{-7})$$

(note that Hg(l) does not appear in the Q expression)

$$E = -0.79 - (0.0296 \log 6.5 + 0.0296 \log 10^{-7})$$

$$E = -0.79 - 0.024 + 0.207$$

$$E = -0.61 \text{ volt}$$

Cell potential, spontaneity, and equilibrium The Nernst equation can also be used to calculate cell potential as a function of concentration. The requirement of a positive potential for spontaneous electron transfer still applies so that spontaneity predictions may now be applied to reactions under conditions other than those of standard state. If a negative potential is found, this indicates that the reaction would proceed spontaneously in the opposite direction from that indicated by the equation used for the Nernst-equation calculation. Prediction of the direction of electron flow for an electrochemical cell is a logical result of Nernst-equation determinations.

When the equilibrium condition is attained, cell potential becomes zero, and the value of Q from the Nernst equation will then be that of the equilibrium constant (K_{equil}) for the cell reaction, i.e., at equilibrium

$$0 = E° - \frac{0.05916}{n} \log K_{equil}$$

from which:

$$\log K_{equil} = \frac{nE°}{0.05916} \qquad (7\text{-}5)$$

It is thus possible to calculate equilibrium constants for oxidation-reduction reactions directly from tabulated values of standard oxidation potentials (Appendix G).

Example 7-8 In Fig. 7-4, the right half of the diagram indicates the direction of initial electron flow for a cell in which $C_{Zn^{2+}} = C_{Cu^{2+}} = 1.0$ M. Predict the direction of initial electron flow for such a cell when $C_{Zn^{2+}} = 1.0$ M and $C_{Cu^{2+}} = 10^{-40}$ M.

Solution

If electron flow is still from zinc to copper, then the cell reaction is described as:

$$\mathbf{Zn}(s) + Cu^{2+}(aq) \rightarrow Zn^{2+}(aq) + \mathbf{Cu}(s)$$

For which (from Appendix G),

$$E° = 0.76 + 0.34 = 1.10 \text{ volts}$$

then

$$E = 1.10 - \frac{0.05916}{2} \log \frac{C_{Zn^{2+}}}{C_{Cu^{2+}}}$$

$$E = 1.10 - 0.0296 \log \frac{1}{10^{-40}}$$

$$E = 1.10 - 0.0296(40) = -0.08 \text{ volt}$$

A negative value for E indicates that the reaction will proceed in the direction opposite to that shown. Hence, for this cell the initial direction of electron flow is from copper to zinc.

Example 7-9 What is the value of the equilibrium constant at 25°C for the reaction indicated by the equation

$$\mathbf{Cu}(s) + 2Ag^{+}(aq) \rightarrow Cu^{2+}(aq) + 2\mathbf{Ag}(s)$$

Solution

$$K_{equil} = \frac{C_{Cu^{2+}}}{C^2_{Ag^+}}$$

From Appendix G:

$$E°_{cell} = -0.34 + 0.80 = +0.46$$

Then, in Eq. (7-5)

$$\log K_{equil} = \frac{(2)(0.46)}{0.05916} = 15.6$$

thus

$$K_{equil} = 4 \times 10^{15} \qquad \text{(approximately)}$$

Concentration cells Since electrode potential varies with concentration, it should be possible to construct a cell in which both electrodes are identical but the concentration of reacting ions in the two half-cells is different. Such cells can, indeed, be constructed, and calculations similar to those discussed for more conventional cells can be made. The cell reaction appears a bit unusual. Consider the concentration cell illustrated in Fig. 7-13. Applying the Nernst equation to a Ag/Ag^+ half-cell, using $E° = -0.80$ volt (Appendix G) shows:

$$E = -0.80 - \frac{0.05916}{1} \log C_{Ag^+}$$

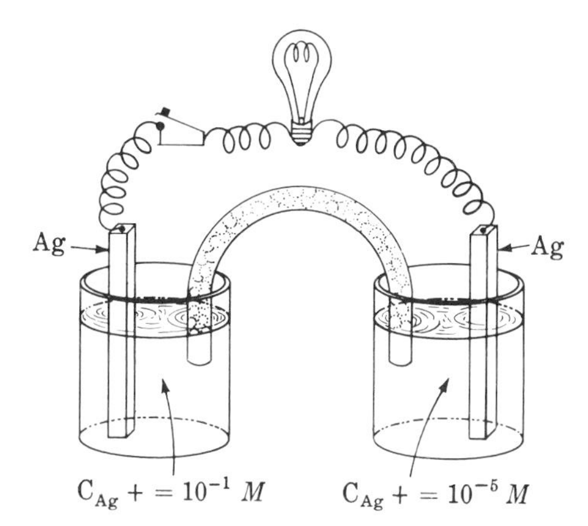

Fig. 7-13 A concentration cell.

This suggests that electrode potential will increase as C_{Ag^+} decreases, i.e., there is a greater tendency for silver to lose electrons when in a more dilute solution of Ag^+ than when in the more concentrated solution. When the cell illustrated discharges, then, initial electron flow will be from right to left as diagrammed. The equation for the cell "reaction" can be written as:

$$\mathbf{Ag}(\text{right}) + Ag^+(\text{left}) \rightarrow Ag^+(\text{right}) + \mathbf{Ag}(\text{left})$$

Since the same electrode is involved in both half-cells,

$$E^\circ_{\text{cell}} = 0$$

thus

$$E = 0 - 0.05916 \log \frac{10^{-5}}{10^{-1}} = +0.24 \text{ volt}$$

Concentration cells may be used for determination of *activities* of ions. In very dilute solution the activity a of an ion is equal to its molar concentration. Since the "exact" form of the Nernst equation includes Q_a, the activity quotient, it is possible to determine the activity of an ion in a concentrated solution by comparing this solution in a concentration cell with a half-cell involving a very dilute solution of the ion. The *activity coefficient* γ, which is defined by the equation

$$\gamma = \frac{a}{c} \tag{7-6}$$

where c is the concentration (molality) based on the amount of solute and solvent used to prepare a given solution, can thus be calculated from the measured cell potential. For dilute solutions molarity and molality are essentially equivalent.

It is important to note that γ is always less than one. The origin of the "activity" expression lies with interparticle forces in solutions which cause particle clusters to form such that there are fewer "effective" solute particles than would have been expected from the "concentration." Since these clusters may contain more than one kind of ion, the "activity" of any particular ion depends on the concentrations of any and all solutes present.

Fortunately, the approximations inherent in the use of molal or molar concentrations rather than activity are usually within the limits of experimental error for all but the most precise measurements, and it is seldom necessary to use activities in normal routine situations.

EXERCISES

7-1. Assign oxidation numbers to each element in each of the following substances: Nitric acid, ammonium sulfate, periodate ion ($H_2IO_6^{3-}$), chloroform ($HCCl_3$), urea ($H_2N\overset{\overset{\displaystyle O}{\|}}{C}NH_2$).

7-2. Which of the following equations describe metathetical reactions?

$2Ag^+(aq) + \mathbf{Zn}(s) \rightarrow 2\mathbf{Ag}(s) + Zn^{2+}$
$2H^+(aq) + \mathbf{Na_2CO_3}(s) \rightarrow 2Na^+(aq) + H_2O(l) + CO_2(g)$
$\mathbf{Mg}(s) + 2MnO_4^-(aq) \rightarrow \mathbf{MgMnO_4}(s) + MnO_4^{2-}(aq)$
$2H^+(aq) + 2CrO_4^{2-}(aq) \rightarrow Cr_2O_7^{2-}(aq) + H_2O(l)$
$\mathbf{LiAlH_4}(s) + 4H^+(aq) \rightarrow Li^+(aq) + Al^{3+}(aq) + 4H_2(g)$

7-3. Write balanced equations for half-reactions for each oxidation-reduction reaction in Exercise 7-2. Label each half-reaction equation as "oxidation" or "reduction."

7-4. When magnesium is dropped into a solution of zinc nitrate, the magnesium begins to disappear and solid zinc precipitates. Diagram a setup that could utilize this reaction for indirect electron transfer. Label all components.

7-5. Permanganate ion reacts with oxalic acid in aqueous solution to form carbon dioxide and manganese(II) ion. (*a*) Write the equilibrium-constant expression for this reaction. (*b*) Explain briefly why this reaction would present problems for an indirect electron-transfer system.

7-6. Chlorine can be manufactured by electrolysis of liquid sodium chloride. What volume of chlorine, measured at STP (273°K, 1 atm), could be produced from excess sodium chloride using a constant current flow of 40.0 amp for 3.00 hours?

7-7. Calculate $E°$ for the cell represented by each of the following equations.

(*a*) $\mathbf{Al}(s) + Fe^{3+}(aq) \rightarrow Al^{3+}(aq) + \mathbf{Fe}(s)$
(*b*) $H_2(g) + Cu^{2+}(aq) \rightarrow 2H^+(aq) + \mathbf{Cu}(s)$
(*c*) $2Br^-(aq) + Cl_2(g) \rightarrow Br_2(l) + 2Cl^-(aq)$
(*d*) $2\mathbf{Au}(s) + 3H_2O_2(aq) + 6H^+(aq) \rightarrow 6H_2O(l) + 2Au^{3+}(aq)$

7-8. Diagram a cell which could be used to determine $E°$ for:

$NH_4^+(aq) + 3H_2O(l) \rightarrow NO_3^-(aq) + 10H^+(aq) + 8e^-$

Label all components.

7-9. Which of the following equations represent reactions that should proceed spontaneously under conditions of standard states?

(*a*) $\mathbf{Cd}(s) + Ni^{2+}(aq) \rightarrow Cd^{2+} + \mathbf{Ni}(s)$

(*b*) $2Cr^{2+}(aq) + Fe^{2+}(aq) \rightarrow 2Cr^{3+}(aq) + \mathbf{Fe}(s)$

(*c*) $H_2(g) + Sn^{2+}(aq) \rightarrow 2H^+(aq) + \mathbf{Sn}(s)$

(*d*) $ClO_4^-(aq) + 2H^+(aq) + 2Br^-(aq) \rightarrow ClO_3^-(aq) + H_2O(l) + Br_2(l)$

7-10. Write balanced equations for:

(*a*) The principal reaction which occurs during recharging of the lead storage battery.

(*b*) The reaction to be expected when hard water [written as $Ca^{2+}(aq)$] is added to the electrolyte concentrate of a lead storage battery.

7-11. Calculate the approximate electrode potential for

$$\mathbf{Zn}(s) \rightarrow Zn^{2+}(aq) + 2e^-$$

when

$$C_{Zn^{2+}} = 10^{-4}\,M$$

7-12. If a fuel cell like that shown in Fig. 7-12 produces a potential of 1.4 volts when $P_{CH_4} = P_{O_2} = 1.0$ atm, $C_{OH^-} = 1.0\ M$ and $C_{CO_3{}^{2-}} = 1.0 \times 10^{-8}\ M$, what potential would be expected after the cell has discharged to the extent that $C_{OH^-} = 0.10\ M$? (Note that gas pressures remain constant.)

7-13. What is the approximate equilibrium constant for

$$Cl_2(g) + 2Br^-(aq) \rightleftharpoons 2Cl^-(aq) + Br_2(l)$$

7-14. Commercial pH meters use rather exotic electrode systems. Sketch an apparatus which could use a hydrogen electrode and any other practical half-cell to determine pH by a measurement of cell voltage. Label all components and indicate the external signs of both electrodes. Derive an equation for pH in terms of the quantities appearing in the Nernst equation.

7-15. Estimate the equilibrium constant for:

$$5H_2S(g) + 6H^+(aq) + 2MnO_4^-(aq) \rightleftharpoons 2Mn^{2+}(aq) + 5S(s) + 8H_2O(l)$$

7-16. Predict whether copper(II) sulfide will dissolve in 6 *M* sulfuric acid according to the equation:

$$\mathbf{CuS}(s) + 4H^+(aq) + SO_4^{2-}(aq) \rightleftharpoons SO_2(g) + Cu^{2+}(aq) + \mathbf{S}(s) + 2H_2O(l)$$

7-17. If a solution is 0.01 *M* in each of the ions $Cr_2O_7^{2-}$, H^+, and Cl^-, will chlorine result according to the equation

$$6Cl^-(aq) + 14H^+(aq) + Cr_2O_7^{2-}(aq) \rightleftharpoons 2Cr^{3+}(aq) + 7H_2O(l) + 3Cl_2(g)$$

7-18. Repeat the calculation of Exercise 7-17, substituting a solution 0.01 *M* in each of the ions $MnO_4^-(aq)$, $H^+(aq)$, and $Cl^-(aq)$. Account for any differences in your conclusions.

RECOMMENDED SUPPLEMENTARY READING

Denaro, A. R.: "Elementary Electrochemistry," Butterworth, Washington, D.C., 1965 (paperback).

Murray, R. E., and Reilley, C. N.: "Electroanalytical Principles," Wiley, New York, 1964 (paperback).

part two

Laboratory Investigation of the Characteristics of the More Common Ionic Species

8
Classification of Ionic Species

Qualitative analysis provides an almost unique medium for the laboratory investigation of the principles of equilibrium and solution chemistry discussed in the preceding chapters. This investigation is accomplished by studying the properties of the ions on a comparative basis, devising techniques based most commonly upon equilibrium processes for their separation, and ultimately utilizing reactions that are specific under the conditions specified for their identification. Fortunately, the objectives of this investigation can be realized by studying in detail the behaviors of only a limited number of common anions and cations, since the properties of these species are to a very large degree characteristic of other and less familiar ions. Thus, the periodic table suggests—and correctly—that knowledge of the properties of the sulfide (S^{2-}) ion enables one to predict with a reasonable degree of certainty the properties of the selenide (Se^{2-}) and telluride (Te^{2-}) ions. Or, if the properties of the sodium (Na^+) and potassium (K^+) ions are known, those of the lithium (Li^+), rubidium (Rb^+), cesium (Cs^+), and francium (Fr^+) ions can be approximated without substantial error. This approach is followed in the laboratory studies outlined in Chaps. 10 and 11, where the behaviors of ions characteristic of a number of the periodic families are investigated.

It is useful to systematize the laboratory approach by first considering together species that exhibit similar reactions with specific and carefully selected reagents. It is the purpose of this chapter both to indicate the ions chosen for laboratory investigation and to outline the systems of classification used for both anionic and cationic species.

ANIONIC SPECIES

The laboratory studies discussed in this textbook are limited to the following anions:

Acetate, $CH_3CO_2^-$
(Ortho)arsenate, AsO_4^{3-}
Bromide, Br^-
Carbonate, CO_3^{2-}
Chloride, Cl^-
Chromate, CrO_4^{2-}
Fluoride, F^-
Iodide, I^-
Nitrate, NO_3^-
Nitrite, NO_2^-
Oxalate, $C_2O_4^{2-}$
Permanganate, MnO_4^-
(Ortho)phosphate, PO_4^{3-}
Sulfate, SO_4^{2-}
Sulfide, S^{2-}
Sulfite, SO_3^{2-}
Thiosulfate, $S_2O_3^{2-}$

In certain cases, changes in pH conditions alter these formulations and permit the inclusion of additional species. Thus, depending on the pH of the system, one can encounter HPO_4^{2-} or $H_2PO_4^-$, as well as PO_4^{3-}; $HAsO_4^{2-}$ or $H_2AsO_4^-$, as well as AsO_4^{3-}; or $Cr_2O_7^{2-}$ as well as CrO_4^{2-}. In each of these instances, however, the characteristic reactions of identification remain unaltered, and the formulation recorded above is adequate to describe the anion in question. The water-derived anions, oxide (O^{2-}) and hydroxide (OH^-), are absent from this list as consequences of (1) the lack of specific reactions for the positive identification of these ions and (2) the formation in aqueous solution of the hydroxide ion by hydrolysis of other basic anions (for example, $CH_3CO_2^-$, CO_3^{2-}, PO_4^{3-}) and the resulting impossibility of drawing a correct conclusion as to the real source of this species. The behaviors of these two anions are described in preceding chapters.

The anions may be conveniently classified into several groups in terms of their reactions with particular cations (for example, Ag^+, Ca^{2+}, Ba^{2+}, Zn^{2+}, H^+) under exactly defined conditions. Inasmuch as the hydrogen and silver ions give the most striking and characteristic reactions, it is convenient to base a classification of this type on these two species. In this way, the anions listed above are divided among four groups, as shown in the following sections.

Anion group I Anions that are decomposed in acidic solution (6 *M* $HClO_4$) with either the evolution of gases of characteristic properties or the formation of precipitates.

Anion group II Anions that are stable in 6 *M* perchloric acid solution and are precipitated as silver salts from this medium.

Anion group III Anions that are stable in 6 *M* perchloric acid solution but are precipitated as silver salts only when the solution is neutralized.

Anion group IV Anions that are stable in 6 *M* perchloric acid solution and are not precipitated by silver ion under either acidic or neutral conditions.

Table 8-1 classifies the anions listed above in terms of these four groups and indicates their behaviors with perchloric acid and silver nitrate as *group reagents.*

Table 8-1 Classification of anions into groups

Anion group	*Group reagent*	*Ion*	*Behavior of ion with group reagent*
I	Dilute (6 *M*) $HClO_4$	CO_3^{2-}	$CO_2\uparrow$ (odorless, colorless)
		NO_2^-	$NO\uparrow + NO_2\uparrow$ (sharp odor, brown)*
		S^{2-}	$H_2S\uparrow$ (decayed egg odor, colorless)*
		SO_3^{2-}	$SO_2\uparrow$ (burning sulfur odor, colorless)
		$S_2O_3^{2-}$	S (white) + $SO_2\uparrow$ (burning sulfur odor, colorless)
II	Dilute (6 *M*) $HClO_4$ + $AgNO_3$	Br^-	**AgBr** (cream)
		Cl^-	**AgCl** (white)
		I^-	**AgI** (pale yellow)
		(S^{2-})	**Ag_2S** (black)
		$(S_2O_3^{2-})$	**$Ag_2S_2O_3$** (white) $\rightarrow$ **Ag_2S** (black)
III	Neutral solution + $AgNO_3$	$(C_2H_3O_2^-)$†	**$AgC_2H_3O_2$** (white)
		AsO_4^{3-}	**Ag_3AsO_4** (red-brown)
		(CO_3^{2-})	**Ag_2CO_3** (yellowish)
		CrO_4^{2-}	**Ag_2CrO_4** (brownish red)
		(NO_2^-)†	**$AgNO_2$** (pale yellow)
		$C_2O_4^{2-}$	**$Ag_2C_2O_4$** (white)
		PO_4^{3-}	**Ag_3PO_4** (yellow)
		(SO_4^{2-})†	**Ag_2SO_4** (white)
		(SO_3^{2-})	**Ag_2SO_3** (white) $\xrightarrow{\Delta}$ **Ag** (black)
		$(S_2O_3^{2-})$	**$Ag_2S_2O_3$** (white) $\xrightarrow{\Delta}$ **Ag_2S** (black)
IV	None	$C_2H_3O_2^-$	
		F^-	
		NO_3^-	
		(NO_2^-)	
		MnO_4^-	
		SO_4^{2-}	

* Poisonous! Danger! Do not inhale!

† Precipitated only if anion concentration exceeds about 5 mg ml^{-1}.

The characteristics of the anions that are responsible for this grouping and their other properties are considered in detail in Chap. 10, where procedures for the identification of these species are given. There are some apparent inconsistencies in Table 8-1 that require comment, particularly with regard to the appearance of some anions in more than one group in terms of the rigorous definitions listed above. The exact grouping of these species depends upon the sequence in which the group tests are applied. If, for example, a solution containing a number of anions is first acidified and then warmed, the group I anions will be destroyed, and silver ion will fail to precipitate any of these anions, which might otherwise have appeared in group II or group III. If, however, these anions had not been destroyed, they would have appeared in one or both of these groups. It is also true that any anion that can be precipitated by silver ion from acidic solutions can be precipitated even more readily from neutral solutions. Table 8-1 includes the formulas for *all* species the behaviors of which satisfy the group definitions. Formulas *not* enclosed in parentheses are for those species that would appear if *one* sample of the original solution were used and the anion groups were removed from it *in the order listed.* Formulas enclosed in parentheses are for those additional species that would appear if a separate sample of solution were employed for *each group test.*

CATIONIC SPECIES

The laboratory studies discussed in this textbook are limited to the following cations:

Aluminum(III), or aluminum, Al^{3+}
Ammonium, NH_4^+
Antimony(III), or antimonyl, SbO^+
Barium(II), or barium, Ba^{2+}
Bismuth(III), or bismuthyl, BiO^+
Cadmium(II), or cadmium, Cd^{2+}
Calcium(II), or calcium, Ca^{2+}
Chromium(III), or chromic, Cr^{3+}
Cobalt(II), or cobaltous, Co^{2+}
Copper(II), or cupric, Cu^{2+}
Iron(III), or ferric, Fe^{3+}
Iron(II), or ferrous, Fe^{2+}
Lead(II), or plumbous, Pb^{2+}
Magnesium(II), or magnesium, Mg^{2+}
Manganese(II), or manganous, Mn^{2+}
Mercury(I), or mercurous, Hg_2^{2+}
Mercury(II), or mercuric, Hg^{2+}
Nickel(II), or nickelous, Ni^{2+}
Potassium(I), or potassium, K^+
Silver(I), or silver, Ag^+
Sodium(I), or sodium, Na^+
Tin(IV), or stannic, Sn^{4+}
Zinc(II), or zinc, Zn^{2+}

In addition, certain complex species are sufficiently unstable to form precipitates with individual cation group reagents and thus give the reactions associated with these cations. Among those involving the cations listed above are the following: chloroantimonate(III) ions, such as $[SbCl_4]^-$ and

$[SbCl_6]^{3-}$; orthoarsenate ion, $[AsO_4]^{3-}$; tetrachlorobismuthate(III) ion, $[BiCl_4]^-$; chloromercurate(II) species, such as $[HgCl_2]$, $[HgCl_3]^-$, and $[HgCl_4]^{2-}$; and hexachlorostannate(IV) ion, $[SnCl_6]^{2-}$. The permanganate and dichromate ions are reduced by sulfide in acidic solution to Mn^{2+} and Cr^{3+}, respectively, which cations are then detected as such.

The cations may be conveniently classified into several groups in terms of their reactions with particular anions (for example, Cl^-, S^{2-}, OH^-, $CO_3{}^{2-}$, $SO_4{}^{2-}$). Of the various groupings proposed over a period of many years, the most satisfactory from the points of view of successful laboratory operation and meaningful application of the principles of chemical equilibrium is that based upon successive precipitations with chloride ion, sulfide ion in acidic solutions, hydroxide ion, sulfide ion in alkaline solutions, and carbonate ion. This system of classification and analysis was apparently devised in the middle of the nineteenth century by K. R. Fresenius, the father of qualitative analysis. The procedure of Fresenius has been extended, modified, and improved upon by countless other investigators. Adverse criticism of this procedure has usually centered in its use of the poisonous and generally obnoxious hydrogen sulfide. The substitution of compounds that hydrolyze to sulfide in aqueous solution (e.g., thioacetamide, page 106) obviates this criticism.

A variant of the sulfide procedure permits classification of the cations listed above into six groups.

Cation group I Cations that are precipitated as chlorides from cold, dilute acidic solution.

Cation group II Cations that are not precipitated as chlorides but are precipitated as sulfides from solutions approximately 0.3 *M* in hydrogen ion.

Cation group III Cations that are not precipitated as chlorides or sulfides from acidic solutions but are precipitated as hydroxides from strongly ammoniacal solutions containing ammonium ion.

Cation group IV Cations that are not precipitated as chlorides, sulfides, or hydroxides under the conditions outlined above but are precipitated as sulfides from ammoniacal solutions containing ammonium ion.

Cation group V Cations that are not precipitated as chlorides, sulfides, or hydroxides under the conditions outlined above but are precipitated as carbonates from ammoniacal solutions containing ethanol (ethyl alcohol).

Cation group VI Cations that are not precipitated under any of the above described conditions. Table 8-2 classifies the cations listed above in terms of these six groups and indicates their behaviors with chloride, sulfide, hydroxide, and carbonate ions as *group reagents.* It is of interest that this grouping also permits the inclusion of cations derived from all of the other metallic elements.

The characteristics of the cations that are responsible for this grouping and their other properties are considered in detail in Chap. 11. It must be pointed out that the reagent for a given cation group will precipitate not only the ions listed for that group but also the ions listed for all the *preceding* groups as well. Table 8-2, as a system of classification, requires that each group of cations be removed in order. On this basis, the only cation that appears in more than one group is Pb^{2+}. The solubility of lead(II) chloride is insufficiently small (page 8) to permit reduction of the concentration of this ion below the level required for sulfide precipitation in cation group II.

Table 8-2 Classification of cations into groups

Cation group	*Group reagent*	*Ion*	*Behavior of ion with group reagent*
I	Cold, dilute (1 *M*) HCl	Pb^{2+}	**$PbCl_2$** (white, crystalline)
		Hg_2^{2+}	**Hg_2Cl_2** (white)
		Ag^+	**AgCl** (white)
II	CH_3CSNH_2 in about 0.3 *M* HCl	SbO^+, $[SbCl_4]^-$, $[SbCl_6]^{3-}$	**Sb_2S_3** (orange)
		AsO_4^{3-}	**As_2S_3** (yellow)
		BiO^+, $[BiCl_4]^-$	**Bi_2S_3** (brown)
		Cd^{2+}	**CdS** (yellow)
		Cu^{2+}	**CuS** (black)
		Pb^{2+}	**PbS** (brown)
		Hg^{2+}, $[HgCl_2]$, $[HgCl_4]^{2-}$	**HgS** (black)
		Sn^{4+}, $[SnCl_6]^{2-}$	**SnS_2** (yellow)
III	Aqueous $NH_3 + NH_4^+$	Al^{3+}	**$Al(OH)_3$** (white)
		Cr^{3+}	**$Cr(OH)_3$** (gray-green)
		Fe^{2+}, Fe^{3+}	**$Fe(OH)_3$** (red-brown)*
		Mn^{2+}	**MnO_2** (black)†
IV	CH_3CSNH_2 in aqueous $NH_3 + NH_4^+$	Co^{2+}	**CoS** (black)
		Ni^{2+}	**NiS** (black)
		Zn^{2+}	**ZnS** (white)
V	$(NH_4)_2CO_3$ in C_2H_5OH solution	Ba^{2+}	**$BaCO_3$** (white)
		Ca^{2+}	**$CaCO_3$** (white)
		Mg^{2+}	**$MgCO_3 \cdot (NH_4)_2CO_3 \cdot 4H_2O$** (white)
VI	None	NH_4^+	
		K^+	
		Na^+	

* Because of oxidation.

† Because of selective prior oxidation with chlorate in acidic solution.

The position of the manganese(II) ion is somewhat anomalous since under the rigorous scheme of classification used this ion should appear in cation group IV. However, oxidation with chlorate ion and removal as manganese-(IV) oxide prior to the precipitation of cation group III (page 253) gives conclusive results and a clean separation and thus permits the inclusion of manganese(II) ion in group III. Certain of the behaviors recorded are influenced by reduction or oxidation processes not directly indicated in Table 8-2. Thus, orthoarsenate ion is conveniently reduced prior to precipitation as sulfide (page 241), and iron(II) ion is oxidized prior to precipitation as hydroxides (page 253).

SYSTEMATIC STUDY OF THE IONS

The identification of a given anion or cation is usually based upon a distinctive reaction that is characteristic of that species. Often separations from interfering species are essential, and in nearly all instances the identifying reaction provides confirmation of evidence based upon reactions carried out prior to that operation. The reactions of the ions in question with a variety of reagents must be known and understood to maximize ease in identification. To this end, equations describing the more characteristic reactions of the anions and cations are given in Appendixes H_a and H_c.

In this textbook, the anions are considered prior to the cations for the following reasons: (1) much of the chemistry of the cations depends upon the effects on their properties produced by the anions; (2) analysis for the anions lends itself less well to complete systematization and thus allows better the development of a logical laboratory approach and the interpretation of observed behaviors; and (3) in the usual sequence of instruction, qualitative analysis follows the material on the nonmetallic or anion-forming elements, and a significant break in instruction is thus avoided.

9 Laboratory Operations and Techniques

The laboratory studies in qualitative analysis have the following primary objectives: (1) to acquaint the student with the reactions of the various anions and cations with particular reagents under selected conditions; (2) to illustrate to the student how these reactions conform to the principles of equilibrium and solution chemistry as outlined in Chaps. 1 to 7; (3) to give the student personal experience with a variety of reactions that can be combined by him into a systematic and logical scheme of analysis; (4) to develop in the student an appreciation of the importance of a carefully planned and logically executed attack upon a laboratory problem in chemistry, and; (5) to teach the student those habits of carefulness, orderliness, cleanliness, and diligence that are so essential to success in the more nearly quantitative laboratory courses in chemistry that follow.

These objectives can be realized by first acquainting the student with the details of laboratory practice through a series of introductory experiments and then having him correlate the results of selected chemical reactions to arrive at the compositions of a number of "unknown" samples. To encourage him to interpret experimental observations and to relate one set of observations to another, he is asked first to analyze samples for anions because the procedures used are not completely systematic and require him constantly to summarize and interpret his results. The analysis of samples for cations, which follows, is simultaneously highly systematic and strongly dependent upon applications of chemical equilibria.

This type of laboratory program requires the development of new and refined techniques and the use of a few types of new apparatus. The general purposes of the present chapter are to discuss these items, to outline the nature and scope of the laboratory investigations, and to introduce the student both to the mechanics of laboratory practice and to the compilation of the written record of his results.

THE SEMIMICRO SCALE OF LABORATORY INVESTIGATION

Laboratory operations can be carried out on various scales, depending upon the size of the sample used. Among these are

Macro scale, 0.1- to 10.0-g sample
Semimicro scale, 0.01- to 0.1-g sample
Micro scale, 0.001- to 0.01-g sample
Ultramicro scale, less than 0.001-g sample

The macro scale involves experiments with solution volumes ranging from about 10 ml upward. It is the scale commonly used in most introductory courses in chemistry. Both the micro and ultramicro scales require the use of extremely small apparatus and considerable instrumentation. The semimicro scale involves operations on roughly one-tenth of the scale practiced in the smallest macro operations, i.e., reagent solutions are added by drop rather than milliliter volumes, and total volumes seldom exceed a few milliliters.

The convenience and rapidity of handling small quantities of matter and the care required of the student in his laboratory manipulations make the semimicro approach particularly well suited to the investigation of solution chemistry. The necessary operations are accomplished also with inexpensive apparatus in a manner that is both new and interesting.

APPARATUS AND TECHNIQUES INVOLVED IN SEMIMICRO STUDIES

With but few exceptions, the apparatus required is the same as that used for investigations on the macro scale. However, smaller articles (e.g., test tubes, beakers, etc.) are used. A complete list of essential items is given on page 351. All of this apparatus is handled in the same way as macro-sized equipment, but with refinements as dictated by the smaller scale of manipulation.

Fig. 9-1 Wash bottle: 16-oz polyethylene bottle.

Fig. 9-2 Stirring rods.

Special semimicro apparatus The following items of apparatus should be assembled before other laboratory work is begun:

1. A *wash bottle* to provide distilled water for rinsing apparatus, washing precipitates, transferring precipitates by washing, or diluting solutions. The polyethylene-bottle type (Fig. 9-1) has the marked advantages of convenience, cleanliness, and indestructibility, but it cannot be heated. This type of wash bottle delivers liquid when squeezed. In use, the wash bottle should never contain anything but distilled water, and the tip of its delivery tube should never contact another container or its contents.
2. Several *stirring rods* of the types shown in Fig. 9-2. These rods are made by bending or flattening and then fire-polishing 3- to 4-mm glass rod. Each should be long enough that either end can be used in a 13 by 100-mm test tube.
3. Several dropper *pipettes* of the type shown in Fig. 9-3. These pipettes are made from 6-mm glass tubing and are equipped with rubber

Fig. 9-3 Pipette with bulb.

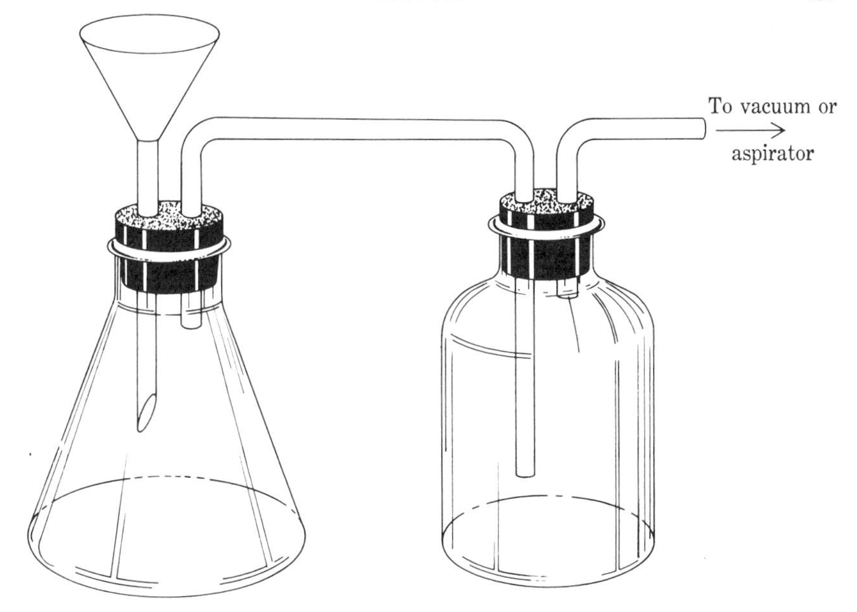

Fig. 9-4 Suction filtration apparatus.

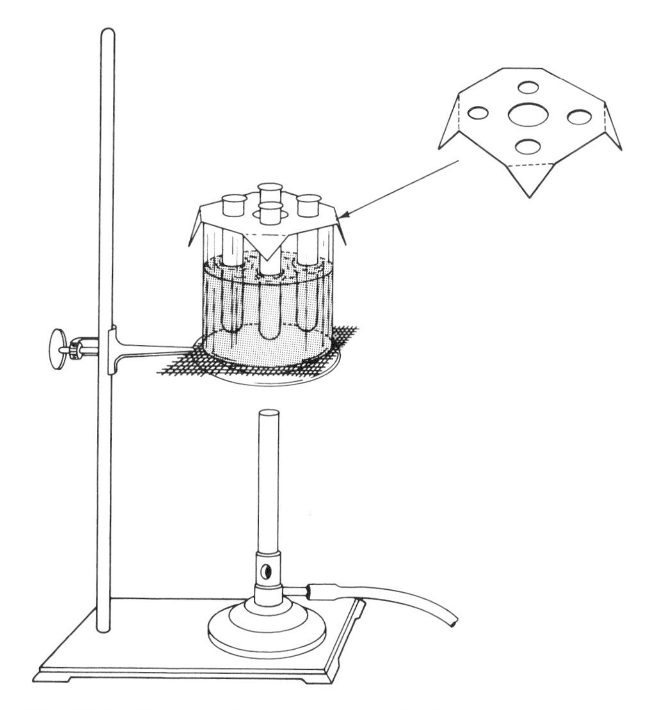

Fig. 9-5 Water-bath arrangement and aluminum bath rack.

dropper bulbs. Each should be calibrated roughly by measuring the volumes of liquid delivered (use a 10-ml graduated cylinder) and marking with file scratches at 0.5-, 1.0-, 1.5-, and 2.0-ml volumes.

4. A *suction filtration apparatus* constructed as shown in Fig. 9-4. To avoid the danger of implosion at reduced pressures, the Erlenmeyer flask used should have a volume no larger than 50 ml. The bottle between the filtration arrangement and the pumping arrangement functions as a trap to protect the filtrate from contamination.
5. A *water bath* constructed from a 150-ml beaker as indicated in Fig. 9-5. This bath is useful either for heating a solution in a test tube or small flask by direct immersion or for heating a material in a casserole placed on top of the beaker.

Reagents and their handling Solid reagents are conveniently stored in wide-mouthed bottles, the hollow glass stoppers of which can be used to remove and dispense the solids as shown in Fig. 9-6. Under no circumstance should a spatula or other object be used to remove a solid from its container. Except as specifically directed for a hygroscopic or strongly oxidizing reagent, a solid can be transferred from the reagent shelf to the student's laboratory bench on a small square of clean (preferably glazed) paper. Although solids must be weighed—on paper squares or watch glasses on the trip balance—in a few cases, quantities can usually be estimated with sufficient accuracy after a little practice.

Solutions are conveniently dispensed from bottles equipped with droppers. The solutions that are required, their concentrations, and the directions for their preparation are listed in Appendix I. Many of these solutions can be placed on the side shelf in about 250-ml bottles for community use, although some instructors may prefer to make sets available to smaller groups of students to reduce the possibilities of contamination. As

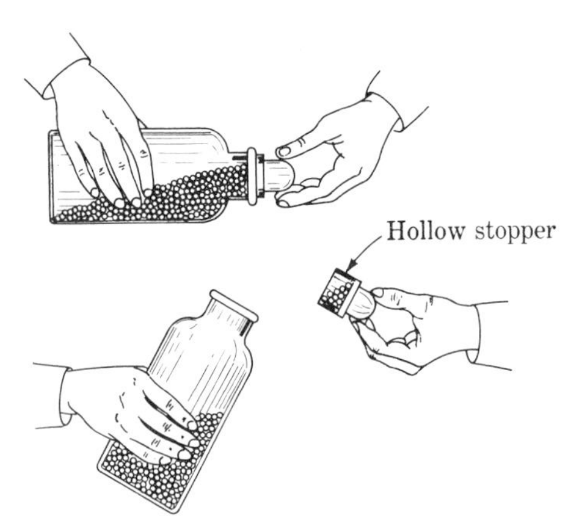

Fig. 9-6 Method of dispensing solid from glass-stoppered reagent bottle.

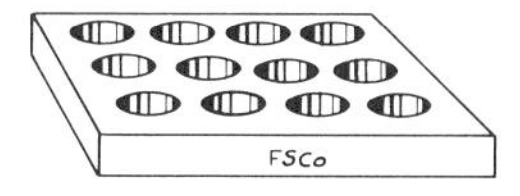

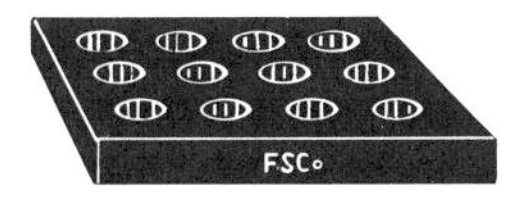

Fig. 9-7 White and black spot plates.

a consequence of frequency of their use, the following solutions may be kept in small (about 15- to 20-ml) dropper bottles in each student's locker:

6 *M* hydrochloric acid	6 *M* acetic acid
12 *M* hydrochloric acid	6 *M* aqueous ammonia
3 *M* sulfuric acid	6 *M* sodium hydroxide
18 *M* sulfuric acid	0.1 *M* barium nitrate
6 *M* nitric acid	0.1 *M* silver(I) nitrate
16 *M* nitric acid	1 *M* thioacetamide

Bottles for these reagents should be washed thoroughly, freshly labeled, and filled from stock bottles before any laboratory exercises are undertaken. When any dropper bottle is used, the delivery tip should never contact the solution being tested or the walls of the container holding it. If more direct measurement of the requisite volume of solution beyond that obtainable by counting drops is necessary, the desired quantity of the solution should be transferred to a clean 10-ml graduated cylinder or a quantity should be added to a clean test tube and then measured into the reaction vessel with one of the prepared pipettes. Each student is held responsible for maintenance of the purity of each shelf reagent.

REACTIONS ON THE SEMIMICRO SCALE

Reactions are carried out in small test tubes, beakers, flasks, or casseroles by dropwise addition of appropriate reagent solutions or by treatment with definite quantities of other reagents. Stirring is conveniently effected by swirling the larger containers or, with test tubes, either by spinning curved stirring rods (Fig. 9-2) between the fingers or by careful tapping with the fingers. Inversion of a test tube and its contents with a finger held over the end is a careless habit leading inevitably to contamination of the sample. Where the characteristics of a solution are to be examined periodically during a reaction, small drops can be withdrawn on clean, straight stirring rods and checked for pH by touching to indicator paper or for other properties by bringing in contact with appropriate reagents in the depressions of a white (for colored products) or a black (for white precipitates) spot plate (Fig. 9-7). Solutions can be heated in beakers or flasks over the open flame, but direct heating in small test tubes is likely to cause spattering because of vaporization in restricted space. Heating by immersion in the water bath (Fig. 9-5) is recommended. Tap water must never be used for dilutions. Distilled or demineralized water is available for this purpose. All equipment must be maintained scrupulously clean. Dirty containers should be washed thoroughly immediately after use and then rinsed with small quantities of distilled water. The student should never dry the inner wall of a container with a towel. A better procedure is to allow all well-rinsed pieces of apparatus to drain on clean paper towels. Stirring rods should never be allowed to touch the desk top, etc., when being used. Crucible tongs or forceps should never be inserted into beakers or flasks to remove them from a heating arrangement. A small piece of folded paper provides a convenient means of handling a hot object without danger of contamination. There is no room for sloppiness or dirty equipment in the analytical laboratory.

Precipitation reactions A precipitate is a solid that forms as a consequence of a chemical reaction in solution. Difficultly soluble sulfides are obtained either by generating sulfide ion within the solution by hydrolysis of thioacetamide (page 106) or by treating the solution with hydrogen sulfide gas. The latter practice is inefficient because of the limited solubility of the gas and dangerous because of its extremely poisonous character.* For this reason, all precipitations of sulfides in subsequently described operations are effected by the hydrolysis of thioacetamide.

Precipitation is perhaps the most important single operation in qualitative analysis. No precipitation reaction should be assumed to be complete until addition of a small quantity of the precipitating reagent to the clear liquid above the precipitate (supernatant liquid) fails to bring down more of the solid. The form in which the precipitate is produced affects its ease of removal from the mother liquor and its subsequent washing. Difficultly soluble materials commonly precipitate in gelatinous form (page 91), and peptization (page 113) often occurs, particularly with sulfides or hydroxides. Digestion, by allowing the solid to stand in contact with the mother liquor on the steam bath, and the presence of added electrolytes (often ammonium salts) improve granular character and reduce peptizability.

Separation of precipitate from mother liquor These separations can be effected by (1) filtration, (2) settling and decantation, or (3) centrifugation. Direct *filtration* is primarily a macro technique, but it can be scaled down to the semimicro scale by using small funnels. As in any filtration, care must be taken to fold the paper and to fit it accurately to the funnel, the paper must first be wet with the solvent present in the solution, and the tip of the funnel

* Hydrogen sulfide, with a minimum lethal dosage even smaller than that of hydrogen cyanide, is one of our most toxic substances. Its odor warns of its presence in lower than lethal concentrations, but larger concentrations so interfere with the olfactory process as to be undetectable in this way.

stem must contact the wall of the receiving container to provide a continuous column of descending liquid. Filtration rate can be increased by using a suction apparatus (Fig. 9-4), but in this case suction should be applied *very carefully* to avoid rupturing the paper. Highly gelatinous precipitates are drawn by suction into mats through which mother liquor passes only very slowly.

Decantation involves allowing the precipitate to settle and pouring off the mother liquor. It is most effective if the particles of the precipitate are sufficiently crystalline to settle rapidly to a compact mass, but, with gelatinous, slow-settling materials, only an incomplete removal of mother liquor can be effected in this fashion. Decantation of much of the mother liquor through a filter paper is a common prelude to removal of the remaining liquid by filtration itself. The rate at which a spherical precipitated particle settles is given ideally by the Stokes relationship

$$R = \frac{2r^2 g}{9\eta}(d - d') \tag{9-1}$$

where R is the rate of settling (cm sec^{-1}), r is the radius of the particle (cm), g is the acceleration of gravity (981 cm sec^{-2}), η is the viscosity of the medium (poises), d is the density of the solid (g cm^{-3}), and d' is the density of the liquid (g ml^{-1}). The limitations upon separation by single decantation are apparent from this equation.

It is also apparent from Eq. 9-1 that for a given system the rate of settling can be increased by increasing the gravitational pull upon the particle. This is done most conveniently by *centrifugation*. A typical centrifuge useful for semimicro separations and the principle of its operations are illustrated in Fig. 9-8. In terms of Fig. 9-8, it can be shown that the centrifugal force f on a particle of mass m is given by

$$f = m\frac{V^2}{x} = m4\pi^2 x n^2 \tag{9-2}$$

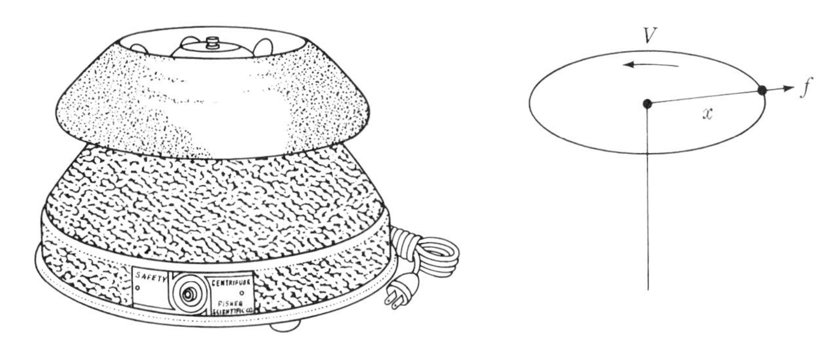

Fig. 9-8 Typical semimicro centrifuge and principle of its operation.

where V is the centrifuge velocity (cm sec^{-1}), x is the distance from the center of the centrifuge to the particle (cm), and n is the number of revolutions per second. For comparison, the normal gravitational force f' upon the same particle would be given by

$$f' = mg \tag{9-3}$$

Division of Eq. 9-2 by Eq. 9-3 and elimination of m then gives

$$\frac{f}{f'} = \frac{4\pi^2 x n^2}{g} \tag{9-4}$$

or

$$f = \frac{4\pi^2 x n^2}{g} f' \tag{9-5}$$

Equation (9-5) thus compares the centrifugal force acting upon a particle with that produced by gravity alone. Thus, for a centrifuge of 10-cm radius operating at 1,000 rpm, substitution gives

$$f = \frac{4 \times 3.1416^2 \times 10\,\text{cm} \times (1{,}000\,\text{rpm})^2}{(981\ \text{cm sec}^{-2}) \times (60\ \text{sec min}^{-1})^2} \times f'$$

$$= 112f'$$

meaning that the force acting to settle the particle is over 100 times that due to gravity alone. Under these conditions, then, sedimentation in the centrifuge would be about 100 times as rapid as simple settling.

Centrifuging most suspensions for a minute or so is sufficient to cause complete settling and produce a sufficiently compact residue to permit complete decantation of the mother liquor. Highly gelatinous or colloidal precipitates may take longer. Centrifugal separations are aided by using constricted tubes, as shown in Fig. 9-9, although for practical purposes 13 by 100-mm test tubes are usually adequate. The student must always remember to insert a second tube carrying essentially the same volume of

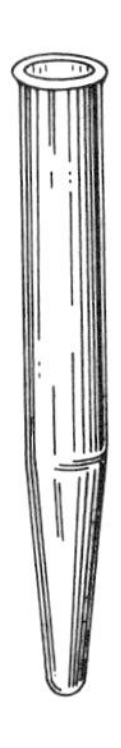

Fig. 9-9 Tapered centrifuge tube.

suspension or solvent in the centrifuge opening opposite from his sample before starting the centrifuge. Otherwise, the lack of balance will cause destructive vibrations within the instrument.

Washing the precipitate Even the most efficient filtration or centrifugation always leaves a certain amount of mother liquor adsorbed on the precipitate. This liquid must be removed in order that either (1) desired ions remaining in solution after precipitation are not lost or (2) the precipitate is not contaminated with ions which might cause interferences in subsequent steps. Consequently, all precipitates are generally washed. The wash liquid is either the pure solvent (usually water) or the solvent to which small quantities of electrolyte have been added to prevent either dissolution or peptization of the precipitate. In certain cases, an organic liquid such as an alcohol may be used.

If the precipitate is on a filter paper, it is generally sufficient to direct a stream of liquid from the wash bottle upon it to break up the precipitate completely, allow all the liquid to drain completely, and then repeat the operation several times. If the precipitate is in a centrifuge tube, the precipitate is either stirred vigorously with the stream of liquid from the wash bottle or agitated with the wash liquid by means of a stirring rod, and the suspension is then recentrifuged. For maximum efficiency, it is necessary to remove all the wash liquid each time before adding a fresh portion. Washing is more effective if carried out with several small portions of wash liquid than if carried out with an equal volume of liquid added in one large portion.

Washing is essentially a process of dilution. Hence, since the total quantity of the species being removed does not change, it may be said

Weight of material present before washing = weight of material present after addition of wash liquid

If C_0 represents the concentration (say, in grams per milliliter) of the impurity present in the residual volume v_r (milliliters) of liquid adhering to the precipitate before the wash liquid is added, then

$$\text{Weight of material present before washing} = C_0 v_r \qquad (9\text{-}6)$$

Correspondingly, if C_1 represents the concentration (grams per milliliter) of the impurity when diluted with a volume v (milliliters) of wash liquid, then

$$\text{Weight of material present after addition of wash liquid} = C_1(v_r + v) \qquad (9\text{-}7)$$

Equating and solving for C_1 then gives

$$C_1 = \frac{C_0 v_r}{v_r + v} \qquad (9\text{-}8)$$

Now, if the precipitate is drained to the same residual volume v_r and washed with a second equal volume v of liquid, the concentration C_2 of the species after addition of the wash liquid is given similarly as

$$C_2 = \frac{C_1 v_r}{v_r + v} \tag{9-9}$$

since $C_1 v_r$ now represents the quantity of substance initially present. Substitution of C_1 from Eq. 9-8 in Eq. 9-9 and simplification then gives for the end of the second washing

$$C_2 = C_0 \left(\frac{v_r}{v_r + v}\right)^2 \tag{9-10}$$

Similarly, after any number n of such washings, each with the same volume of wash liquid, the ultimate concentration C_n is given as

$$C_n = C_0 \left(\frac{v_r}{v_r + v}\right)^n \tag{9-11}$$

The utility of Eq. 9-11 for evaluating the efficiency of a washing process is illustrated by the data summarized in Table 9-1 for a typical case.

It must be emphasized that washing can remove only species which are adsorbed by the precipitate. Foreign ions included in the precipitate as a result of coprecipitation or postprecipitation cannot be so removed.

Table 9-1 Efficiency of washing a $BaSO_4$ precipitate containing Mg^{2+} ion
General conditions: C_0 of Mg^{2+} = 0.003 g ml^{-1}, v_r always = 0.05 ml (about 1 drop)

		Quantity of Mg^{2+} *remaining with* $BaSO_4$	
Case	*Washing conditions*	g ml^{-1}	g
1	One wash with 2 ml of water	7.3×10^{-5} (= C_1)	3.7×10^{-6}
2	Two successive washes, each with 1 ml of water	1.4×10^{-4} (= C_1)	7.0×10^{-6}
		6.7×10^{-6} (= C_2)	3.4×10^{-7}
3	Four successive washes, each with 0.5 ml of water	2.7×10^{-4} (= C_1)	1.4×10^{-5}
		2.6×10^{-5} (= C_2)	1.3×10^{-6}
		2.4×10^{-6} (= C_3)	1.2×10^{-7}
		2.2×10^{-7} (= C_4)	1.1×10^{-8}
4	Eight successive washes, each with 0.25 ml of water	1.8×10^{-9} (= C_n)	9.0×10^{-11}

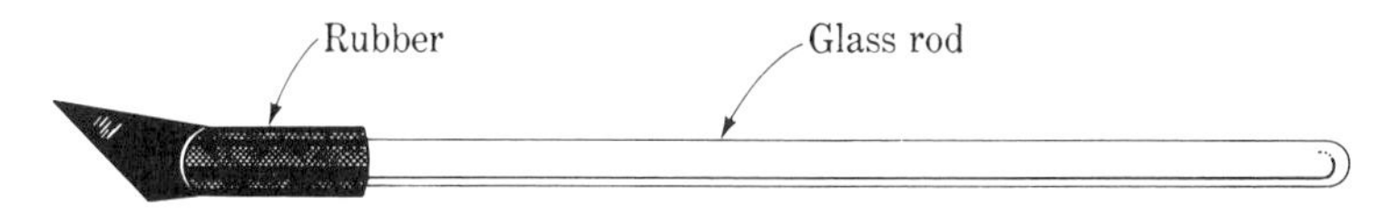

Fig. 9-10 Rubber policeman.

Transferring the precipitate Precipitates collected in centrifuge tubes can usually be dissolved or otherwise treated directly in those tubes. Precipitates collected on filters can be washed off the paper with the wash bottle and liquid removed by centrifuging, or they can be removed by scraping with a rubber policeman (Fig. 9-10). Precipitates should never be allowed to dry before being handled.

Solvent Extraction Molecular or complex species commonly dissolve to different extents in different liquids. If two liquids, which we may designate as A and B, that are either immiscible or only partially miscible are mixed in contact with a species of this type and the system then allowed to stand until the two liquid layers separate, the solute distributes itself between the two liquids in terms of the relationship

$$\lambda = \frac{C_A}{C_B} \qquad (9\text{-}12)$$

where λ is a distribution coefficient or ratio and C_A and C_B are the concentrations in the two liquid phases. If the distribution ratio is large enough, essentially all of the solute concentrates in one liquid layer, and the solute is thus removed from the other liquid by solvent extraction. Thus for iodine in a carbon disulfide–water system at room temperature,

$$\lambda = \frac{C_{CS_2}}{C_{H_2O}} = 625$$

It follows by simple calculation that if the concentration of iodine in the carbon layer after equilibration is 0.1 mole liter^{-1}, its concentration in the aqueous layer cannot exceed 0.00016 mole liter^{-1}.

Applications of solvent extraction that are important in the laboratory work described in Chaps. 10 and 11 include detection of iodine by extraction into carbon tetrachloride (page 222), detection of chromium(VI) by extraction of the blue peroxo compound CrO_5 into ether (page 227), and detection of cobalt(II) by extraction of the blue complex species $[Co(NCS)_4]^{2-}$ into amyl alcohol-ether or benzyl alcohol (page 260).

Ion-exchange chromatography If the distribution ratio between two solvents [Eq. (9-12)] is small, the removal of the solute in question from one liquid phase requires many separate batchwise solvent extraction steps. It is apparent that a process of this type would be slow and inefficient for the separation of two solutes from each other. Efficiency can be increased markedly by adsorbing the solvent containing the solute species upon rather finely divided inert particles held in a vertical column, allowing the other solvent to flow slowly through the column, and collecting the eluate. Partition occurs at each particle, thereby enhancing the separation. This type of process, involving a stationary and a moving phase, is an example of column chromatography.

The solid support can itself act as the stationary phase if its molecular structure is such that it can contain exchangeable cations or anions. Separation of species by ion exchange then results by the displacement of one ion by another as a solution moves down a column on which the ions in question have been adsorbed. A practical case in point is the "softening" of hard water by exchanging the calcium ions for sodium ions when the water is passed through a column of a cation exchanger containing sodium ions. If the insoluble anionic exchanger is represented by R^-, the process can be described by the equilibrium equation

$$2\mathbf{NaR}(s) + Ca^{2+}(aq) \rightleftharpoons \mathbf{CaR_2}(s) + 2Na^+(aq)$$

The more highly charged Ca^{2+} ion bonds the more firmly to the exchanger and thus displaces the Na^+ ion. Inasmuch as the process is reversible, the Ca^{2+} ion can be replaced by Na^+ ion in a regeneration cycle by treating with a concentrated sodium salt solution.

Anion exchange is particularly useful in the separation of cobalt(II) from nickel(II) (page 203). Both ions form tetrachloro anions in the presence of sizeable quantities of chloride ion,

$$M^{2+}(aq) + 4Cl^-(aq) \rightleftharpoons [MCl_4]^{2-}(aq)$$

If a solution containing the two anions $[CoCl_4]^{2-}$ (blue to green) and $[NiCl_4]^{2-}$ (pale greenish) is passed through an anion exchange column containing chloride ion, these ions are adsorbed by displacement of the chloride. The two chloro anions have different formation constants (Chap. 6, Appendix F), however, and elution with hydrochloric acid of the correct concentration removes the nickel ion more readily as a consequence of the greater ease of dissociation of its chloro complex.

Anion and cation exchangers that are available commercially are most commonly synthetic resins of appropriate composition. The experimental arrangement for effecting ion-exchange chromatographic separations is shown in Fig. 9-11.

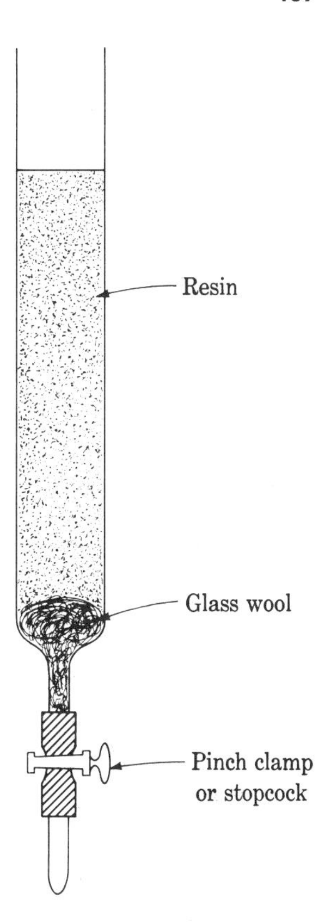

Fig. 9-11 Ion-exchange column.

OUTLINE OF LABORATORY OPERATIONS

After the student has familiarized himself with new items of apparatus and equipment, has prepared these additional pieces of apparatus as suggested in earlier sections of this chapter, and has been completely oriented to the laboratory, its rules, and its objectives, he should then be required to carry out the introductory preliminary experiments which appear at the end of this chapter. These experiments are designed to acquaint him with various types of manipulation, to give him some experimental background, and to indicate to him the scope, limitations, and sensitivities of the laboratory operations. These studies are then followed by the identification of individual anions in selected unknowns. Systematic identification of the members of the various cation groups (Chap. 11) is introduced through analyses of "knowns." Unknowns for the individual cation groups are employed to test the accuracy of the knowledge so obtained, and finally general unknowns are analyzed to summarize information for anions and cations combined. Details of all such assignments are left to the individual instructor.

THE WRITTEN LABORATORY RECORD

No person can obtain the maximum benefit from laboratory investigation unless he sets down his observations in a complete written record. This record should be kept in a permanently bound notebook of convenient size and should be compiled in a neat and legible form which is equally intelligible to both student and instructor. It is imperative that all records be made in the laboratory and concurrently with the laboratory studies. A record made at odd times after an experiment is completed can never fix the details of that exercise as firmly in the mind of the student as a record made when the observation is still fresh and the actual result is before him. The written record should be sufficiently detailed and clear so that the student can use the information obtained for subsequent study. This item is of particular importance in regard to unknowns, where the final conclusion as to the presence or absence of various ions must be based upon the combined results of a number of experiments—experiments which may have been carried out over a considerable period of time.

Records for preliminary experiments Notes for a preliminary experiment can be written in outline or semioutline form. All operations carried out, all chemical reactions investigated, and all results obtained should be included. Conclusions based upon the experiment must be summarized. The liberal use of ionic equations to describe reactions is encouraged. In these equations, it is desirable to indicate clearly the solid substances [e.g., by a boldface formula, as $\mathbf{PbCrO_4}$; or designating by a small (*s*), as $PbCrO_4(s)$] and the gaseous substances [e.g., by inserting a vertical arrow or a (*g*) after the formula, as $Cl_2\uparrow$ or $Cl_2(g)$]. An indication of the color, general appearance, or other interesting property of a substance written under its formula provides useful information for subsequent study. These points are illustrated for an exercise on precipitation (Expt. I-1) as follows:

A yellow precipitate forms,

$$2Pb^{2+} + Cr_2O_7^{2-} + H_2O \rightleftharpoons \underset{\text{yellow}}{\mathbf{PbCrO_4}} + 2H^+$$

Completeness is determined by either allowing the precipitate to settle or centrifuging and then adding a drop of the dichromate solution to the clear supernatant liquid.

Records for knowns and unknowns These records are conveniently written in outline form. A type of record which combines simplicity with the systematic arrangement of extensive detail is indicated below for the typical use of a cation group I sample.

Unknown No. ______ Date received: ______
Materials present: Cation Group I Date completed: ______
Nature of sample: Colorless solution Ions found: Pb^{2+}, Hg_2^{2+}, Ag^+

No.	*Material treated*	*Reagent or treatment*	*Observation*	*Conclusion*
1	Unknown solution	Cold, 1 *M* HCl	White, curdy ppt.	Cation group I present
2	Precipitate 1	H_2O, boil	White residue	AgCl or Hg_2Cl_2 probable
3	Centrifugate 2	K_2CrO_4	Yellow ppt.	Pb^{2+} present
4	Residue 2	aq NH_3	Black residue	Hg_2^{2+} present
5	Centrifugate 4	Dil HNO_3 to acid	White ppt.	Ag^+ present

The student should note that the material treated is designated by a number that refers to the line from which that material was obtained. Some instructors may prefer to add another column in which equations describing the observations are recorded.

Records for additional items The laboratory record should contain answers to selected exercises that appear at the ends of Chaps. 10 and 11. At the discretion of the instructor, appropriate flow sheets or other schematic outlines may be inserted in the notebook prior to the records on unknowns.

Both student and instructor will find it advantageous for the latter to check the laboratory record carefully at the conclusion of each phase of the laboratory work.

INTRODUCTORY PRELIMINARY EXPERIMENTS

The laboratory work is conveniently introduced by means of general preliminary experiments of the following categories: (1) laboratory manipulations, (2) separations, (3) tests for various species, and (4) applications of solution theory. The results of each experiment should be recorded in the laboratory notebook.

Laboratory manipulations The commonest operations are concerned with the formation and handling of precipitates.

Experiment I–1 Precipitation or crystallization

By temperature change

Place sufficient solid lead(II) chloride in a test tube to fill the rounded bottom, fill the test tube half full of water, and boil gently for 1 to 2 min. Decant the clear liquid into another test tube, and cool by placing this tube in a beaker of cold water. Record your observations, and explain in terms of numerical data for the solubility of lead(II) chloride.

By chemical reaction

Place 4 drops of 0.1 *M* lead(II) nitrate solution in a test tube, dilute to 1 ml, and add 4 drops of 0.1 *M* potassium dichromate solution, shaking after each drop. Record your observations, and give an appropriate equation. How could you determine whether or not precipitation had been complete? Try your method.

Place 4 drops of 0.1 *M* copper(II) nitrate solution in a clean 50-ml flask, add 1 drop of 1 *M* hydrochloric acid solution, dilute to 2 ml, add several drops of 1 *M* thioacetamide solution, and heat to boiling. Note the rate of sulfide formation, the appearance of the precipitate, and the rate of settling of the precipitate. Record your observations, and write the necessary equation.

By change of solvent

Place 1 ml of saturated sodium chloride solution in a test tube, and add ethyl alcohol drop by drop (up to 2 to 3 ml). Account for your observations.

Experiment I-2 Removal of mother liquor from a precipitate Dilute 10 drops of 0.1 *M* lead(II) nitrate solution to 10 ml in a small beaker, and add 3 *M* sulfuric acid solution drop by drop until precipitation is apparently complete and excess sulfate ion is present. Stir the suspension to prevent settling, and divide into three equal portions. Allow one portion to stand, and remove the mother liquor by decantation. Centrifuge the second (page 192) for 1 to 2 min, again removing the mother liquor by decantation. Pour the third through a filter paper contained in a small funnel, and collect the filtrate, allowing the liquor to drain completely. Retain all three samples of precipitate for Expt. I-3. Compare the times necessary for complete removal of mother liquor. Discuss briefly the relationship between ease of removal of mother liquor and crystal form.

Experiment I-3 Washing the precipitate Use the precipitates retained from Expt. I-2. To the sample freed from mother liquor by simple decantation, add 2 ml of water, shake thoroughly, allow to settle, and decant the supernatant liquid into a clean test tube. Repeat the operation three more times, preserving the washings each time in a separate test tube. To each tube, add 2 drops of 0.1 *M* barium chloride solution. Record the time necessary to complete these washing operations, and indicate roughly the relative amount of sulfate ion remaining after each washing from the quantity of precipitate obtained.

Repeat these operations with the sample freed from mother liquor by centrifuging, using 2-ml volumes of wash water in each case, but centrifuging instead of using gravity sedimentation. Record your results similarly.

Repeat again with the sample freed from mother liquor by filtration by pouring 2-ml volumes of water on the precipitate on the filter paper and collecting each washing separately. Again record your results similarly. Compare the relative efficiencies of the three washing procedures in terms of your observations.

Repeat the precipitation outlined in Expt. I-2, stir vigorously, and divide the suspension into two *equal* portions in test tubes. Centrifuge both, and discard the centrifugates. Wash one precipitate by centrifugation, using *five* separate 2-ml volumes of water. Check each wash liquor for sulfate-ion content as before. Wash the second precipitate with *two* separate 5-ml volumes of water, again checking for sulfate-ion content. Reserve the residues for Expt. I-4. Compare the efficiencies of the two techniques of washing. What qualitative support do these results give to Eq. 9-11?

Experiment I-4 Transferring the precipitate Suspend one residue from Expt. I-3 in 5 ml of water by directing a stream of water from the wash bottle directly into the centrifuge tube. Quickly decant the suspension into a small beaker, using a small volume of water from the wash bottle to transfer any remaining precipitate to the beaker. Stir and decant into a second beaker. This time, remove any precipitate adhering to the interior of the first beaker by adding water, scraping with the rubber policeman (Fig. 9-10), decanting, and ultimately washing with a fine stream of water from the wash bottle.

Suspend the second residue from Expt. I-3 in water, and decant through a filter paper. Allow to drain. Carefully remove the paper containing the solid from the funnel, unfold partially, and while holding over a small beaker, wash the solid into the beaker, using the wash bottle. Refilter, and this time remove the precipitate by unfolding the paper, placing it on a clean glass plate or watch glass, and scraping with the rubber policeman. Indicate clearly in your notes the influence of the physical character of the precipitate upon the mode and ease of its transfer.

Separations Separations of importance in qualitative analysis may involve selective processes such as precipitation, solvent extraction, volatilization, dissolution, oxidation or reduction, complex-ion formation, amphoterism, or ion exchange.

Experiment II-1 Precipitation Dilute a mixture of 2 to 3 drops of 0.1 *M* sodium chloride solution and 2 to 3 drops of 0.1 *M* sodium orthophosphate solution to 2 ml, and acidify with 6 *M* nitric acid solution. Add 0.1 *M* silver(I) nitrate solution drop by drop until precipitation is complete. Identify the precipitate. Centrifuge. Neutralize the centrifugate by adding 6 *M* aqueous ammonia solution drop by drop (avoid excess), and identify the precipitate. In terms of the properties of the anions involved (Appendix

H_a), indicate why a separation is effected. Write ionic equations for all reactions involved.

Experiment II-2 Solvent extraction Dilute 2 to 3 drops of 0.1 *M* potassium iodide solution to 2 ml, and add several drops of 0.2 *M* potassium nitrite solution. Acidify with 3 *M* sulfuric acid solution. Note and account for the observed color change. Add 1 ml of carbon tetrachloride, and shake. Note and account for your observations in terms of numerical data from some suitable reference on the solubility of iodine in water and carbon tetrachloride at room temperature. Prepare a filter paper moistened with *water* in a funnel, and decant the two-phase liquid system through it. Repeat the experiment, but this time use a dry filter paper in a *dry* funnel, and wet the paper thoroughly with *carbon tetrachloride* just before decanting the mixture into it. Account for the difference, and suggest a general method for separating immiscible liquids. What other techniques can be employed? Describe all chemical reactions with balanced equations.

Experiment II-3 Volatilization Carefully evaporate just to dryness in a casserole over the water bath a solution prepared by mixing 5 drops of 0.1 *M* ammonium chloride solution with 5 drops of 0.1 *M* potassium chloride solution. Scrape a small portion of the residue into a small beaker, and apply a flame test to it (page 270), using a clean platinum wire. Then add 1 ml of 6 *M* sodium hydroxide solution, and warm (do not boil) gently, testing the vapors evolved with a piece of moist red litmus or other indicator paper (page 267). Interpret the results of these two tests. Next, carefully heat the casserole and its remaining contents over an open flame until evolution of white fumes ceases. Allow to cool, and then apply the flame and sodium hydroxide tests to the residue in the casserole. Interpret your results, and account clearly for differences noted between the two sets of results. Write pertinent ionic equations.

Experiment II-4 Dissolution of precipitate Repeat the initial precipitation in Expt. II-1, but omit the addition of the nitric acid solution. Centrifuge and wash once with 1 ml of water, discarding the centrifugate and washings. Add to the residue in the centrifuge tube 1 ml of water and 1 ml of 6 *M* nitric acid solution. Shake thoroughly, recentrifuge, and decant the centrifugate. Note and interpret the color change in the solid. Add 6 *M* aqueous ammonia solution drop by drop (avoid excess) to the centrifugate. Account for your observations. Discuss clearly the principles illustrated by this separation.

Experiment II-5 Oxidation To a solution prepared by mixing in a small beaker 2 to 3 drops of 0.1 *M* iron(III) chloride solution and 2 to 3 drops of 0.1 *M* chromium(III) chloride solution and diluting to 1 ml, add 6 *M* sodium hydroxide solution until the resulting suspension is strongly alkaline. Note chemical and color changes. Add 3 percent hydrogen peroxide solution drop by drop until no further reaction (beyond oxygen evolution) takes place. Centrifuge, and note colors of both residue and centrifugate. Indicate the separation which has occurred and its chemical basis (Appendixes H_a, H_c). Record appropriate ionic equations for all reactions that have occurred.

Experiment II-6 Reduction Dilute a mixture of 2 to 3 drops of 0.1 *M* copper(II) nitrate solution and 2 to 3 drops of 0.1 *M* cadmium nitrate solution to 5 ml. Remove 1 ml of this solution, add several drops of 1 *M* thioacetamide solution, and heat to boiling. Add a few drops of 3 *M* sulfuric acid solution to the remaining solution, and then some iron filings. Boil gently until the color of the copper(II) ion disappears. Centrifuge or allow to settle. If the clear decantate is not acidic, make just acidic with the sulfuric acid solution. Then add several drops of 1 *M* thioacetamide solution and heat to boiling.

Interpret the observed differences between the two sulfide precipitates. Write ionic equations for all reactions which occurred. Explain clearly why iron acts as a selective reducing agent in this system.

Experiment II-7 Complex-ion formation In separate test tubes, dilute 2 to 3 drops each of 0.1 *M* bismuth(III), aluminum, copper(II), and zinc nitrate solutions to 2-ml total volumes. To each, add 6 *M* aqueous ammonia solution drop by drop, shaking after each drop, until no further changes are apparent. Record observed changes, and describe each by means of an ionic equation (Chap. 6). Predict, on the bases of these observations, pairs of these cations which can be separated by use of excess aqueous ammonia and pairs which cannot. Check experimentally the validity of your prediction for each type of pair, and record your observations.

Experiment II-8 Amphoterism Repeat Expt. II-7, but substitute 6 *M* sodium hydroxide solution for the aqueous ammonia solution (Chap. 6). Record your results similarly, and test your predictions in the same fashion.

Experiment II-9 Ion-exchange chromatography Prepare an ion-exchange column (Fig. 9-11) from a 15-cm length of 10-mm [outside diameter (OD)] tubing by pouring into the tube a slurry of 100-mesh Dowex 1-X8 resin in distilled water to a resin height of 7 to 8 cm. Permit the column to drain until the liquid level is at the level of the top of the resin. Allow about 10 ml of 10 *M* hydrochloric acid, prepared by dilution of the 12 *M* acid, to pass through the column by adjusting the pinch clamp to a flow rate of 1 to 2 drops every 2 sec. Do not permit the liquid level to drop below the top of the resin. Prepare a solution by evaporating 2 ml each of 0.1 *M* cobalt(II) and nickel(II) chloride solutions to dryness and dissolving the residue in 2 ml of 10 *M* hydrochloric acid. Pour this solution on the column and drain until the liquid level reaches the top of the resin. Arrange a series of ten test tubes to collect the eluate in 5-ml fractions. Pour 10 *M* hydrochloric acid through the column, adjust the pinch clamp to a constant flow rate of 5 to 10 drops per second, and collect the fractions. Note color changes on the column as flow continues and the colors of the eluted fractions. Test separate portions of each fraction for nickel(II) by barely neutralizing with 6 *M* aqueous ammonia, saturating with solid sodium acetate, and adding 1 percent dimethylglyoxime solution (*Procedure C*-26, page 261) and for cobalt(II) by adding amyl alcohol-ether mixture or benzyl alcohol, 1 to 2 g solid ammonium thiocyanate, and shaking (*Procedure C*-25, page 260). Record your observations, account for the separation observed, and indicate whether the procedure is quantitative or not. Write ionic equations for all reactions observed.

Tests A *test* is a laboratory operation which is designed to show definitely the presence or absence of a particular species. As such, tests may be *positive* or *negative*. A test that is applied for final proof of the presence of an ion after separations have been completed is termed a *confirmatory* test. A test that is given only by the species in question and by no other species which may be present is called a *specific* test. It is often desirable to compare the results of a particular test, either with the behavior of the reagents alone under comparable conditions (*blank* test) or with the behavior of a known sample of the species in question (*control* test). Tests commonly involve (1) formation or dissolution of a precipitate, (2) formation or reaction of a gas, (3) development of a characteristic odor, or (4) formation or disappearance of a color. Each such test has its own *sensitivity*, i.e., the smallest quantity or concentration of the particular species detectable by that test.

Experiment III-1 Precipitation Dilute 2 to 3 drops of 0.1 *M* silver(I) nitrate solution to 1 ml, and add 0.1 *M* sodium chloride solution drop by drop until reaction is complete. Write the appropriate ionic equation. Centrifuge and wash once with 1 ml of water, discarding the centrifugate and the washings. Suspend the residue in 0.5 ml of water, and add 6 *M* aqueous ammonia solution drop by drop until a clear solution results. Write an ionic equation. Add 6 *M* nitric acid until an acidic reaction results. Write an ionic equation. Indicate, on the basis of your results, how one would test for chloride ion or for silver(I) ion. In consultation with your instructor and by reference to Appendix H, indicate clearly why the initial precipitation reaction does not constitute a completely positive test for either species.

Experiment III-2 Gas formation To 0.1 g of solid sodium carbonate in a small test tube, add 2 to 3 drops of 3 *M* sulfuric acid solution. Carefully check the evolved gas for odor. Add more of the sulfuric acid solution, and this time observe what happens when a drop of saturated barium hydroxide solution held on the end of a glass tube is brought in contact with the evolved gas. Write appropriate ionic equations, and outline a suitable test for carbonate ion. In consultation with your instructor and by reference to Appendix H_a, indicate whether or not there may be interferences with this test as outlined.

Experiment III-3 Odor To 5 drops of 0.1 *M* sodium acetate solution in a small test tube, add 5 drops of ethyl alcohol and 3 to 4 drops of 18 *M* sulfuric acid solution. Warm gently, and carefully note the odor. Carry out a *blank* test in the same fashion, omitting the sodium acetate, and compare the two odors.* Write an appropriate equation.

Experiment III-4 Color reactions Dilute 2 to 3 drops of 0.1 *M* potassium chromate or dichromate solution to 2 ml in a small test tube and make definitely acidic with 3 *M* sulfuric acid solution. Add 0.5 ml of ether (*Caution.* Keep away from flames!) and 1 to 2 drops of 3 percent hydrogen peroxide solution. Shake *once*, and allow to stand. The blue peroxochromic acid (page 227) in the ether layer is useful as a test for either chromium(VI) or peroxide.

Dilute 2 to 3 drops of 0.1 *M* iron(III) chloride solution to 2 ml, and add 1 to 2 drops of 0.1 *M* ammonium thiocyanate solution. Note the reaction, and write an appropriate equation. Then add 1 ml of ether, shake thoroughly, and observe. Is the test rendered more obvious?

Experiment III-5 Sensitivity of a test From the iron(III) chloride test solution (concentration = 5 mg of $Fe^{3+}ml^{-1}$), prepare *two* 5-ml volumes each of solutions of the following concentrations: (*a*) 1 mg $Fe^{3+}ml^{-1}$, (*b*) 0.1 mg $Fe^{3+}ml^{-1}$, (*c*) 0.01 mg $Fe^{3+}ml^{-1}$, (*d*) 0.001 mg $Fe^{3+}ml^{-1}$, (*e*) 0.0001 mg $Fe^{3+}ml^{-1}$, and (*f*) 0.00001 mg $Fe^{3+}ml^{-1}$. To do this, measure exactly 4 ml of the initial test solution into a graduated cylinder, and dilute to exactly 20 ml. The concentration of this solution will be 1 mg $Fe^{3+}ml^{-1}$. Save two 5-ml volumes of this solution. Discard all but exactly 2 ml of the remaining solution, and dilute this volume to exactly 20 ml. The concentration of this solution will be 0.1 mg $Fe^{3+}ml^{-1}$. Again save two 5-ml volumes. Continue this dilution procedure to produce the other listed concentrations. *Label* each set of solutions to avoid confusion, and place at the end of the series two additional test tubes, each containing 5 ml of water.

To each test tube, add 2 drops of 6 *M* hydrochloric acid solution and exactly 1 ml of 1 *M* potassium or ammonium thiocyanate solution. Shake each tube, avoiding contact of the fingers with the contents. Allow one series of tests (1 mg $Fe^{3+}ml^{-1}$ through distilled water) to stand. To each

* In checking for odor, sniff cautiously, but *never* inhale deeply. Gases may produce toxic effects if inhaled. Furthermore, large quantities of gases may dull the sense of smell and lead to false conclusions.

tube in the second series, add 1 ml ether, and shake again. Note in each series the smallest concentration of iron(III) ion giving a detectable color, and describe the *sensitivity* of the test in milligrams Fe^{3+} per milliliter and in gram ions (moles) per milliliter. Account for any differences between the two series.

Applications of solution theory Quantitative illustrations of the validity of equilibrium-constant expressions and their exact application to solution chemistry must be deferred until the more precise techniques of quantitative analysis and physical chemistry have been studied. It is possible, however, to illustrate in a general way some of the qualitative aspects of relationships such as (1) acid-base equilibria, (2) equilibria involving precipitates, and (3) equilibria involving complex ions.

Experiment IV-1 Acid-base equilibria Measure into separate containers 10-ml volumes of 6 *M* acetic acid solution and 6 *M* sodium acetate solution. Determine the approximate pH (page 72) of each solution with an indicator paper (such as "pHydrion"). Mix the two solutions thoroughly, and again determine the approximate pH. Account for any observed change in hydronium-ion concentration. Divide the solution into ten 2-ml portions, saving one portion for comparison. To four of the remaining tubes, add the following volumes of 6 *M* hydrochloric acid solution: 0.5 ml, 1.0 ml, 1.5 ml, 2.0 ml. Where necessary, add distilled water to make the total volume in each case the same (= 4 ml). Repeat with four more of the tubes, substituting 6 *M* sodium hydroxide solution for the hydrochloric acid solution. Dilute the contents of the remaining tube to 4 ml with distilled water. Determine the pH of each solution as before. In separate test tubes, dilute 0.5 ml each of 6 *M* sodium hydroxide solution and 6 *M* hydrochloric acid solution to 4.0 ml with distilled water, and determine the approximate pH values. Record your results, and account qualitatively for your observations in terms of the equilibrium involving acetic acid and its ions.

Experiment IV-2 Precipitation equilibria Add 35 ml of distilled water to 1 g of solid silver(I) acetate, and heat the suspension nearly to boiling, stirring vigorously. Disregard any darkening that may occur. Filter while hot, and allow the filtrate to cool to room temperature. Some white crystals of silver acetate should form, showing the solution to be saturated. To 10 ml of this saturated silver acetate solution, add 0.5 to 1.0 ml of saturated silver(I) nitrate solution. To another 10-ml portion, add 0.5 to 1.0 ml of saturated sodium acetate solution. What is the precipitate in each case? Account for its formation in terms of the equilibrium existing in the saturated silver acetate solution. To a third 10-ml portion, add 0.5 to 1.0 ml of saturated sodium nitrate solution. Account for your observation. To the suspension from the silver nitrate addition, add 6 *M* nitric acid solution until the precipitate dissolves. Treat the other suspension from the sodium acetate addition similarly with 6 *M* aqueous ammonia solution. Account for your observations in terms of the equilibria involved.

Experiment IV-3 Complex-ion equilibria Precipitate silver(I) chloride, and dissolve in aqueous ammonia as in Expt. III-1. To the resulting solution, add 0.1 *M* potassium iodide solution drop by drop until precipitation is complete. Add to the alkaline suspension with constant shaking 0.1 *M* potassium cyanide solution (*Caution.* Poison! Avoid contact with body, mouth, or nose!) until dissolution is complete. Saturate the final solution with hydrogen sulfide. Describe each reaction by an appropriate ionic equation. Give an appropriate equilibrium expression for each silver-containing species present, and give the corresponding equilibrium constant (Appendixes E and F). Account for the sequence of reactions observed in terms of these data. Could the reactions be made to take place in the reverse order? Why?

10
Identification of Anions

The identification of each anion present in a given sample can be accomplished either by means of a systematic procedure involving its prior separation from other anions or by means of a series of characteristic tests that give meaningful results without requiring detailed, systematic separations. For routine analyses where the emphasis is upon the result rather than upon the instructional value of obtaining it, the first approach is advantageous. However, this approach does deemphasize the importance of careful observation and thoughtful interpretation of the reactions of the ions. Experimentally, each anion can be identified in terms of its reactions with a number of selected reagents. Although for each reagent interfering reactions by other anions commonly exist, identification can be effected in terms of the summed-up behavior of the sample with several reagents. This type of approach emphasizes the importance of painstaking observation in the laboratory, the necessity for correlating the results of a number of reactions, and the fact that the procedures of qualitative analysis need not be completely "cookbookish" in nature. It offers, also, continuing instruction in the chemical characteristics of the anions.

As such, then, the classification of the anions into four groups, as outlined in Chap. 8, is more an arrangement of convenience than of completely practical utility. However, this classification does permit the student to make rather broad group eliminations when he analyzes a particular sample (Procedure A-7, page 213). It is also a classification that could be adapted to systematic separations, for groups of anions can be removed successively from a sample as (1) gases or precipitates through distillation with dilute perchloric acid, (2) silver(I) salts through precipitation from acidic medium, and (3) silver(I) salts through precipitation from neutral or slightly alkaline medium.

The identification of the anions, as outlined in this chapter, involves:

1. A detailed *preliminary examination* of the sample through the application of tests that give extensive information about the presence or absence of the various anions.
2. *Specific tests* for individual anions to confirm the interpretations made in the preliminary examination and to prove definitely the presence or absence in the sample of a given anion for which the preliminary examination does not provide definitive information.

Each step in the preliminary examination contributes to the final result. It is particularly important, therefore, that the result of each experiment be made a part of the permanent written record (pages 198 and 199) and that all possible identifications and eliminations be made for each step. When the preliminary examination has been completed, all of the information obtained must be summarized to avoid conflicting conclusions and to suggest those specific tests that must be carried out for final confirmation.

PREPARATION OF THE SOLUTION FOR DETECTION OF THE ANIONS

Samples to be used for analysis may be obtained either as solutions or as solids. Although it is usually possible to use a solution, as obtained, for most of the required tests, only a few of these tests can be applied directly to a solid sample. For most purposes, therefore, a solid must be dissolved. Whether or not a given solid is soluble in water can be determined by shaking a small quantity (<0.1 g) with 5 ml of cold water and warming if necessary. If the sample is not soluble in water, one may be tempted to dissolve it in an acid. Although this procedure is adequate for the ultimate detection of the cations (page 238), certain anions are destroyed by acids, and the oxidizing and reducing strengths of other anions are sufficiently enhanced in acidic media that these species may mutually destroy each other. In alkaline media, however, the anions are quite generally stable both with respect to decomposition to gaseous products and with respect to destruction by oxidation-reduction reactions. Dissolution of the sample under alkaline conditions is, therefore, desirable. A further complication is that cations other than sodium, potassium, and ammonium interfere with certain of the necessary tests. Elimination of heavy-metal cations from solutions to be analyzed for the anions is also essential.

Dissolution of the sample under alkaline conditions and removal of heavy-metal cations can be accomplished simultaneously by boiling the solid sample with saturated sodium carbonate solution. When a difficultly soluble salt is heated with this solution, an insoluble heavy-metal carbonate or its hydrolysis product precipitates and the anion is released to the solution, as

$$\mathbf{MA}(s) + CO_3^{2-}(aq) \rightleftharpoons \mathbf{MCO_3}(s) + A^{2-}(aq)$$

or

$$\mathbf{MA}(s) + CO_3^{2-}(aq) + H_2O(l) \rightleftharpoons \mathbf{M(OH)_2}(s) + CO_2(g) + A^{2-}(aq)$$

Whether a reaction of this type can take place depends upon the relative solubilities of the original salt and the corresponding carbonate, basic carbonate, or hydroxide (see Procedure A-2, Note 3, page 211). However, in all but a few cases, the conversion as indicated is favored.

Similarly, solutions received as such or obtained by dissolving samples in water can be rendered alkaline and freed from interfering cations by treating with sodium carbonate solution.

Treatment with sodium carbonate does add carbonate ion to the solution that is then to be tested for the anions and thus requires that the carbonate ion always be detected in a separate portion of the original sample. The solution obtained by this treatment is referred to in subsequent procedures as the "sodium carbonate prepared solution."

Procedure A-1 Treatment of solutions or water-soluble samples Use a 10-ml volume of the solution as submitted, or dissolve 0.10 to 0.15 g of the solid sample by warming with 5 ml of water, and dilute to 10 ml (Note 1). Test the solution with indicator paper, and, if acidic, add 6 *M* sodium hydroxide solution until faintly alkaline. To the alkaline solution, add 1 to 2 drops of saturated sodium carbonate solution (Note 2). If no precipitate forms, use the solution directly for indicated anion tests. If a precipitate forms, add 2 ml of saturated sodium carbonate solution, heat to boiling in a casserole, and boil gently for at least 10 min, adding water to replace that lost by evaporation. Centrifuge, and save the centrifugate for anion tests. Test the solubility of a small portion of the residue in warm 6 *M* acetic acid solution. If the residue dissolves, it consists entirely of carbonates or hydroxides and conversion of all anions to soluble sodium salts may be assumed to be complete. If the residue is only partially soluble or is completely insoluble in acetic acid solution, re-treat it with 2 ml of saturated sodium carbonate solution, and add the centrifugate to the anion solution. Dilute the sodium carbonate prepared solution to 10 ml before using for anion tests.

Procedure A-2 Treatment of water-insoluble samples To 0.10 to 0.15 g of the finely ground sample in a casserole, add 2 ml of saturated sodium carbonate solution (Note 2). Cover with a clean watch glass, and heat to boiling. Boil gently for 10 min., adding small quantities of water periodically to prevent evaporation to dryness. Wash the residue and the solution into a small test tube with 3 to 4 ml of water. Cool and centrifuge. Wash the residue with 2 ml of water, adding the washings to the initial centrifugate. If the residue is not soluble in 6 *M* acetic acid solution (Procedure A-1), reboil with 2 ml of saturated sodium carbonate solution, centrifuge, and add the centrifugate to the solution from the initial sodium carbonate digestion. Any final residue that is insoluble in acetic acid should be discarded, reserved for indicated specific tests, or treated as outlined in Procedure C-41, page 284 (Note 3). Dilute the sodium carbonate prepared solution to 10 ml and reserve for anion tests.

Note 1 The quantity of solid should be within the limits suggested. Rough or reasonably accurate weighing is indicated.

Note 2 The sodium carbonate used must be chemically pure. Samples containing chloride ion or sulfate ion cannot be used. These anions can be detected by diluting 2 to 3 drops of the saturated sodium carbonate solution to 1 ml, acidifying with 6 *M* nitric or perchloric acid solution, dividing into two portions, and adding 0.1 *M* silver(I) nitrate solution to one and 0.1 *M* barium nitrate solution to the other.

Note 3 Inasmuch as carbonates, basic carbonates, or hydroxides are less soluble in alkaline medium than the majority of the other salts, solubilization of salts of many of the anions is effected. With most sulfides, a number of phosphates and arsenates, and silver bromide or iodide, however, the reverse is true. With these substances, tests must be run either upon original samples or upon the residue from the sodium carbonate treatment. Even if solubility differences are moderately unfavorable, increased concentration of carbonate ion may assure adequate conversion. Thus, conversion of barium chromate ($K_{sp} = 8.5 \times 10^{-11}$ mole^2 liter^{-2}) to barium carbonate ($K_{sp} = 1.6 \times 10^{-9}$ mole^2 liter^{-2}) might not appear possible. At equilibrium, the concentration of barium ion in contact with each precipitate is the same. Hence,

$$\frac{\cancel{C_{Ba^{2+}}} \times C_{CrO_4^{2-}}}{\cancel{C_{Ba^{2+}}} \times C_{CO_3^{2-}}} = \frac{8.5 \times 10^{-11}\ \cancel{\text{mole}^2\ \text{liter}^{-2}}}{1.6 \times 10^{-9}\ \cancel{\text{mole}^2\ \text{liter}^{-2}}}$$

$$\frac{C_{CrO_4^{2-}}}{C_{CO_3^{2-}}} = 5.3 \times 10^{-2}$$

or

$$C_{CrO_4^{2-}} = 5.3 \times 10^{-2} C_{CO_3^{2-}}$$

emphasizing this point. However, if the concentration of carbonate ion present is 1.5 *M* (saturated sodium carbonate solution), the concentration of chromate ion in the resulting solution is

$$C_{CrO_4{}^{2-}} = 5.3 \times 10^{-2} \times 1.5 \text{ mole liter}^{-1} = 0.08 \text{ mole liter}^{-1}$$

which is sufficient to detect.

PRELIMINARY EXAMINATION FOR ANIONS

The following procedures may or may not all be applicable, depending upon the nature of the sample as received. For general unknowns, these procedures should be supplemented by information on the solubility of the sample as related to its cation content (pages 7–9).

Procedure A-3 General examination of the sample Solid samples should be checked for color, crystalline form, and homogeneity (Notes 1, 2, 3). Solutions should be checked for color.

Note 1 Color can arise from colored anions, colored cations, or a high degree of covalency in solid salts containing colorless ions. It is therefore not an absolute criterion of ion content, but it may be indicative. White solid samples or colorless solutions obviously cannot contain colored ionic species. Colored anions are chromate (yellow), dichromate (orange), and permanganate (purple).

Note 2 Crystal form can never serve as more than a qualitative indication of solid composition since all compounds crystallize in a limited number of geometrical arrangements and since perfection in crystals as obtained is commonly the exception rather than the rule. On occasion, this type of observation can be helpful, but routine comparison by the student of his samples with the reagents on the side shelf is a waste of time.

Note 3 Complete homogeneity in a solid commonly indicates the presence of but a single salt in the sample, but complete homogeneity or its absence is not an absolute criterion. Thus, mixtures of potassium chloride and bromide might appear homogeneous, whereas mixtures of anhydrous sodium carbonate and one of its hydrates would be heterogeneous and still contain a single anion.

Procedure A-4 Presence of free acids in solutions This test is applicable only to samples obtained as solutions. Place a drop of the solution upon a piece of litmus paper or other suitable indicator paper, or place a drop of methyl orange in a depression on a white spot plate, and add 3 to 4 drops of water and 1 drop of the solution in question. A definitely acidic reaction indicates the probable absence of the following species: $CO_3{}^{2-}$, $SO_3{}^{2-}$, $S_2O_3{}^{2-}$, S^{2-}, and $NO_2{}^-$ (Note 1).

Note 1 Sulfurous acid or thiosulfuric acid certainly can exist in dilute solution; so the results of this observation are not absolute. The test has no real significance when applied to solutions obtained by dissolving solid samples, nor is it applicable to the sodium carbonate prepared solution.

Procedure A-5 Presence of oxidizing agents To 1.0 ml of a 0.1 *M* solution of manganese(II) chloride in 12 *M* hydrochloric acid solution, add 1 to 2 drops either of the unknown solution or of the sodium carbonate prepared solution. Development of a brown to black color (manganese(III) chloro complex) indicates the presence of oxidizing agents. Anions reacting as oxidizing agents: are NO_2^-, $Cr_2O_7^{2-}$, NO_3^-, and MnO_4^- (Notes 1, 2).

Note 1 The test is effective only in strongly acidic solution. Sufficient acid is present in the reagent to neutralize the sodium carbonate prepared solution.

Note 2 A positive test for oxidizing agents means the probable absence of reducing anions such as S^{2-}, SO_3^{2-}, $S_2O_3^{2-}$, and I^-; but this conclusion is valid only when the solution tested is initially acidic. Normally, oxidizing and reducing anions can coexist in the alkaline sodium carbonate solution.

Procedure A-6 Presence of reducing agents Mix 1 drop of 0.1 *M* iron(III) chloride solution, 1 drop of freshly prepared 0.1 *M* potassium hexacyanoferrate(III) solution, and 2 drops of 6 *M* hydrochloric acid solution. Dilute to 0.5 ml, and add 1 drop of either the unknown solution or the sodium carbonate prepared solution. A dark-blue color ($\mathbf{KFe^{III}[Fe^{II}(CN)_6]}$) indicates the presence of reducing agents. Anions reacting as reducing agents are: SO_3^{2-}, $S_2O_3^{2-}$, S^{2-}, NO_2^-, and I^- (Notes 1, 2, 3).

Note 1 Sufficient hydrochloric acid solution must be present both to neutralize the sodium carbonate and to maintain an acidic medium.

Note 2 A positive test for reducing agents means the probable absence of oxidizing anions such as $Cr_2O_7^{2-}$, NO_3^-, and MnO_4^-, but again this conclusion is valid only when the solution tested is initially acidic.

Note 3 The inclusion of nitrite ion with both oxidizing and reducing agents is due to its dualistic behavior. It is doubtful that either the presence or absence of nitrite ion can be established with absolute certainty by these tests alone.

Procedure A-7 Detection of anion groups Dilute 3 to 4 drops of the sodium carbonate prepared solution to 1 ml, just acidify with 6 *M* perchloric acid solution, and add 1 to 2 drops in excess. Note the evolution of gases (Note 1), and attempt to identify them (presence of anion group I). Heat carefully to boiling, and boil gently for 1 min to remove gases. Cool and centrifuge if necessary (Note 2). To the *acidic* solution, add 0.1 *M* silver(I) nitrate solution drop by drop until precipitation (presence of anion group II) is complete and a small excess of silver ion is present (Note 3). Centrifuge if a precipitate forms. To the clear centrifugate (or solution if no precipitate has formed), add 6 *M* aqueous ammonia solution drop by drop with vigorous shaking until a faintly alkaline reaction results (Note 4). The formation of a precipitate indicates the presence of anion group III (Note 5). No test can be applied to detect anion group IV.

Note 1 Carbon dioxide gas will of course be released, but gases such as sulfur dioxide and hydrogen sulfide are easily detected in its presence. Perchloric acid is added as a strong, nonoxidizing (when dilute) acid which does not add any of the anions here studied to the system. Dilute nitric or sulfuric acid may be substituted, since their anions appear in group IV.

Note 2 Thiosulfate ion gives sulfur.

Note 3 The color of the precipitate in this and the next step should be carefully noted and interpreted as clearly as possible since it is commonly very characteristic (Chap. 8, page 179). Undestroyed group I anions may precipitate either at this point or in the subsequent test.

Note 4 Excess aqueous ammonia solution must be carefully avoided because of the ease of conversion of the precipitated silver salts into the soluble diammine complexes by ammonia. If excess ammonia is inadvertently added, 6 *M* acetic acid solution should be added drop by drop to destroy the complex ion. A neutral or faintly alkaline solution is desirable for precipitation at this point. Sodium or potassium hydroxide solution cannot be employed because of precipitation of black silver(I) oxide.

Note 5 Although the formation of a precipitate at this point is ordinarily a conclusive indication of the presence of group III anions, concentrated acetate or sulfate solution may yield a precipitate as well (page 179).

Procedure A-8 Reaction with 18 *M* sulfuric acid solution To a pinch of the *original* solid sample (or, if the sample is received as a solution, to the residue remaining after evaporation of 3 to 4 drops of the solution to dryness) contained in a clean, *dry* test tube, add 1 to 2 drops of 18 *M* sulfuric acid solution. Observe. Warm gently, and finally heat more strongly, but not to boiling. Then transfer the flame to the wall of the test tube, and heat the escaping vapors strongly. Observe carefully and interpret your results in terms of the data in Table 10-1 (Notes 1–6).

Note 1 No other single preliminary test is capable of giving as much useful information about anions as this one. Extreme care in applying this procedure and in interpreting its results will be most rewarding.

Note 2 Although the reactions of the individual anions are definite and easily interpreted, mixtures may give obscuring or even misleading information. Thus, elemental iodine liberated from an iodide may hide or mask other liberated gases.

Note 3 Small quantities of many salts (e.g., acetates, nitrates) often fail to give completely positive results. Furthermore, insoluble substances (e.g., sulfides, silver halides) or highly covalent or complexed materials [e.g., cadmium iodide, mercury(II) chloride] react so slightly that positive information is not obtained. The results of the sulfuric acid treatment cannot always be considered as completely positive, but they are usually very helpful.

Note 4 Hot 18 *M* sulfuric acid solution is a corrosive and dangerous chemical. Never look down into the test tube during this test, and never point the tube at yourself or any other person. Because of its highly exothermic reaction with water, the sulfuric acid

Table 10-1 Behavior of anions with 18 *M* sulfuric acid solution

	Specific examples	
Observed behavior	*Cold*	*Hot*
No apparent change	PO_4^{3-}, AsO_4^{3-}, AsO_2^-, $C_2O_4^{2-}$, $C_2H_3O_2^-$, NO_3^-, SO_4^{2-}	PO_4^{3-}, AsO_4^{3-}, SO_4^{2-}
Color change	CrO_4^{2-} (yellow) $\rightarrow$ $Cr_2O_7^{2-}$ (orange); $Cr_2O_7^{2-}$ (orange) $\rightarrow$ **CrO_3** (red)	Same
Colorless, odorless gas evolved*	$CO_3^{2-} \rightarrow CO_2\uparrow$	$CO_3^{2-} \rightarrow CO_2\uparrow$ $C_2O_4^{2-} \rightarrow CO_2\uparrow + CO\uparrow$
Colorless gas with odor evolved*	$SO_3^{2-} \rightarrow SO_2\uparrow$ $S_2O_3^{2-} \rightarrow SO_2\uparrow + S$ $S^{2-} \rightarrow H_2S\uparrow$ $Cl^- \rightarrow HCl\uparrow$ $F^- \rightarrow HF\uparrow$	Same, and in addition $C_2H_3O_2^- \rightarrow HC_2H_3O_2\uparrow$
Colored gas evolved*	$NO_2^- \rightarrow NO_2\uparrow$ $Br^- \rightarrow Br_2\uparrow$ $I^- \rightarrow I_2\uparrow$ $Cl^- + CrO_4^{2-}$ or $Cr_2O_7^{2-} \rightarrow CrO_2Cl_2\uparrow$	Same, and in addition $NO_3^- \rightarrow NO_2\uparrow$ (if vapors heated)
Violent reaction	MnO_4^-	Same

* With or without effervescence.

solution liberates steam with moist samples and with some hydrates. Spattering may thus result. Test tubes from this procedure should be allowed to cool before being washed.

Note 5 If the sample is dark in color and yields purple solutions (presence of MnO_4^- ion), *do not apply this test! Danger!*

Note 6 Nitrate ion yields colorless nitric acid vapor as the system is heated. Brown vapors of nitrogen(IV) oxide result only when the escaping gas is heated, as

$$\underset{\text{Colorless}}{4HNO_3(g)} \rightarrow \underset{\text{Brown}}{4NO_2(g)} + 2H_2O(g) + O_2(g)$$

Procedure A-9 Reaction with 6 *M* hydrochloric or perchloric acid solution To a pinch of the original solid (or residue from evaporation of unknown solution), add 1 to 2 drops of 6 *M* hydrochloric or perchloric acid solution. Of the anions which readily evolve gases with cold 18 *M* sulfuric acid solutions, only the following react under these conditions: CO_3^{2-}, SO_3^{2-}, $S_2O_3^{2-}$, S^{2-}, and NO_2^- (Note 1).

Note 1 This test is, of course, the same as that in the first part of Procedure A-7, and the results are generally similar. It is convenient as a logical supplement to the sulfuric acid treatment, however.

SPECIFIC TESTS FOR ANIONS

Equations describing the reactions that are used in the specific tests are summarized in Appendix H_a. If either the chemistry or the expected observations for a test are in doubt, the student should first apply the test to a sodium or potassium salt containing the anion in question (control test, page 204) and compare the result with that given by his sample or by the sodium carbonate prepared solution.

Group I anions

The general scheme followed in identifying the members of this group is summarized in Table 10-2.

Procedure A-10 Sulfide To a pinch of the original solid sample or to the acetic acid-insoluble residue from preparation of the sodium carbonate solution (Procedure A-1 or 2) contained in a test tube, add 2 to 3 drops of 6 *M* hydrochloric acid solution (Note 1). Warm on the water bath, and test the evolved gases with filter paper moistened with freshly prepared sodium plumbite solution (Note 2). Formation of a silvery to dark-brown precipitate of lead(II) sulfide on the paper proves the presence of *sulfide* ion (Note 3). If no reaction is noted, add a pinch of zinc dust to the test tube, warm, and retest the gases (Note 4).

Note 1 Few sulfides are sufficiently soluble to dissolve in water or to decompose when treated with sodium carbonate solution. Tests made upon the original solid or upon the residue from the sodium carbonate treatment are more nearly valid than those made on the sodium carbonate prepared solution.

Note 2 The plumbite solution is prepared by diluting 2 drops of 0.1 *M* lead(II) nitrate solution to 1 ml and adding, with shaking, 6 *M* sodium hydroxide solution until the precipitate that forms first disappears *permanently* and the resulting *clear* solution is strongly alkaline.

Note 3 Although massive lead(II) sulfide is dark brown in color, a silvery deposit often appears under these conditions. Lead acetate solution can be substituted for the plumbite solution, but it is less sensitive.

Note 4 Zinc may reduce sulfite or thiosulfate ion to hydrogen sulfide. These species are not probable in the residue from the sodium carbonate treatment.

Procedure A-11 Thiosulfate and sulfite The presence of one or the other of these species will have been suggested by the evolution of sulfur dioxide in Procedures A-7, A-8, and A-9, and the presence of thiosulfate will have been indicated by the simultaneous precipitation of sulfur (Note 1).

Table 10-2 Schematic outline for analysis of anion group I

Separate portions of original solid sample		*Separate portions of sodium carbonate prepared solution*			
A-10. Add 6 *M* HCl alone or plus **Zn**. Warm	A-12. Add **Zn**, 3% H_2O_2, 3 *M* H_2SO_4. Warm	A-11. Add 3 *M* H_2SO_4. Warm	A-11. Add saturated $Sr(NO_3)_2$. Let stand		A-13. Add 3 *M* H_2SO_4, 0.1 *M* $FeSO_4$
Gas: H_2S, etc. A-10. Pass gas onto paper moistened with $NaHPbO_2$ solution	Gas: CO_2 A-10. Pass gas into saturated $Ba(OH)_2$	Gas: SO_2, etc. A-11. Pass gas into HNO_3-$KMnO_4$-$Ba(NO_3)_2$	Precipitate: **$SrSO_3$** (white) A-11. Add 6 *M* HCl, 0.1 *M* $BaCl_2$, 3% H_2O_2	Centrifugate: $S_2O_3^{2-}$, etc. A-11. Add 6 *M* HCl. Warm	Solution: $[Fe(NO)]^{2+}$ (dark brown) (from NO_2^-)
Precipitate: **PbS** (silvery to dark brown) (from S^{2-})	Precipitate: **$BaCO_3$** (white)	Precipitate: **$BaSO_4$** (white) (from SO_3^{2-} or $S_2O_3^{2-}$)	Precipitate: **$BaSO_4$** (white) (from SO_3^{2-})	Precipitate: **S** (white) + SO_2 (from $S_2O_3^{2-}$)	Or use thiourea alternative—A-13
			Or use ethylenediamine alternative—A-11 If sulfide present, remove first with **$CdCO_3$** suspension		

Sulfite or thiosulfate alone Assemble the gas evolution and absorption apparatus (Fig. 10-1), placing in the trap or open test tube a solution prepared by adding 1 drop of 0.1 *M* potassium permanganate solution, 1 drop of 6 *M* nitric acid solution, and 1 drop of 0.1 *M* barium nitrate solution to 2 ml of water. Place 2 to 3 drops of the sodium carbonate prepared solution (or a pinch of the *original* solid sample) in the test tube, add 2 to 3 drops of 3 *M* sulfuric or 6 *M* perchloric acid solution, close the tube immediately with the stopper carrying the gas trap or delivery tube, and warm the test tube in the water bath. Destruction of the permanganate color and simultaneous precipitation of white barium sulfate in the trap indicates the presence of *sulfite* ion or *thiosulfate* ion in the sample (Note 2).

Sulfite and thiosulfate in the presence of each other. Use either of the following procedures:

1. Dilute 2 to 3 drops of the sodium carbonate prepared solution to no more than 0.5 ml, and add an equal volume of saturated strontium nitrate solution. A white precipitate, forming slowly, may indicate the presence of sulfite ion (Note 3). Let stand 15 to 30 min to ensure complete precipitation, and centrifuge if necessary. Save the centrifugate. To the residue in the centrifuge tube, add 1 ml of saturated strontium nitrate solution. Shake or stir vigorously, centrifuge again, and discard the centrifugate. Suspend the residue in the centrifuge tube in 1 ml of cold water. Stir or shake with 2 to 3 drops of 6 *M* hydrochloric or perchloric acid solution, and then add 1 to 2 drops of 0.1 *M* barium chloride solution. Centrifuge, and discard the

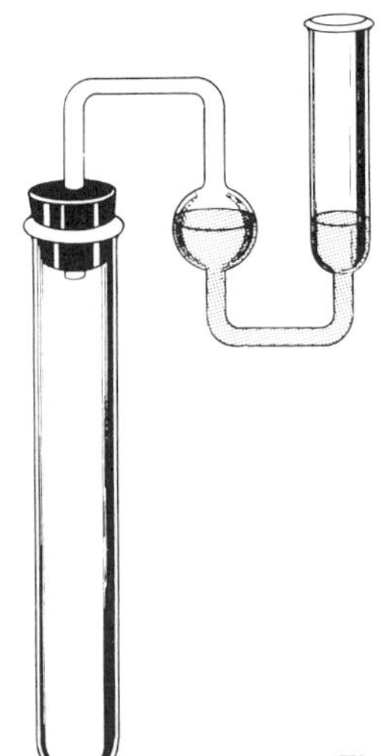

Fig. 10-1 Gas-absorption apparatus.

residue. To the clear centrifugate, add 1 to 2 drops of 3 percent hydrogen peroxide solution. The formation of a white precipitate of barium sulfate proves the presence of *sulfite* ion. Acidify the centrifugate from the strontium nitrate precipitation with 6 *M* hydrochloric acid solution, and warm. The formation of a white precipitate (or turbidity) of sulfur and the simultaneous evolution of sulfur dioxide prove the presence of *thiosulfate* ion.

2. To 1 ml of 0.1 *M* nickel nitrate solution, add 10 percent ethylenediamine solution drop by drop until a clear violet-colored solution results (Note 4). Dilute 2 to 3 drops of the sodium carbonate prepared solution to 1 ml, and add the nickel-ethylenediamine reagent solution drop by drop until precipitation is complete or the solution has a violet color. The formation of a violet-colored crystalline precipitate of composition $\mathbf{[Ni(en)_3]S_2O_3}$ proves the presence of *thiosulfate* ion. The clear centrifugate can then be tested for sulfite ion with peroxide and barium ion, as above.

Sulfite and thiosulfate in the presence of sulfide ion If sulfide ion is present (Procedure A-10), it can be removed as follows: Dilute 2 to 3 drops of 0.1 *M* cadmium nitrate solution to 1 ml, and add 0.1 *M* sodium carbonate solution in excess. Centrifuge. To the residue in the centrifuge tube, add the solution obtained by diluting 2 to 3 drops of the sodium carbonate prepared solution to 1 ml. Shake thoroughly, centrifuge, and test the centrifugate for sulfite and thiosulfate ions as outlined above. Sulfide ion is removed as yellow cadmium sulfide (Note 5).

Note 1 Solutions containing sulfide ion and sulfite ion may behave similarly because of the formation of sulfur as a consequence of oxidation of hydrogen sulfide by sulfur dioxide.

Note 2 Hydrogen sulfide liberated from sulfides also reduces permanganate ion and gives a white precipitate (of sulfur) whether or not barium ion is present. Reduction of permanganate ion by liberated hydrogen iodide or bromide or nitrogen(IV) oxide is possible, but a precipitate does not form under the conditions used.

Note 3 Any ion yielding an insoluble strontium salt will precipitate at this point; so formation of a precipitate is not positive proof of the presence of sulfite ion. Strontium sulfite precipitates very slowly.

Note 4 Nickel hydroxide may precipitate first because of the presence of hydroxide ion formed as

$$H_2NC_2H_4NH_2 + H_2O \rightleftharpoons H_2NC_2H_4NH_3^+ + OH^-$$

Excess ethylenediamine (en) redissolves this precipitate to give the violet complex ion, $[Ni(en)_3]^{2+}$.

Note 5 Formation of a yellow precipitate provides additional confirmation of the presence of sulfide ion.

Procedure A-12 Carbonate Prepare the gas evolution and absorption apparatus with saturated barium hydroxide solution in the trap or open tube. To a pinch of the finely powdered *original solid sample* (Note 1) in the test tube, add a small granule or two of zinc (Note 2), 3 to 4 drops of 3 percent hydrogen peroxide solution (Note 3), and several drops of 3 *M* sulfuric acid solution or 6 *M* perchloric acid solution. Promptly stopper with the gas-trap assembly, and warm on the water bath. A white precipitate of barium carbonate in the absorption tube proves the presence of the *carbonate* ion (Note 4).

Note 1 Obviously the sodium carbonate prepared solution cannot be used for this test.

Note 2 Hydrogen liberated by reaction of zinc with hydrogen ion aids in sweeping liberated carbon dioxide into the gas trap. The zinc has no other function.

Note 3 Sulfite and thiosulfate ions are oxidized to sulfate ion, which does not liberate sulfur dioxide with acids. In the absence of these ions, addition of hydrogen peroxide may be omitted, although oxygen liberated by its decomposition also aids in carrying carbon dioxide into the trap.

Note 4 Excess carbon dioxide dissolves the precipitate, owing to the formation of hydrogen carbonate ion.

Procedure A-13 Nitrite The presence or absence of nitrite ion will have been suggested by the evolution of brown nitrogen(IV) oxide in Procedures A-7, A-8, and A-9. Use either of the following specific tests:

1. Dilute 3 to 4 drops of the sodium carbonate prepared solution to 0.5 ml and just acidify with 3 *M* sulfuric acid solution or 6 *M* acetic acid solution (Note 1). Cool and add a few drops of freshly prepared 0.1 *M* iron(II) sulfate solution. A brown color ($[Fe(NO)]^{2+}$ ion) throughout the solution proves the presence of *nitrite* ion (Notes 2, 3).
2. Dilute 3 to 4 drops of the sodium carbonate prepared solution to 1 ml, acidify with 6 *M* acetic acid solution (Note 1), and add 0.1 *M* silver(I) nitrate solution drop by drop until precipitation is complete (Note 4). Centrifuge and add to the acidic centrifugate a pinch of solid thiourea. Warm and add 1 drop of 0.1 *M* iron(III) chloride solution. A red color ($[Fe(NCS)]^{2+}$ ion) proves the presence of the *nitrite* ion (Notes 3, 5).

Note 1 Although nitrous acid decomposes in strongly acidic solution, it is sufficiently stable under these conditions to be detected by the subsequent tests. Effervescence due to the evolution of carbon dioxide causes no difficulty.

Note 2 Nitrate ion causes no interference at low hydrogen-ion concentrations. Interference from iodide ion may occur if oxidizing anions are present, because of the liberation of iodine upon acidification. If a dark color appears prior to the addition of the iron(II) sulfate solution, extract with 1 ml of carbon tetrachloride, and then test the aqueous layer. Other anions do not interfere.

Note 3 Because of oxidation, nitrate ion is always present in nitrite-containing samples.

Note 4 Iodide ion interferes and is removed as the silver(I) salt. Some silver nitrite may precipitate, but sufficient nitrite ion remains to give the test. Added nitrate ion causes no interference.

Note 5 If the red color is not readily apparent, extract with 1 ml of ether to intensify it. Thiocyanate ion is formed in terms of the equation

$$SC(NH_2)_2 + HNO_2 \rightarrow N_2(g) + H^+ + 2H_2O + NCS^-$$

Nitrate ion does not interfere.

Group II anions

The presence or absence of members of this group will have been indicated by the results of Procedure A-7. The general scheme followed in identifying the members of the group is summarized in Table 10-3.

Procedure A-14 Iodide, bromide, and chloride The following procedures should be employed in the order listed:

Iodide and bromide in the presence of chloride Use either of the following procedures:

1. Dilute 2 to 3 drops of the sodium carbonate prepared solution to 1 ml, acidify with 6 M hydrochloric acid solution, and add a few drops in excess. Add 1 ml of carbon tetrachloride and then 1 drop of chlorine

Table 10-3 Schematic outline for analysis of anion group II

Separate portions of sodium carbonate prepared solution		
A-14. Add 6 M HCl, CCl_4, Cl_2 water. Shake	A-14. Add solid $\mathbf{(NH_4)_2S_2O_8}$ or $\mathbf{K_2S_2O_8}$, 3 M H_2SO_4. Warm. Decant	
CCl_4 layer: I_2 (violet) (from I^-)	Decantate: I_2, Br_2, Cl^-, etc. A-14. Add CCl_4, shake. Repeat	
Aqueous layer: Br^- A-14. Add more Cl_2 water	CCl_4 layer: I_2, Br_2 (dark)	Aqueous layer: Cl^- A-14. Add 6 M HNO_3, 0.1 M $AgNO_3$
CCl_4 layer: Br_2 (yellow to brown) (from Br^-)		Precipitate: **AgCl** (white)
Or use nitrite alternative—A-14	Or add $\mathbf{K_2Cr_2O_7}$, and remove CrO_2Cl_2—A-14 If Br^-, I^- absent, precipitate **AgCl**—A-14	

water or acidified sodium hypochlorite solution. Shake. A violet to purple color in the carbon tetrachloride layer proves the presence of iodide ion (Note 1). Continue addition of chlorine water drop by drop with constant shaking until the violet color disappears (Note 2). A yellowish to reddish-brown color in the carbon tetrachloride then proves the presence of *bromide* ion (Note 3).

2. Dilute 2 to 3 drops of the sodium carbonate prepared solution to 1 ml, acidify with 6 *M* hydrochloric acid solution, and add 1 ml of carbon tetrachloride. Add 0.1 *M* sodium or potassium nitrite solution drop by drop with shaking (Note 4). A violet carbon tetrachloride layer proves the presence of *iodide* ion. Remove the carbon tetrachloride layer, add 2 drops of the nitrite solution, and boil gently to expel free iodine (Note 5). When the solution is colorless, cool, add 1 ml of carbon tetrachloride, and then add chlorine water drop by drop with shaking. A yellow to brownish-red carbon tetrachloride layer proves the presence of *bromide* ion.

Chloride in the presence of bromide or iodide or both Use either of the following procedures if bromide or iodide is present:

1. Dilute 2 to 3 drops of the sodium carbonate prepared solution to 1 ml, add a small excess of either solid ammonium peroxodisulfate or potassium peroxodisulfate, acidify with 3 *M* sulfuric acid solution, and warm at 80°C for 5 min (Note 6). Decant the liquid from any residue, and add to the liquid 1 ml of carbon tetrachloride. Shake and remove the carbon tetrachloride layer. Repeat, using another small portion of the peroxodisulfate and a fresh portion of carbon tetrachloride (Note 7). Dilute the remaining liquid to 2 ml, and add 1 drop of 6 *M* nitric acid solution and 1 drop of 0.1 *M* silver(I) nitrate solution. A white cloudiness or precipitate, readily soluble in 6 *M* aqueous ammonia solution, proves the presence of *chloride* ion (Note 8).

2. Evaporate 2 to 3 drops of the sodium carbonate prepared solution to dryness, and transfer the residue to a small test tube. Fit this tube with a gas-absorption trap or tube (Fig. 10-1) containing 6 *M* sodium hydroxide solution. To the solid in the test tube, add an equal volume of finely powdered solid potassium dichromate. Then add several drops of 18 *M* sulfuric acid solution, and stopper immediately with the trap. Warm the test tube in the water bath. Escaping brown vapors (Notes 9, 10) which yield a *yellow* solution in the trap indicate the presence of chloride ion (Note 11). Test this yellow solution for chromate ion with hydrogen peroxide according to Procedure A-17. A positive test proves the presence of *chloride* ion (Note 12).

Chloride in the absence of bromide or iodide The results of Procedure A-7 should provide definite proof of the presence of *chloride* ion.

Note 1 Iodide ion is more readily oxidized to free halogen than bromide ion; so careful addition of a small quantity of the oxidant permits detection of iodide in the presence of bromide.

Note 2 Elemental iodine is oxidized further to colorless iodate ion, which then appears in the aqueous layer.

Note 3 Ultimate oxidation to elemental bromine or bromine monochloride permits detection of bromide ion.

Note 4 Nitrous acid oxidizes only iodide ion, leaving bromide and chloride ions unaffected.

Note 5 Complete removal of all iodide ion is desirable, although addition of excess chlorine water will obviate any difficulties imposed by its incomplete removal.

Note 6 Bromide and iodide ions are selectively oxidized to the elemental halogens.

Note 7 Complete removal of both iodide and bromide ions is essential for the success of this procedure.

Note 8 Dilution obviates precipitation of silver(I) sulfate. Any silver bromide or iodide would not dissolve in small quantities of aqueous ammonia.

Note 9 Brown fumes may come from nitrate or bromide ions, as well as from dichromate ion plus chloride ion (Table 10-1). Iodide ion gives purple iodine.

Note 10 Carbon dioxide from the excess carbonate ion aids in carrying liberated gases into the trap.

Note 11 Chromyl chloride hydrolyzes to yellow chromate ion. Other liberated gases are absorbed to give colorless solutions and do not carry chromium into the trap. Pertinent reactions are summarized by the equations

$$4Cl^- + 6H^+ + Cr_2O_7^{2-} \rightarrow \underset{\text{Brown}}{2CrO_2Cl_2(g)} + 3H_2O$$

$$CrO_2Cl_2(g) + 4OH^- \rightarrow CrO_4^{2-} + 2H_2O + 2Cl^-$$

Note 12 A positive test for chromium(VI) results only if chloride ion is present in the initial sample.

Group III anions

The presence or absence of anions of this group will have been indicated by the results of Procedure A-7. The general scheme followed in identifying the members of the group is summarized in Table 10-4.

Procedure A-15 Orthoarsenate The presence of arsenic-containing species is established as follows: Dilute 1 to 2 drops of the sodium carbonate prepared solution to 1 ml with 12 *M* hydrochloric acid solution, add 0.5 ml

Table 10-4 Schematic outline for analysis of anion group III

Separate portions of sodium carbonate prepared solution					
A-15. Add 12 *M* HCl, saturated $SnCl_2$. Warm	A-15. Add 6 *M* HCl, $CH_3C(S)NH_2$. Heat	A-15. Add 6 *M* $HC_2H_3O_2$, magnesia mixture. Let stand	A-16. Add 15 *M* HNO_3, boil, add $(NH_4)_2MoO_4$-HNO_3, warm to 40°C	A-17. Add 6 *M* $HC_2H_3O_2$, boil. Cool, add 0.1 *M* $Ba(C_2H_3O_2)_2$. Centrifuge	A-18. Add 3 *M* $HC_2H_3O_2$, boil. Cool, add 0.1 *M* $Ca(NO_3)_2$. Centrifuge
Precipitate: **As** (brown to black) (from AsO_4^{3-})	Precipitate: **As_2S_3** (yellow) (from AsO_4^{3-})	Precipitate: **$MgNH_4AsO_4 \cdot 6H_2O$** or **$MgNH_4PO_4 \cdot 6H_2O$** (white, crystn.) A-15. Wash with 1 *M* aq NH_3. Add 6 *M* $HC_2H_3O_2$, 0.1 *M* $AgNO_3$	Precipitate: **$(NH_4)_3[P(Mo_{12}O_{40})]$** or perhaps **$(NH_4)_3[As(Mo_{12}O_{40})]$** (yellow)	Precipitate: **$BaCrO_4$** (yellowish) plus other salts A-17. Add 6 *M* HNO_3, cool. Add ether, 3% H_2O_2	Precipitate: **CaC_2O_4** (white) plus other salts A-18. Add 1 *M* HNO_3. Heat to 60–70°C, add 0.01 *M* $KMnO_4$
		Residue: **Ag_3AsO_4** (reddish) or **Ag_3PO_4** (yellow)	If AsO_4^{3-} present, remove with $CH_3C(S)NH_2$ and then apply above test for PO_4^{3-}—A-16	Ether layer: CrO_5 (deep blue) (from CrO_4^{2-} or $Cr_2O_7^{2-}$)	Solution: colorless plus CO_2 gas (due to $C_2O_4^{2-}$)

of saturated tin(II) chloride solution, and warm. The development of first a dark-brown color and then a black color proves the presence of *arsenic species* (Note 1). If the test is positive, the color of the precipitate in Procedure A-7 may aid in confirming the species as arsenate ion.

Further confirmation is obtained by either of the following procedures:

1. Dilute 2 drops of the sodium carbonate prepared solution to 1 ml, acidify with 6 *M* hydrochloric acid solution, and add 2 to 3 drops in excess. Add several drops of 1 *M* thioacetamide solution and heat to boiling. A bright-yellow precipitate indicates the presence of *orthoarsenate* ion (Note 2). The precipitate should dissolve readily in cold 6 *M* sodium hydroxide solution (Note 3).
2. Dilute 1 to 2 drops of the sodium carbonate prepared solution to 1 ml, and just neutralize with 6 *M* acetic acid solution. Add 1 ml of magnesia mixture, stir vigorously, and allow to stand for at least 10 min. If no precipitate appears, scratch the inner walls of the tube with a stirring rod, and then allow to stand again (Note 4). A white crystalline precipitate may indicate the presence of orthoarsenate ion (Note 5). If a precipitate forms, centrifuge, and wash once with 1 ml of 1 *M* aqueous ammonia solution (Note 6). Moisten the residue in the tube with a few drops of 6 *M* acetic acid solution (Note 7), and add a few drops of 0.1 *M* silver(I) nitrate solution. Conversion of the residue to a reddish precipitate proves the presence of *orthoarsenate* ion (Note 8). Formation of a yellow silver salt proves the presence of orthophosphate ion (Note 9).

Note 1 Arsenic and arsenous acids are reduced to elemental arsenic. This test does no more than establish the presence or absence of arsenic, since any arsenic-containing species would react in this way.

Note 2 Thioacetamide first reduces arsenic(V) to arsenic(III), which then reacts with the extra sulfide ion generated by the hydrolysis of the reagent to give insoluble yellow arsenic(III) sulfide, As_2S_3 (page 242).

Note 3 Oxidation of sulfide ion or decomposition of thiosulfate ion may cause finely divided sulfur to precipitate. This material is normally white or pale yellow in color and dissolves very slowly in cold alkaline solutions.

Note 4 Supersaturation is common at this point and is minimized by these techniques.

Note 5 Magnesium ammonium orthophosphate 6-hydrate is indistinguishable in appearance from the corresponding arsenate. Gelatinous precipitates may form with other anions, but these cause no interference in subsequent tests.

Note 6 The presence of aqueous ammonia is necessary to reduce solubility and inhibit hydrolysis. Prepare a 1 *M* solution by diluting 1 ml of the 6 *M* solution to 6 ml.

Note 7 Excess aqueous ammonia must be neutralized to avoid complex formation with silver(I) ion.

Note 8 The silver salts are less soluble than the magnesium ammonium salts. The color of silver(I) orthoarsenate is apparent in the presence of yellow silver(I) orthophosphate and masks the color of the latter.

Note 9 In the absence of orthoarsenate, this test is positive for orthophosphate.

Procedure A-16 Orthophosphate Depending upon the results of Procedure A-15, use one of the following:

In the absence of orthoarsenate Dilute 1 to 2 drops of the sodium carbonate prepared solution to 0.5 ml, acidify with 15 *M* nitric acid solution, add 1 to 2 drops in excess, heat to boiling (Note 1), and add 5 to 10 drops of ammonium molybdate–nitric acid reagent solution. Warm to not above 40°C (Note 2) if necessary. A bright-yellow precipitate of ammonium molybdophosphate proves the presence of *orthophosphate* ion (Note 3).

In the presence of orthoarsenate Dilute 2 drops of the sodium carbonate prepared solution to 2 ml with 12 *M* hydrochloric acid solution, heat to boiling, add several drops of 1 *M* thioacetamide solution, and again heat to boiling (Note 4). Centrifuge, and test the centrifugate for completeness of precipitation. When precipitation is complete, carefully evaporate the solution almost to dryness in a casserole. Add 0.5 ml of 15 *M* nitric acid solution, and again evaporate nearly to dryness (Note 5). Take up the residue in 5 drops of 15 *M* nitric acid solution and 5 drops of water. Remove any sulfur residue with a stirring rod, and then add 5 to 10 drops of ammonium molybdate–nitric acid reagent solution. Warm to 40°C if necessary. A bright-yellow precipitate proves the presence of *orthophosphate* ion.

Note 1 Reducing anions, e.g., sulfite, thiosulfate, sulfide, and iodide, interfere by giving blue precipitates or colors with molybdate ion. Boiling with nitric acid effects their oxidation.

Note 2 Precipitation normally occurs without heating, but the rate of precipitation is increased by raising the temperature.

Note 3 Orthoarsenate ion gives a similar precipitate. Interference from this anion is minimized by not allowing the temperature to rise above 40°C.

Note 4 The condition of high acidity is necessary to effect complete conversion of orthoarsenate ion to sulfide. Iodide ion (as 1 drop of 1 *M* ammonium iodide solution) may be added to reduce arsenate ion and facilitate precipitation.

Note 5 The second evaporation with nitric acid removes chloride and sulfide ions, which may interfere by reducing molybdate ion.

Procedure A-17 Chromate and dichromate The presence of chromate ion is indicated by a yellow-colored sodium carbonate prepared solution (Note 1) and by the results of the preliminary examination (Procedures A-5, A-7, A-8). If doubt exists, the following test can be applied: Dilute 2 drops of the sodium carbonate prepared solution to 1 ml, acidify with 6 *M* acetic acid solution, heat to boiling (Note 2), cool, and add several drops of 0.1 *M* barium acetate solution. A yellowish precipitate indicates the presence of chromate ion (Note 3). Centrifuge, and wash with 1 ml of water. To the residue in the test tube, add 2 to 3 drops of 6 *M* nitric acid solution and 1 ml of water. Warm to ensure dissolution of the precipitate (Note 4). Cool in ice water. Then add 0.5 ml of ether or amyl alcohol (Note 5) and 1 drop of 3 percent hydrogen peroxide solution. Shake once. A deep blue color, collecting in the nonaqueous layer, proves the presence of *chromium(VI)* (Notes 6, 7).

Note 1 The only other colored anionic species considered here is permanganate ion. Polysulfide solutions resulting from partial oxidation of sulfide ion are yellow also. These solutions yield sulfur upon acidification, whereas yellow chromate ion gives orange dichromate ion.

Note 2 Boiling removes carbon dioxide and may decompose other precipitatable anions.

Note 3 Chromate ion is precipitated as the barium salt to separate it from reducing species such as iodide ion which could interfere with the peroxochromic acid test. Other barium salts insoluble in acetic acid are white. Their presence lightens the color of barium chromate.

Note 4 Barium sulfate, if present, will remain undissolved.

Note 5 Because of the flammability of ether, amyl alcohol may be preferred. Ether often contains sufficient peroxide to develop a color when it is added.

Note 6 The stability of the peroxochromic acid is so dependent upon conditions that its existence may be too fleeting to permit its extraction into the nonaqueous layer. The production of a blue color, however fleeting it may be, is proof of the presence of chromium(VI).

Note 7 Because of the interconversion of chromate and dichromate ions as a result of pH changes, distinction between the two is normally not made in the report.

Procedure A-18 Oxalate ion Dilute 2 drops of the sodium carbonate prepared solution to 0.5 ml, and just acidify to litmus with 3 *M* acetic acid solution, counting the drops added. Add the same number of drops of the acid in excess. Heat to boiling to expel carbon dioxide. Cool, and add 3 to 4 drops of 0.1 *M* calcium nitrate solution. A white precipitate, forming immediately, strongly indicates the presence of oxalate ion (Notes 1, 2). Centrifuge and wash the precipitate with two 1-ml portions of water. Dissolve the residue in 1 to 2 drops of 6 *M* nitric acid solution, and add 1 ml of water. Heat the solution to 60 to 70°C in the water bath, and add 0.01 *M* potassium permanganate solution drop by drop, noting the number of drops required to give a permanent pink color (Note 3). Discharge of the color of the first several drops accompanied by the evolution of carbon dioxide gas (Note 4) proves the presence of *oxalate* ion. If only a single drop of the permanganate solution is needed to give a color, oxalate ion is absent.

Note 1 Calcium oxalate is insoluble in neutral solutions and in dilute acetic acid but is soluble in strongly acidic solutions.

Note 2 Other ions, especially the fluoride, may give precipitates also. These precipitates are seldom crystalline. None of these species reduces permanganate ion.

Note 3 The reduction of permanganate ion is catalyzed by manganese(II) ion formed in the process. Decolorization of the first drop of permanganate solution by oxalate may require a moment until sufficient catalyst is generated. Reaction with subsequent drops is instantaneous.

Note 4 Carbon dioxide can be detected as described in Procedure A-12.

Group IV anions

Except for permanganate ion (colored, oxidizing), information obtained on the members of this group by means of the preliminary examination is limited to that from Procedure A-8. The general scheme followed in identifying the members of the group is summarized in Table 10-5.

Procedure A-19 Acetate To 3 to 4 drops of the sodium carbonate prepared solution or to a pinch of the original solid sample, add 3 to 4 drops of 18 *M* sulfuric acid solution. Warm carefully on the water bath until gas evolution ceases (Note 1). Then add 4 drops of ethyl alcohol, and heat for 1 min in the boiling-water bath. Carefully note the odor of the escaping gas. A pleasant, fruitlike odor (due to ethyl acetate) proves the presence of *acetate* ion (Note 2). If the odor is not clearly identifiable, carry out in parallel a *blank* test using only the reagents and a *control* test using sodium acetate (Expt. III-3, page 205). Compare the odors.

Note 1 Anions, such as sulfide, sulfite, and nitrite, that give gases the odors of which could hide that of ethyl acetate, are thus eliminated.

Note 2 The odor is distinctive, but some workers have difficulty in detecting it unless they carry out control and blank tests. The odors of the laboratory may dull the sense of smell somewhat.

Procedure A-20 Nitrate Depending upon the composition of the sample, use one of the following procedures:

In the absence of nitrite, bromide, iodide, or chromate ions In a test tube, dilute 2 to 3 drops of the sodium carbonate prepared solution to 1 ml, and acidify with 3 *M* sulfuric acid solution. Add 2 to 3 drops of freshly prepared 0.1 *M* iron(II) sulfate solution, and cool the mixture (Note 1).

Table 10-5 Schematic outline for analysis of anion group IV

Portion of original solid or of residue from evaporation of sodium carbonate prepared solution	*Separate portions of sodium carbonate prepared solution*			
A-22. Add $\mathbf{SiO_2}$, 18 *M* H_2SO_4. Warm	A-19. Add 18 *M* H_2SO_4, warm. Add C_2H_5OH. Heat	A-20. Add 3 *M* H_2SO_4, 0.1 *M* $FeSO_4$. Cool. Add 18 *M* H_2SO_4 under solution	A-21. Add 3 *M* H_2SO_4, warm. Add 3% H_2O_2 or 0.1 *M* $Na_2C_2O_4$	A-23. Add 6 *M* HCl, 0.1 *M* $BaCl_2$
Gas: SiF_4 (colorless) Gives turbidity with drop of water	Gas: $CH_3CO_2C_2H_5$ (colorless), gives fruitlike odor	Brown ring: $[Fe(NO)]^{2+}$—(from NO_3^- or NO_2^-)	Solution: colorless from violet or purple (due to MnO_4^-)	Precipitate: $\mathbf{BaSO_4}$ (white)
		If NO_2^- present, remove with $CO(NH_2)_2$ or HSO_3NH_2. Then use **Zn** dust-NaOH alternative—A-20. (Also in presence of Br^-, I^-, or CrO_4^{2-})		

By means of a pipette, carefully introduce 0.5 ml of 18 *M* sulfuric acid solution into the bottom of the test tube so that it forms a separate layer under the other solution. Allow to stand undisturbed for a few moments. A brown ring at the interface between the two liquids proves the presence of *nitrate* ion (Note 2).

In the presence of the above-mentioned species Remove nitrite ion, if present, by diluting 2 to 3 drops of the sodium carbonate prepared solution to 1 ml, adding 2 to 3 drops of 3 *M* sulfuric acid solution beyond that necessary for acidification, adding either a few crystals of urea or of sulfamic acid (Note 3), and warming until evolution of gas ceases (Note 4). Then transfer the mixture to a small casserole, make strongly alkaline with 6 *M* sodium hydroxide solution, and boil gently until any evolution of ammonia ceases (Note 5). Add carefully a pinch of zinc dust or aluminum powder, and cover immediately with a watch glass to the bottom side of which a piece of moist red litmus paper has been stuck (Note 6). Warm gently, but do not boil. A blue color, developing uniformly across the litmus paper, proves the presence of nitrate ion (Note 7).

Note 1 The brown $[Fe(NO)]^{2+}$ ion decomposes in hot solutions.

Note 2 Nitrite ion would give the same color, but throughout the solution, upon addition of iron(II) sulfate solution. Bromide and iodide ions develop confusing rings of the brownish halogens because of oxidation by sulfuric acid. Reduction of dichromate ion to dark-green chromium(III) ion by iron(II) ion gives an obscuring color. All these species except nitrite can be removed by prior precipitation with silver(I) sulfate solution, but the limited solubility of the latter compound is a complicating factor. Samples containing nitrite ion invariably contain nitrate ion because of oxidation.

Note 3 Sulfamic acid is preferred since it reacts very rapidly and completely. Urea reacts more slowly and always gives some nitrate ion.

Note 4 The gas is nitrogen, formed as

$$SO_3NH_2^- + HNO_2 \rightarrow N_2(g) + H^+ + H_2O + SO_4^{2-}$$

Note 5 Ammonia comes from ammonium salts. Its removal must be complete to render the remainder of the test valid.

Note 6 A universal indicator paper or a strip of filter paper moistened with mercury(I) nitrate solution may be substituted. Ammonia converts mercury(I) nitrate to **Hg** (black) and $\mathbf{Hg_2N(NO_3){\cdot}H_2O}$(white). A dark stain on the paper indicates the presence of ammonia.

Note 7 Boiling may spatter droplets of the alkaline solution and thus produce blue spots on the paper. Ammonia reacts uniformly to give an over-all color. The test is valid only if extreme care has been used to eliminate nitrite and ammonium ions. The other species mentioned above do not interfere.

Procedure A-21 Permanganate The presence of permanganate ion is indicated by a purple to black original sample and a violet to purple sodium carbonate prepared solution. If the latter solution has no violet color, permanganate ion is absent. If the solution has this color, dilute 2 to 3 drops to 0.5 ml, and acidify with 3 *M* sulfuric acid solution. Either add 3 percent hydrogen peroxide solution drop by drop, or warm and add 2 to 3 drops of 0.1 *M* sodium oxalate solution (Note 1). Conversion of the violet solution to a colorless one proves the presence of *permanganate* ion.

Note 1 Dichromate ion interferes both by its own color and the obscuring color of its reduction product. If present, it should be removed first by precipitation with barium ion from acetate solution.

Procedure A-22 Fluoride To a pinch of the original solid sample or to the residue obtained by evaporating 5 drops of the sodium carbonate prepared solution to dryness, add an equal volume of finely ground silicon dioxide (Note 1). Mix thoroughly, and transfer to a dry test tube, which may be equipped for the drop-of-water test (Fig. 10-2). Add sufficient 18 *M* sulfuric acid solution to equal twice the volume of the solid mixture, insert the stopper bearing the tube holding the drop of water, and warm the test tube gently over a small flame for 2 to 3 min. A white turbidity developing in the water drop proves the presence of *fluoride* ion (Note 2).

Note 1 Quartz is preferred to precipitated silica because of reaction of the latter with liberated silicon tetrafluoride to give nonvolatile products.

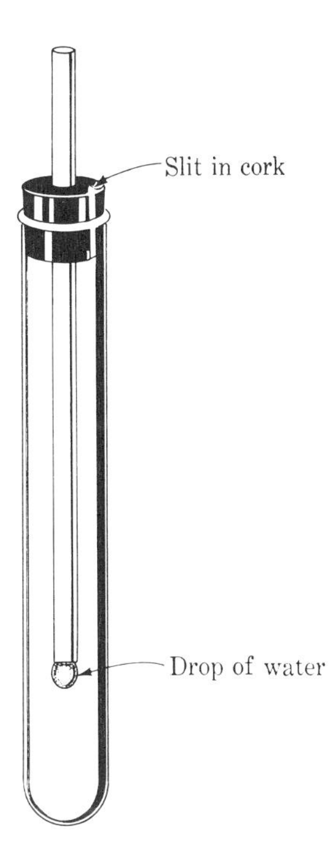

Fig. 10-2 Apparatus for drop-of-water test.

Note 2 Since the reaction depends upon the hydrolysis of gaseous silicon tetrafluoride, there is no interference among the species here studied.

Procedure A-23 Sulfate Dilute 2 drops of the sodium carbonate prepared solution to 1 ml, and acidify with 6 *M* hydrochloric acid solution. Add 2 to 3 drops of the acid solution in excess, and then add 1 to 2 drops of 0.1 *M* barium chloride solution. A white precipitate, insoluble in acidic solution, proves the presence of the *sulfate* ion (Notes 1, 2, 3).

Note 1 Other barium salts are not precipitated under strongly acidic conditions. In the presence of fluoride ion, doubling the quantity of hydrochloric acid solution is recommended.

Note 2 Samples containing sulfite, thiosulfate, or sulfide ion always contain traces of sulfate ion because of oxidation. These quantities are shown by a faint turbidity upon addition of barium ion. Sulfate ion present by intention gives a definitely milky suspension or a precipitate.

Note 3 The presence of sulfate ion can be confirmed by mixing the above obtained precipitate with an equal volume of powdered charcoal in a crucible, igniting strongly, and then adding a few drops of 6 *M* hydrochloric acid solution to the cooled residue. Evolution of hydrogen sulfide gas from the sulfide obtained by reduction indicates the presence of sulfate ion.

EXERCISES

10-1. A white unknown sample containing Zn^{2+}, Ba^{2+}, and K^+ is completely soluble in water. List the anions which cannot be present and the anions which could be present.

10-2. A mixture of the two water-soluble salts K_2HPO_4 and $Co(NO_3)_2$ is only partially soluble in water. Explain. What general principle of the solubility of the sample in relationship to its anion and cation content does this example illustrate?

10-3. The solubility-product constants of the compounds CaC_2O_4 and PbC_2O_4 are, respectively, 1.3×10^{-9} mole^2 liter^{-2} and 8.3×10^{-12} mole^2 liter^{-2}. Determine the concentration of oxalate ion obtainable in each case by treating the solid with excess 1.5 *M* sodium carbonate solution.

10-4. A solid sample gives a brownish gas with either 18 *M* sulfuric acid solution or 6 *M* hydrochloric acid solution. What anion is indicated to be present? If only the sulfuric acid solution had given a brown gas, what anion would have been indicated? Account for the difference.

10-5. A solid sample gives a brownish gas with cold 18 *M* sulfuric acid solution but no fumes with 6 *M* perchloric acid solution. A water solution of this sample, when acidified and shaken with chlorine water and carbon tetrachloride, gives only a colorless carbon tetrachloride layer. Interpret these results.

10-6. A colorless solution darkens a manganese(II) chloride–hydrochloric acid solution but gives no change when treated with 3 *M* sulfuric acid solution and 0.1 *M* iron(II) sulfate solution. What anion is most probably present, and why?

10-7. Although either a mixture of solid KI and solid KNO_2 or a sodium carbonate solution containing iodide and nitrite ions is stable, the addition of acid to each effects its decomposition. Explain.

10-8. Give a concise reason why the original solid sample is preferred for specific tests for each of the following anions: sulfide, carbonate, fluoride.

10-9. Describe by means of an ionic equation the behavior of each anion of group I with a strong acid. Indicate clearly why each of these reactions takes place.

10-10. Outline in as few steps as you can a possible procedure for identifying positively each anion in each of the following combinations: (*a*) SO_3^{2-}, $S_2O_3^{2-}$, S^{2-}; (*b*) CO_3^{2-}, NO_2^-; (*c*) CO_3^{2-}, $S_2O_3^{2-}$.

10-11. Give a group I anion described by each of the following statements:

(*a*) Gives with silver(I) ion a white precipitate that changes through orange to brown or black.

(*b*) Has both oxidizing and reducing properties.

(*c*) Solid sodium salt liberates brown fumes when treated with cold hydrochloric acid solution.

10-12. The ions Fe^{3+} and Al^{3+} are precipitated as hydroxides rather than as carbonates by potassium carbonate solution. Explain.

10-13. Why is it easier to dissolve copper(II) sulfide in hot dilute nitric acid solution than it is to dissolve mercury(II) sulfide?

10-14. Silver(I) carbonate and sulfite dissolve readily in 6 *M* perchloric acid solution, but silver(I) bromide does not. Give a reason for this difference.

10-15. Write a balanced ionic equation to illustrate the behavior of hydrogen sulfide with each of the following reagents: (*a*) dichromate ion in dilute sulfuric acid solution; (*b*) sulfurous acid solution; (*c*) aqueous sodium hydroxide solution.

10-16. The solubility-product constants for **PbS** and **$PbCO_3$** are 7×10^{-29} mole^2 liter^{-2} and 1.5×10^{-13} mole^2 liter^{-2} respectively. Calculate how many milligrams of lead(II) sulfide can be converted into the carbonate by boiling an excess of the solid with 100 ml of 3 *N* sodium carbonate solution.

10-17. From appropriate oxidation-potential data, calculate equilibrium constants for

$$2X^- + 2H^+ + 2HNO_2 \rightleftharpoons 2NO + 2H_2O + X_2$$

where X = I and Br. Of what significance are your results?

10-18. The presence of carbonate ion is not clearly indicated by treatment of a solid sample with 3 *M* sulfuric acid solution to liberate a colorless gas which gives with barium hydroxide solution a white precipitate. Explain.

10-19. Indicate clearly how each anion can be identified in a sample containing the the species NH_4^+, I^-, Na^+, NO_2^-, and NO_3^-.

10-20. Indicate clearly, using the fewest steps possible, how each of the following can be identified:

(*a*) Chloride ion in a sample containing fluoride ion

(*b*) Orthoarsenate ion in a sample containing orthophosphate ion

(*c*) Sulfite ion in a sample containing thiosulfate and sulfide ions

(*d*) Fluoride ion in the presence of oxalate ion

10-21. Give the symbol for a cation of which:

(*a*) The sulfide is insoluble in either 6 *M* or 12 *M* hydrochloric acid solution.

(*b*) The sulfide is completely hydrolyzed in contact with water.

(*c*) The orthophosphate and iodide are both yellow.

(*d*) The thiosulfate rapidly changes color in contact with water.

(*e*) The carbonate is insoluble in water but is soluble in water saturated with carbon dioxide.

10-22. Suggest the simplest and most suitable solvent to dissolve each of the following: $\mathbf{PbCO_3}$, **AgBr**, **CdS**, $\mathbf{BiCl_3}$, $\mathbf{Na_3PO_4}$, $\mathbf{NiSO_4}$, $\mathbf{CaF_2}$.

10-23. Five unlabeled bottles contain the solid salts **NaCl**, $\mathbf{K_2SO_4}$, $\mathbf{NH_4NO_3}$, **KI**, and $\mathbf{Ca(NO_2)_2}$. Indicate clearly how treatment with 18 *M* sulfuric acid solution could provide information necessary to the correct labeling of these bottles.

10-24. A water-soluble sample is shown to contain only the Cu^{2+} and Ni^{2+} ions by cation analysis. What anions can be present?

10-25. Account clearly for the observation that solid sodium bromide reacts readily with cold 18 *M* sulfuric acid solution but does not react with hot or cold 12 *M* hydrochloric acid solution.

10-26. Offer an explanation for the increase in color in the series **AgCl-AgBr-AgI.**

10-27. Why does reducing strength increase in the series Cl^--Br^--I^-?

10-28. Why are the silver(I) salts of the group II anions insoluble in dilute acids?

10-29. The solubility-product constants for AgNCS and AgBr are, respectively, 1×10^{-12} mole^2 liter^{-2} and 5×10^{-13} mole^2 liter^{-2}. Could thiocyanate and bromide ions be separated by selective precipitation with silver(I) ion? Explain.

10-30. Four unlabeled bottles are known to contain the solid salts **NaCl**, **NaBr**, $\mathbf{Na_2C_2O_4}$, and **NaI**, but it is not known specifically what is in each. Select *one* reagent which will permit positive identification of each, and give equations illustrating the behavior of each salt with this reagent.

10-31. By means of simple tests, indicate how one could distinguish between the solids in each of the following pairs:

(*a*) $\mathbf{BaSO_4}$ and $\mathbf{BaCO_3}$
(*b*) **KBr** and **KI**
(*c*) $\mathbf{Na_2S_2O_3}$ and $\mathbf{Na_2SO_3}$
(*d*) $\mathbf{CaC_2O_4}$ and $\mathbf{CaCO_3}$
(*e*) **KF** and $\mathbf{K_2SO_4}$

10-32. A single salt, when treated with cold 18 *M* sulfuric acid solution, effervesced, giving a colorless, fuming gas with a sharp odor. Identify the anion present.

10-33. Using as a sample a combination of the ions $CO_3{}^{2-}$, Cl^-, $CrO_4{}^{2-}$, and $SO_4{}^{2-}$, show clearly how the classification of anion groups employed in this chapter can be made the basis of a systematic scheme of separation and identification.

10-34. For each of the following combinations of anions, outline a possible procedure by means of which each component can be identified: (*a*) $HPO_4{}^{2-}$, $HAsO_4{}^{2-}$, $C_2O_4{}^{2-}$; (*b*) $CO_3{}^{2-}$, $C_2O_4{}^{2-}$; (*c*) $CrO_4{}^{2-}$, $AsO_4{}^{3-}$, $C_2O_4{}^{2-}$.

10-35. Discuss clearly the circumstances under which a solution might contain each of the following species: $PO_4{}^{3-}$, $HPO_4{}^{2-}$, $H_2PO_4{}^-$.

10-36. Compare the oxidizing strengths of orthophosphoric acid and orthoarsenic acid (Appendix G). What effect does pH have upon oxidation-reduction behavior in the arsenite-arsenate system?

10-37. Discuss the "chromate-dichromate" equilibrium, and indicate how it can be displaced. Criticize the statement "Dichromate ion is an oxidation product of chromate ion."

10-38. Which of the following solutions would be stable, and which would be unstable: (*a*) $Cr_2O_7{}^{2-}$, $C_2O_4{}^{2-}$, H^+; (*b*) $HPO_4{}^{2-}$, $HAsO_4{}^{2-}$; (*c*) $CrO_4{}^{2-}$, $C_2O_4{}^{2-}$, OH^-; (*d*) I^-, $Cr_2O_7{}^{2-}$, H^+? Explain.

10-39. Calculate the volume of 0.125 *N* potassium permanganate solution needed to react completely with 0.36 g of the salt $Na_2C_2O_4$.

10-40. Why is it possible to precipitate chromates by addition of alkali-metal dichromate solutions? Calculate the chromate-ion concentration in 0.1 *M* dichromate solution (Appendix C).

10-41. Why are lanthanum and thorium oxalates difficultly soluble in acidic solutions?

10-42. Outline clearly the chemical basis for each of the following:

(*a*) Use of peroxodisulfate to remove bromide and iodide prior to testing for chloride

(*b*) Use of iron(III) chloride–hexacyanoferrate(III) solutions to test for reducing agents

(*c*) Use of hydrogen peroxide to eliminate sulfite ion interference in the carbonate test

(*d*) Use of sodium carbonate to obtain solutions for anion detection

(*e*) Use of silicon dioxide in identifying fluoride ion

10-43. Explain clearly why an acetate solution containing iron(III) or aluminum(III) may precipitate upon boiling, whereas one containing calcium or barium ion does not.

10-44. Nitrite solutions commonly contain nitrate ion, but the unintentional presence of nitrite ion in nitrate solutions is unlikely. Explain.

10-45. Explain clearly why solutions containing acetic acid and alkali-metal acetates act as buffers.

10-46. Compare the oxidizing strengths of nitrate ion in acidic and alkaline media (Appendix G). By means of balanced ionic equations, formulate reactions in which four different reduction products are obtained from nitrate ion.

10-47. Indicate how each anion in each of the following combinations can be positively identified: (*a*) NO_2^-, Br^-, NO_3^-; (*b*) $C_2H_3O_2^-$, F^-, SO_4^{2-}; (*c*) MnO_4^-, CO_3^{2-}, $C_2O_4^{2-}$; (*d*) NO_3^-, I^-, NH_4^+.

10-48. Under what circumstances might each of the following exist as a stable combination: (*a*) MnO_4^-, I^-, SO_3^{2-}; (*b*) MnO_4^-, $C_2O_4^{2-}$; (*c*) NO_2^-, $C_2H_3O_2^-$, F^-? In every case, what single reagent might be added to destroy part or all of the combination? Explain.

10-49. List the anions which can coexist with permanganate ion in acidified solution. To what extent does this list suggest an experimental approach to ascertaining the anion content of samples containing permanganate ion.

10-50. Give the anion described by each of the following items:

(*a*) Solid sodium salt shows no reaction with cold 18 *M* sulfuric acid solution but yields brown fumes when this mixture is heated strongly.

(*b*) Solid sodium salt gives a colorless fuming gas with cold 18 *M* sulfuric acid solution, but an aqueous solution of the sodium salt gives no precipitate with silver(I) ion.

10-51. From the solubility-product constants for barium fluoride and barium carbonate (Appendix E), calculate the concentration of fluoride ion which might be expected in the solution resulting from treating excess solid barium fluoride with 50 ml of 3 *N* sodium carbonate solution.

10-52. 25 ml of 0.1 *M* $KMnO_4$ solution is added to an acidified solution containing 0.26 g of $NaHSO_3$. Determine the volume of 0.2 *M* oxalic acid solution needed to complete the reduction of the MnO_4^- ion.

10-53. Calculate the total volume of 0.1 *M* Ag_2SO_4 solution that should be added to a solution containing 4 ml each of 0.5 *M* K_2CrO_4, 0.5 *M* KI, and 0.5 *M* KNO_3 solutions to remove interferences with the "brown-ring" test for $NO^-{}_3$ ion.

10-54. To 0.24 mole of Na_3AsO_4 are added an excess of NaI and dilute H_2SO_4 solutions. Calculate the volume of 0.3 *M* $Na_2S_2O_3$ solution required to decolorize the mixture.

10-55. Determine the composition of the aqueous solution formed by adding 5 ml of 1 *M* $NaNO_2$ solution to a mixture of 10 ml of 0.3 *M* KI and 2 ml of 6 *M* CH_3COOH solutions and extracting completely with carbon tetrachloride.

10-56. Calculate the weight of barium chromate theoretically obtainable from the chromyl chloride produced from 0.585 g of sodium chloride.

10-57. A precipitated magnesium "carbonate" contains water and 48.15 percent MgO and 36.13 percent CO_2 by analysis. Determine the empirical formula of the compound.

10-58. A solution is prepared by dissolving 0.53 mole of NaH_2PO_4 and 1.66 moles of Na_3PO_4 in water and diluting to 500 ml. Calculate (1) the pH of the solution and (2) the molar concentrations of the species H_3PO_4, $H_2PO_4^-$, HPO_4^{2-}, and PO_4^{3-}.

10-59. Calculate the molar concentrations of H_3O^+, A^-, and HA_2^- ions in a 1 *M* solution of a weak acid HA if the pK_a for HA is 8.0 and if the equilibrium constant for the reaction described as

$$HA + A^- \rightleftharpoons HA_2^-$$

is 4 liters $mole^{-1}$.

10-60. Solid $SrSO_4$ and solid $BaSO_4$ are shaken with 1 liter of water until complete equilibrium is established. Calculate the concentrations of the ions Sr^{2+}, Ba^{2+}, and SO_4^{2-} in the resulting solution.

11
Identification of Cations

The individual cations present in a sample can be identified following the same general type of procedure used with the anions, by applying a series of specific tests that do not require the prior separation of the various ions from each other. However, unless organic reagents (page 145) are used very extensively, it is difficult to find sufficient reactions of this type that are free from interferences by other cations to allow for convenient identification of all of the important species. On the other hand, the system of classification outlined in Chap. 8 does permit systematic separation into major groups of cations. The separation and identification of individual species within each group are then simplified and can be carried out equally systematically. This approach has the advantages of convenience and logic, and it permits the direct application of many of the principles of solution chemistry, as outlined in Part I of this book, to a greater degree than any other approach. It is unfortunately true, however, that unless the student has a thorough understanding of what he seeks to do and appreciates thoroughly what he can learn from this approach, the procedure followed can become quite "cook-bookish."

The order in which the procedures for the cations are presented is the regular order of groups I through VI, as outlined in Chap. 8. This order is thus directly applicable to the analysis of a general unknown since it effects the systematic removal of each species before it can interfere with tests for additional species. Unfortunately, this sequence, based as it is solely on the solubilities of selected classes of compounds, does not develop the chemistry of the cations in the simplest possible way nor with any direct relationship to the periodic classification. In terms of developing the chemistry of the cations logically from the simplest stages to the more complex ones and of showing increasingly the application of principles, it is better to consider the cation groups in the order groups VI, V, I, II, III, and IV. In the discussion that follows, each group is discussed independently of the others and can thus be studied in this recommended order.

Inasmuch as many of the reactions used in the procedures for cations may be unfamiliar to the student, it is recommended that the investigation of each cation group begin with the study of a known sample for that group. Each cationic species is thus both carried through all of its important reactions and compared in behavior with all the other cationic species of that group before it is encountered in an unknown sample. Equations for the reactions of the cations are summarized in Appendix H_c.

PREPARATION OF THE SAMPLE FOR ANALYSIS

The procedures that follow are all based upon the use of solutions. Samples obtained as solutions can be used directly, since, by contrast with the behavior of anions, procedures for the cations are seldom interfered with by the anions which can be present. Solid samples must be dissolved. The following procedure is adequate for most samples. For those which are not suitable for this treatment, Procedure C-41 (page 284) should be applied.

Procedure C-1 Dissolution of the sample If the solid is not finely powdered, grind a sample of it in a clean mortar. Determine a suitable solvent by vigorously shaking pinch quantities of the finely divided solid with 1-ml volumes of (1) water, cold and hot; (2) 6 *M* nitric acid solution, cold and hot; (3) 16 *M* nitric acid solution, cold and hot; (4) 6 *M* hydrochloric acid solution, cold and hot; (5) 12 *M* hydrochloric acid solution, cold and hot; and (6) aqua regia (3 volumes of 6 *M* hydrochloric acid solution plus 1 volume of 6 *M* nitric acid solution). Try the behavior of the solvents *in the order listed.* In each case, allow time for reaction to occur before concluding that the sample is insoluble.

When a suitable solvent is found, dissolve 0.10 to 0.15 g of the finely powdered solid in 2 to 3 ml of that solvent, and use the solution for the appropriate cation-group procedure (Note 1).

Table 11-1 Schematic outline for analysis of cation group I

<table>
<tr><td colspan="5">Solution containing cations of group I only or of groups I–VI
C-2. Add cold, 6 *M* HCl. Centrifuge</td></tr>
<tr><td colspan="4">Precipitate: $\mathbf{PbCl_2}$, $\mathbf{Hg_2Cl_2}$, **AgCl** (all white)
C-3. Suspend in H_2O and boil. Filter</td><td rowspan="3">Centrifugate: Cations of groups II–VI (see Table 11-2)</td></tr>
<tr><td colspan="2">Residue: $\mathbf{Hg_2Cl_2}$, **AgCl**
C-4. Add 6 *M* aq NH_3</td><td colspan="2">Filtrate: Pb^{2+}
C-3. Divide into two portions</td></tr>
<tr><td>Residue: **Hg** + $\mathbf{HgNH_2Cl}$ (black to dark gray)</td><td>Solution: $[Ag(NH_3)_2]^+$ + Cl^-
C-5. Add 6 *M* HNO_3 until acidic
Precipitate: **AgCl** (white)</td><td>Portion I
C-3. Add 3 *M* H_2SO_4
Precipitate: $\mathbf{PbSO_4}$ (white)</td><td>Portion II
C-3. Add 0.1 *M* K_2CrO_4
Precipitate: $\mathbf{PbCrO_4}$ (yellow)</td></tr>
</table>

Note 1. Organic matter, if present, might interfere with the precipitation of hydroxides in cation group III because of complex-ion formation. However, evaporation with 16 *M* nitric acid solution in Procedure C-17 for the removal of manganese effectively eliminates this possible interference.

Cation group I

The general scheme followed in separating this group and in identifying its members is summarized in Table 11-1.

Procedure C-2 Precipitation of group Dilute the solution (Note 1) to 2 ml with cold water, and add 6 *M* hydrochloric acid solution drop by drop with vigorous shaking until precipitation appears to be complete (Note 2). Centrifuge, and test for completeness of precipitation by adding 1 drop of 6 *M* hydrochloric acid solution to the clear liquor above the residue. If a precipitate forms, add 2 more drops of the acid and recentrifuge. Wash the precipitate twice with 0.5-ml volumes of cold water, each containing 1 drop of 6 *M* hydrochloric acid solution (Note 3). Add the washings to the centrifugate. Save this solution for Procedure C-6 if the other cation groups are to be detected. Otherwise, discard it. Analyze the precipitate by Procedure C-3.

Note 1 Either a solution submitted directly as a known or unknown or one prepared according to Procedure C-1 can be used. If it is desired to analyze a known solution containing only Group I cations, mix 2 drops each of 0.1 *M* silver(I) nitrate, 0.1 M mercury(I) nitrate, and 0.1 *M* lead(II) nitrate solutions, dilute to 2 ml, and use directly.

Note 2 Cold solutions are used to minimize the solubility of lead(II) chloride. Care must be taken to avoid a large excess of hydrochloric acid to prevent dissolution of lead(II) and silver(I) chlorides as chloro-complex ions, $[PbCl_4]^{2-}$ and $[AgCl_2]^-$.

Note 3 Washing frees the precipitate from cations of subsequent groups. A small quantity of acid is needed to prevent hydrolysis of the ions Bi^{3+}, Sb^{3+}, and Sn^{4+} to water-insoluble basic salts.

Procedure C-3. Separation and identification of lead(II) ion Suspend the precipitate from Procedure C-2 in 2 ml of water, and transfer the suspension to a small beaker. Heat carefully to boiling and boil for 1 min (Note 1). Filter rapidly while hot. Save the residue on the filter paper for Procedure C-4. Divide the filtrate into two equal portions (Note 2). To one, add several drops of 3 *M* sulfuric acid solution; to the other, add several drops of 0.1 *M* potassium chromate solution. A white precipitate of lead(II) sulfate and a yellow precipitate of lead(II) chromate confirm the presence of *lead*(II) ion (Note 3).

Note 1 Boiling the precipitate with water increases the rate of dissolution of lead(II) chloride. Removal by merely washing with hot water is seldom complete.

Note 2 If crystals of the chloride separate as the filtrate cools, reheat until they dissolve before carrying out the confirmatory tests.

Note 3 Since lead(II) chromate is less soluble than the sulfate, the chromate test may give positive results when the sulfate test fails. Lead(II) sulfate may precipitate slowly.

Procedure C-4 Removal of silver(I) ion and identification of mercury(I) ion If lead(II) ion was found in Procedure C-3, first wash the residue on the filter paper with 1-ml portions of boiling water until the washings no longer give a precipitate with chromate ion (Note 1). Otherwise, proceed directly as follows: Treat the residue on the filter with 1 ml of 6 *M* aqueous ammonia solution, saving the filtrate for Procedure C-5. A dark-gray to black residue on the filter paper proves the presence of *mercury*(I) ion (Notes 2, 3).

Note 1 Any lead(II) chloride remaining will be converted to a finely divided basic salt by aqueous ammonia. This compound gives a turbid solution for the silver-ion test. Since the basic salt dissolves in the nitric acid solution used in this test, no interference results. However, the turbidity is misleading.

Note 2 Aqueous ammonia simultaneously detects mercury(I) ion through formation of insoluble materials (black **Hg** and white $\mathbf{HgNH_2Cl}$) and removes silver(I) ion by converting its chloride to the soluble diammine complex, $[Ag(NH_3)_2]^+$.

Note 3 If the mercury(I) to silver(I) ion ratio is large, released elemental mercury may itself reduce so much silver ion,

$$\mathbf{Hg} + 2Ag^+ \rightarrow 2\mathbf{Ag} + Hg^{2+}$$

that the latter cannot be detected. Dilute aqua regia converts this residue of silver and mercury to white silver(I) chloride and a solution of mercury(II) chloride, after which the usual test for silver ion can be applied.

Procedure C-5 Identification of silver(I) ion To the filtrate from Procedure C-4, add 6 *M* nitric acid solution until the system is definitely acidic (Note 1). A white precipitate of silver(I) chloride, insoluble in nitric acid, proves the presence of *silver*(I) ion (Note 2).

Note 1 Precipitation depends upon destruction of the $[Ag(NH_3)_2]^+$ ion, which cannot be complete until the solution is rendered acidic. If silver is present, the first drop of acid usually gives a precipitate which redissolves in the ammoniacal solution upon shaking. A permanent precipitate results only in acidic solution.

Note 2 Necessary chloride ion comes from the original silver(I) chloride.

Cation group II

The general scheme of analysis followed in separating this group and in identifying its members is summarized in Table 11-2.

Procedure C-6 Precipitation of group II Transfer the centrifugate reserved from Procedure C-2, or 1 ml of a solution containing only group II cations (Note 1), to a 25-ml beaker. If the solution is not already acidic to litmus (Note 2), add sufficient 6 *M* hydrochloric acid solution to make it so. Then add 4 to 5 drops of 3 percent hydrogen peroxide solution, heat to boiling, and boil carefully for 1 to 2 min (Note 3). Allow the solution to cool, and then add 6 *M* aqueous ammonia solution drop by drop until an alkaline reaction results, disregarding any precipitate which forms (Notes 4, 5).

Add 6 *M* hydrochloric acid solution drop by drop with stirring until the solution is just acidic to litmus, and then add 1.0 ml in excess. Pour the solution into a graduated cylinder and dilute to exactly 19 ml with cold water. Add 6 *M* hydrochloric acid solution drop by drop with stirring until a drop of the resulting solution or suspension gives the same color with methyl violet paper or with methyl violet solution on a white spot plate as does a solution prepared by diluting exactly 2 ml of 6 *M* hydrochloric acid solution to exactly 20 ml with water (0.6 *M* in hydronium ion, pH = about 0.5) (Note 6). Add 1 ml of 1 *M* thioacetamide solution (Notes 7, 8). Mix thoroughly, and transfer back to a small beaker. Heat on the water bath for 10 min (Note 9). Transfer the suspension to two test tubes, and centrifuge. To the clear centrifugate, add 0.5 ml of 1 *M* thioacetamide solution. Reheat. (Note 10). If an additional precipitate forms, recentrifuge, reserving the centrifugate for analysis for groups III to VI (Procedure C-17). If no precipitate forms, reserve the *clear* solution for Procedure C-17. Wash all residues into a single test tube with cold water and centrifuge. Add the washings to the centrifugate saved for groups III to VI. Suspend the residue in 1 ml of cold water and analyze by Procedure C-7 (Note 11).

Table 11-2 Schematic outline for analysis of cation group II

<table>
<tr><td colspan="6">Centrifugate from cation group I (Procedure C-2) or solution containing only group II cations $CH_3C(S)NH_2$, and heat. Centrifuge</td></tr>
<tr><td colspan="6">Precipitate: **HgS** (black), **PbS** (brown), **Bi_2S_3** (brown), **CuS** (black), **CdS** (yellow), **As_2S_3** (yellow),
C-7. Add 15 *M* aq NH_3. Heat. Add $CH_3C(S)NH_2$, and heat. Centrifuge</td></tr>
<tr><td colspan="6">Residue: **HgS, PbS, Bi_2S_3, CuS, CdS**
C-8. Add H_2O and 6 *M* HNO_3. Warm. Centrifuge</td></tr>
<tr><td rowspan="4">Residue: **HgS, S**
C-8. Add 16 *M* HNO_3, 12 *M* HCl
Evaporate
Test by either:
(*a*) Add 3 *M* $NaC_2H_3O_2$, 0.1 *M* $Co(NO_3)_2$, 0.1 *M* NH_4NCS
Precipitate:
$Co[Hg(NCS)_4]$
(deep blue)
(*b*) Add 0.1 *M* $SnCl_2$ in excess
Precipitate:
Hg_2Cl_2 + Hg
(gray to black)</td><td colspan="5">Centrifugate: Pb^{2+}, BiO^+, Cu^{2+} (blue), Cd^{2+}
C-9. Add 3 *M* H_2SO_4. Evaporate to white SO_3. Cool. Add H_2O.
Centrifuge</td></tr>
<tr><td colspan="2">Precipitate: **$PbSO_4$, $(BiO)_2SO_4$(?)** (white)
C-9. Add H_2O, 6 *M* NaOH. Centrifuge</td><td colspan="3">Centrifugate: BiO^+, Cu^{2+} Cd^{2+}
C-10. Add 6 *M* aq NH_3. Centrifuge</td></tr>
<tr><td rowspan="2">Precipitate:
BiOOH
(discard)</td><td rowspan="2">Centrifugate:
$HPbO_2^-$
C-9. Add 6 *M* $HC_2H_3O_2$; 0.1 *M* K_2CrO_4
Precipitate:
$PbCrO_4$
(yellow)</td><td rowspan="2">Precipitate:
BiOOH
(white)
C-10. Add $NaHSnO_2$
Precipitate:
Bi (jet black)</td><td colspan="2">Centrifugate: $[Cu(NH_3)_4]^{2+}$ (deep blue), $[Cd(NH_3)_4]^{2+}$ (colorless)
C-11. Divide in two portions</td></tr>
<tr><td>Portion I
C-11. Add 6 *M* $HC_2H_3O_2$
Evaporate
Test by either:
(*a*) Add 0.1 *M* $K_4[Fe(CN)_6]$
Precipitate:
$Cu_2[Fe(CN)]_6$
(reddish)
(*b*) Add C_5H_5N, 0.1 *M* NH_4NCS.
Precipitate:
green
$[Cu(py)_2(NCS)_2]$
(gives green $CHCl_3$ solution)</td><td>Portion II
C-12. Add 1 *M* NaCN to colorless. Add $CH_3C(S)$-NH_2, boil.
Precipitate:
CdS
(yellow)</td></tr>
</table>

Note 1 To prepare a known solution for this group, mix 2-drop volumes of 0.1 *M* mercury(II) nitrate, 0.1 *M* lead(II) nitrate, 0.1 *M* bismuth(III) chloride, 0.1 *M* copper(II) nitrate, 0.1 *M* cadmium nitrate, 0.1 *M* sodium arsenate, 0.1 *M* antimony(III) chloride, and 0.1 *M* tin(IV) chloride solutions, and dilute to 1 ml. If a precipitate forms, disregard it, but be certain the precipitate is suspended uniformly in the sample used for analysis.

C-6. Add 6 *M* HCl, 3% H_2O_2. Boil. Add 6 *M* aq NH_3. Adjust acidity with HCl. Add

---

Sb_2S_3 (orange), **SnS_2** (yellow)

Centrifugate: Cations of group III-VI (see Table 11-3)

Centrifugate: $[AsS_2]^-$, $[AsO_2]^-$, $[SbS_2]^-$, $[SbO_2]^-$, $[SnS_3]^{2-}$, $[Sn(OH)_6]^{2-}$, S_x^{2-}
C-13. Add 6 *M* $HC_2H_3O_2$. Warm. Centrifuge

Precipitate: **As_2S_3, Sb_2S_3, SnS_2, S_8**
C-14. Add H_2O, 12 *M* HCl. Heat. Centrifuge

Centrifugate: (discard)

Residue: **As_2S_3, S_8**
C-14. Add 6 *M* NaOH. Heat

Centrifugate: $[SbCl_4]^-$, $[SnCl_6]^{2-}$
C-15. Divide in two equal portions

Solution: $[AsS_2]^-$, $[AsO_2]^-$. Divide in two portions

Portion I
C-14. Add Al. Heat, test gas with $AgNO_3$
Precipitate: **Ag** (gray to black)

Portion II
C-14. Add 3% H_2O_2, boil. Add 6 *M* $HC_2H_3O_2$, 0.1 *M* $AgNO_3$
Precipitate: **Ag_3AsO_4** (reddish)

Portion I
C-15. Heat. Add 6 *M* aq NH_3, 6 *M* $HC_2H_3O_2$. Add solid **$Na_2S_2O_3$**
Precipitate: **Sb_2OS_2** (orange)

Portion II
C-16. Boil, add Al. Filter into saturated $HgCl_2$
Precipitate: **Hg_2Cl_2** (white)

Note 2 The solution will normally be acidic, because of acid added in the precipitation of group I, acid added to prevent hydrolysis, or acid formed during hydrolysis. The quantity of acid present is unknown.

Note 3 Hydrogen peroxide oxidizes tin(II) to tin(IV). Tin(IV) sulfide is less soluble in hydrogen ion than its tin(II) analogue and much more soluble in sulfide ion. Thus, both the precipitation of the tin and its separation into the appropriate subgroup are aided by having the material in its higher oxidation state. Excess peroxide is destroyed by boiling. Arsenic(III) is also converted into nonvolatile arsenate ion. Tin(II) and arsenic(III) are not listed as species for this group, but they may be present in some samples obtained for analysis. The procedures outlined will detect tin or arsenic irrespective of its original oxidation state or type of complexation.

Note 4 Ammonia is added to neutralize the undetermined quantity of free acid present in the solution so that a careful acidity control can be effected.

Note 5 Any precipitated hydroxide or basic salt can be ignored since it will be either dissolved or converted to less soluble sulfide in subsequent steps.

Note 6 The quantities of solution, water, and acid must be so chosen that the solution will be 0.25 to 0.3 *M* in hydrogen ion after the final dilution. In the presence of basic anions such as acetate, more than the given quantity of 6 *M* hydrochloric acid solution will be needed to produce this hydrogen-ion concentration. The procedure given allows for this adjustment by a colorimetric comparison with a solution of the correct acidity.

Note 7 The solution at this point is ca. 0.6 *M* in hydrogen ion, the higher hydrogen-ion concentration than in the hydrogen sulfide procedure being necessitated by the buffering action of acetate liberated in the hydrolysis of thioacetamide (page 106). If the solution is initially only 0.3 *M* in hydrogen ion, the acidity will be sufficiently decreased to allow precipitation of sulfides of group IV cations. Under the conditions outlined, the final hydrogen-ion concentration approaches that needed for complete separation.

Note 8 Thioacetamide acts as its own reducing agent toward arsenic acid, thereby eliminating problems of precipitating arsenic(V) sulfide. Tin must be in the tetrapositive state to be precipitated quantitatively.

Note 9 The forms and colors of the precipitated sulfides may give indications as to the cations present (Table 8-2, page 182).

Note 10 Precipitation must be complete to avoid complications in suceeding groups.

Note 11 The precipitate should be yellow to brown or black if group II cations are present. A light-colored precipitate is probably sulfur resulting from oxidation of hydrogen sulfide by nitrate, iron(III), dichromate, etc., ions.

Procedure C-7 Separation into copper and arsenic subgroups

To the suspension reserved from Procedure C-6, add 2 ml of 15 *M* aqueous ammonia solution, and warm in the water bath. Add 1 ml of 1 *M* thioacetamide solution. Heat on the water bath for 10 min (Note 1). Centrifuge, reserving the centrifugate for Procedure C-13. To the residue, add 2 to 3 ml of water plus 3 to 5 drops of 4 *M* ammonium nitrate solution (Note 3). Shake, centrifuge, and discard the centrifugate. Suspend the residue in 2 ml of water, and analyze by Procedure C-8 (Note 4).

Note 1 Reaction is much more rapid at elevated temperatures, but the suspension cannot be boiled without causing decomposition. Arsenic(III) sulfide dissolves readily, even in the absence of sulfide ion. Antimony(III) and tin(IV) sulfides dissolve more slowly. Tin must be in the tetrapositive state to dissolve under these conditions.

Note 2 Under the conditions used, no dissolution of mercury(II) sulfide or peptization of copper(II) sulfide should result. The centrifugates should be clear, with no more than a yellow color. The presence of some polysulfide ion (yellow) in the centrifugate is probably inevitable because of slight oxidation of sulfide ion by atmospheric oxygen. If the combined centrifugate is not to be analyzed until the following laboratory period it should be stored in a stoppered container.

Note 3 The ammonium ion prevents peptization by excess sulfide ion.

Note 4 If ions from the copper subgroup are present, the residue will be yellow to dark brown or black. Suspending in water limits air oxidation of sulfide ion to sulfate ion. The latter might cause loss of some lead(II) ion in the next procedure. Precipitated sulfides should never be allowed to dry in air.

Procedure C-8 Separation and identification of mercury(II) ion

To the suspension reserved from Procedure C-7, add 1 ml of 6 *M* nitric acid solution (Note 1). Heat almost to boiling, and maintain the sample at this temperature until reaction appears to be complete (Note 2). A dark-colored residue is probably mercury(II) sulfide (Note 3). Cool and centrifuge, reserving the centrifugate for Procedure C-9 (Note 4). Wash the residue with 1 ml of water, and add the washings to the centrifugate above. To the residue in the test tube, add a mixture of 3 drops of 6 *M* nitric acid solution and 11 drops of 12 *M* hydrochloric acid solution. Heat on the water bath until reaction is complete, and remove the small globule of liberated sulfur with a stirring rod. Evaporate the solution to 0.5 ml (Note 5). Apply either of the following tests: (*a*) To 2 drops of this solution on a white spot plate, add 3 drops of 3 *M* sodium acetate solution, 1 drop of 0.1 *M* zinc nitrate solution, and 1 drop of 0.1 *M* cobalt(II) nitrate solution. Then add 0.1 *M* ammonium thiocyanate solution drop by drop. A deep-blue precipitate confirms the presence of *mercury*(II) ion. (*b*) To the solution from the aqua regia treatment, add 0.1 *M* tin(II) chloride solution drop by drop. A white precipitate of mercury(I) chloride, turning gray to black (elemental mercury) with excess tin(II) ion, confirms the presence of *mercury*(II) ion (Note 6).

Note 1 The separation depends upon the insolubility of mercury(II) sulfide in 2 *M* nitric acid solution. More concentrated acid causes dissolution. The procedure given yields a 2 *M* solution.

Note 2 Boiling may dissolve some mercury(II) sulfide.

Note 3 Positive conclusions are not warranted because either (*a*) sulfur may enclose and protect from the acid small quantities of other sulfides or (*b*) mercury(II) sulfide may be converted to light-colored compounds of the composition $Hg(NO_3)_2 \cdot xHgS$.

Note 4 Small quantities of solid floating on this solution [sulfur plus some mercury(II) sulfide] should be removed with a stirring rod and returned to the undissolved residue. A blue color in this solution is due to copper(II) ion, but the color is not usually sufficiently intense to permit positive identification.

Note 5 Excess chloride ion is removed to reduce interference with the tin(II) chloride reduction test.

Note 6 Copper(II) ion, if present, would reduce to white copper(I) chloride, but no further. Hence, conversion to dark elemental mercury is a necessary part of the test. Excess tin(II) chloride is needed because of oxidation of tin(II) ion by any remaining aqua regia.

Procedure C-9 Separation and identification of lead(II) ion Transfer the centrifugate reserved from Procedure C-8 to a small casserole, and add 6 to 8 drops of 3 *M* sulfuric acid solution. *Under the hood*, carefully evaporate until *dense white* fumes of sulfur trioxide are liberated (Note 1). Allow to cool *completely*, and *cautiously* pour the contents of the casserole into 1 ml of water in a test tube (Note 2). Add 1 ml of water to the casserole, stir carefully, and decant back into the test tube. Finally rinse with a few drops of water, and add this to the material in the test tube. Cool. A finely divided white precipitate indicates the presence of lead(II) ion (Notes 3, 4). Centrifuge, carefully decant the liquid from the residue, and reserve the centrifugate for Procedure C-10. To the residue in the test tube, add 0.5 ml of water and 5 to 6 drops of 6 *M* sodium hydroxide solution (Note 5). Shake thoroughly. If the solid dissolves completely (no BiO^+ ion present), add 2 to 3 drops of 0.1 *M* potassium chromate (or dichromate) solution, and acidify the solution with 6 *M* acetic acid. A yellow precipitate of lead(II) chromate confirms the presence of *lead*(II) ion (Note 6). If the solid is not completely soluble in sodium hydroxide solution, centrifuge, and test the centrifugate as outlined above.

Note 1 These fumes are persistent, heavy, and densely white. They are not to be confused with steam. Their formation from the decomposition of sulfuric acid serves as an indicator for the complete removal of the more volatile nitric acid. The latter solubilizes lead(II) sulfate through formation of HSO_4^- ions.

Note 2 Concentrated sulfuric acid liberates sufficient heat when diluted to cause spattering unless care is taken. Water should *never* be poured into the concentrated acid.

Note 3 Dilution is necessary to effect precipitation because of the solubility of lead(II) sulfate in concentrated sulfuric acid solution. The precipitate is seldom large, particularly if the original sample was a general unknown and lead(II) ion was detected in cation group I. It is best seen against the dark background of the desk top or a black spot plate. In the casserole, it is probably invisible.

Note 4 Large quantities of bismuth(III) ion may precipitate as white $(BiO)_2SO_4$. A confirmatory test for lead(II) ion is always essential.

Note 5 Lead(II) sulfate is dissolved as plumbite ion, whereas insoluble bismuth(III) hydroxide is formed. Hot 3 *M* ammonium acetate solution can be used also to dissolve lead(II) sulfate selectively.

Note 6 No precipitate results in alkaline solutions, but, in neutral solutions, orange to reddish precipitates of basic chromates may form. Centrifuging renders the precipitate more readily visible.

Procedure C-10 Separation and identification of bismuth(III) ion To the centrifugate reserved from Procedure C-9, add 6 *M* aqueous ammonia solution until an excess is present and a strongly alkaline reaction results (Note 1). A white precipitate (Note 2) indicates the presence of bismuth(III) ion and a deep-blue color the presence of copper(II) ion (Note 3). Centrifuge, and reserve the centrifugate for Procedure C-11. Wash the residue with 1 ml of water, and add the washings to the centrifugate for Procedure C-11. To the precipitate in the test tube, add a few drops of a cold sodium stannite solution (freshly prepared as in Note 4). An immediate conversion of the white precipitate to jet-black elemental bismuth confirms the presence of *bismuth*(III) ion (Note 5).

Note 1 Excess ammonia is needed to ensure dissolution of initially precipitated copper(II) and cadmium hydroxides. The separation depends upon lack of ammine complex-ion formation by bismuth(III).

Note 2 The whiteness of the precipitate may become apparent only upon centrifuging because of the blue of the copper(II) complex.

Note 3 Although the blue color is a sufficient test for copper (II) ion, it is often not sufficiently intense to permit detection of small quantities of this ion.

Note 4 Dilute 2 drops of 0.1 *M* tin(II) chloride solution to 1 ml, cool under the cold-water tap, and add 6 *M* sodium hydroxide solution drop by drop until the precipitate which first forms redissolves and the solution is strongly alkaline. Cool and use immediately.

Note 5 Since other hydroxides may precipitate with aqueous ammonia, the confirmatory test is necessary. Reduction of the bismuth compound is usually instantaneous. Any darkening after 1 to 2 min is probably due to reduction of other hydroxides or to decomposition of the reagent in terms of the equations

$$2HSnO_2^- + 2H_2O \rightleftharpoons \underset{\text{(black)}}{\mathbf{Sn}} + [Sn(OH)_6]^{2-} \qquad \text{(excess } OH^-\text{)}$$

and

$$HSnO_2^- \rightleftharpoons \underset{\text{(black)}}{\mathbf{SnO}} + OH^- \qquad \text{(insufficient } OH^-\text{)}$$

Each of these interfering reactions is favored by heat. The reagent must be cold and freshly prepared.

Procedure C-11 Identification of copper(II) ion Use half the centrifugate reserved from Procedure C-10 for the following tests, and save the remainder for Procedure C-12 (Note 1). Add 6 *M* acetic acid solution drop by drop until the solution is acidic (Note 2), and then evaporate to 0.5 ml. Apply either of the following tests: (*a*) To several drops of the resulting solution on a white spot plate, add 2 to 3 drops of 0.1 *M* potassium hexacyanoferrate(II) solution. A reddish coloration or precipitate of $\mathbf{Cu_2[Fe(CN)_6]}$ confirms the presence of *copper*(II) ion (Note 3). (*b*) To the solution, add pyridine until an alkaline reaction results (Note 4), and then add several drops of 0.1 *M* ammonium thiocyanate solution. A green precipitate of composition **$[Cu(py)_2(NCS)_2]$**, which dissolves to a green solution when shaken with 0.5 ml of chloroform, confirms the presence of *copper*(II) ion (Note 5).

Note 1 In the absence of copper(II) ion, the tests outlined may give indication of the presence of cadmium ion and substitute for Procedure C-12.

Note 2 The tetraammine complex ion must be destroyed to give maximum sensitivity to the test used.

Note 3 In the absence of copper(II) ion, the yellowish cadmium salt precipitates. The copper(II) compound is the less soluble and forms preferentially. Its color is sufficiently intense that even small quantities of copper(II) ion can be detected in the presence of sizable quantities of cadmium ion.

Note 4 Pyridine resembles ammonia in first precipitating hydroxides and then giving soluble complex species. The ion $[Cu(py)_4]^{2+}$ is dark blue, whereas the ion $[Cd(py)_4]^{2+}$ is colorless.

Note 5 In the absence of copper(II) ion, white **$[Cd(py)_2(NCS)_2]$** precipitates and dissolves in chloroform to give a colorless solution. The copper compound again forms preferentially and masks the cadmium compound. The color of the copper complex is best seen against a white background.

Procedure C-12 Identification of cadmium ion If the tests used in Procedure C-11 indicate that the copper(II) ion is absent, add several drops of 1 *M* thioacetamide solution to the solution reserved from Procedure C-11 and heat to boiling. A yellow to orange precipitate of **CdS** confirms the presence of *cadmium* ion (Note 1). If copper(II) ion is present, add 1 *M* sodium cyanide solution drop by drop (*Hood!*) to the *ammoniacal* solution (*Test with litmus!*) from Procedure C-11 until the solution is colorless (Notes 2, 3). Add several drops of 1 *M* thioacetamide solution and heat to boiling. A yellow precipitate of cadmium sulfide confirms the presence of *cadmium* ion (Note 4).

Note 1 The precipitate is normally yellow, but orange products result under certain conditions.

Note 2 The solution *must be alkaline* when the cyanide is added to prevent release of hydrogen cyanide gas and provide for absorption of liberated cyanogen. If the litmus test does not show an alkaline solution, add 6 *M* ammonia solution to alkalinity before adding cyanide. Reasonable care in keeping the nose and mouth away from cyanide-containing solutions and in washing the hands is necessary.

Note 3 Excess cyanide ion reduces copper(II) to a copper(I) complex

$$2[Cu(NH_3)_4]^{2+} + 8CN^- \rightleftharpoons 2[Cu(CN)_3]^{2-} + 8NH_3 + (CN)_2\uparrow$$

The liberated cyanogen is absorbed in the alkaline solution,

$$(CN)_2 + 2OH^- \rightleftharpoons CN^- + OCN^- + H_2O$$

The copper(I) complex gives too small a concentration of copper(I) ions to form a precipitate with sulfide ion. Cadmium ion yields a relatively unstable tetracyano ion.

Note 4 If the precipitate is dark in color, indicating a faulty procedure, centrifuge, wash the residue with water containing a few drops of 1 *M* ammonium sulfate solution, and discard the washings. Treat the residue with 1 ml of 3 *M* sulfuric acid solution, and heat. Centrifuge, dilute the centrifugate with 5 ml of water, and heat with thioacetamide to give yellow cadmium sulfide.

Procedure C-13 Reprecipitation of arsenic subgroup sulfides To the centrifugate reserved from Procedure C-7, add 6 *M* acetic acid solution drop by drop with constant stirring (*Hood!*) until a slightly acidic reaction results (Note 1). Then add 1 ml in excess. Warm on the water bath to coagulate the precipitate. Centrifuge, and wash the solid with 1 ml of water, discarding both centrifugate and washings. Reserve the precipitate for Procedure C-14 (Note 2).

Note 1 An acidic solution is necessary to decompose the mixed oxo- and sulfoanions and reprecipitate the sulfides, but excess hydrogen ion would dissolve tin(IV) sulfide. Acetic acid is too weak to dissolve the precipitated sulfides.

Note 2 The precipitate is yellow or orange if members of this subgroup are present. A light-colored precipitate is probably sulfur.

Procedure C-14 Separation and identification of arsenic To the precipitate reserved from Procedure C-13, add 1 ml of water and 2 ml of 12 *M* hydrochloric acid solution. Heat the mixture to 70°C on the water bath for 3 min while stirring the contents of the tube with a glass rod to break up lumps of precipitate (Note 1). Centrifuge. Treat the residue with 0.5 ml of water and 1 ml of 12 *M* hydrochloric acid in the same fashion. Centrifuge again. Combine the two centrifugates, and reserve them for Procedure C-15. Wash the yellow residue (Note 2) with water, added in 2-ml portions, until the washings give no more than a faint test for chloride ion with 0.1 *M* silver(I) nitrate solution (Note 3). Dissolve the precipitate in 2 ml of 6 *M* sodium hydroxide solution, and divide the resulting solution into two equal portions. To *one* portion, contained in a small test tube, add a few pieces of granulated aluminum. Insert a small wad of loosely packed cotton well into the tube, and place over the mouth of the tube a piece of filter paper moistened with 1 to 2 drops of 0.1 *M* silver(I) nitrate solution. Warm the test tube in the water bath. A gray to black precipitate of elemental silver on the underside of the filter paper confirms the presence of *arsenic* in the sample (Notes 4, 5). To the other portion of the sodium hydroxide solution, add 1 ml of 3 percent hydrogen peroxide solution. Heat to boiling, and boil gently until oxygen evolution from decomposition of the excess peroxide ceases (Notes 6, 7). Add 6 *M* acetic acid solution drop by drop until the solution is just acidic, and then add 2 to 3 drops of 0.1 *M* silver(I) nitrate solution. A reddish precipitate of silver(I) orthoarsenate confirms the presence of *arsenic* (Note 8). If no precipitate results, add 6 *M* aqueous ammonia solution drop by drop, shaking after each drop. If arsenate ion is present, a precipitate will form before the solution becomes alkaline.

Note 1 The separation depends upon the greater solubility of antimony(III) and tin(IV) sulfides in acidic solutions, but if the solution is boiled or the treatment prolonged, appreciable quantities of arsenic(III) sulfide will dissolve. Incomplete treatment leaves antimony and tin materials, but these species do not interfere with the tests outlined.

Note 2 If arsenic is absent, no more than a light-colored residue of sulfur should remain.

Note 3 Chloride ion must be removed because of preferential precipitation of silver(I) chloride in the silver-ion test that follows.

Note 4 The test is specific for arsenic(III) as carried out in alkaline medium. Under acidic conditions, both arsenic(V) and antimony(III) give similar results. The reducing agent is liberated arsine.

Note 5 Hydrogen sulfide, even in traces, darkens the paper by reaction with silver(I) ion. The test should not be carried out close to any source of hydrogen sulfide.

Note 6 Hydrogen peroxide oxidizes arsenite and sulfoarsenite ions to orthoarsenate, converting all sulfide species to sulfate. Orthoarsenate ion gives a more sensitive and distinctive test with silver(I) ion than arsenite ion.

Note 7 Removal of excess peroxide eliminates any possible side reaction with silver ion.

Note 8 Silver(I) orthoarsenate precipitates only at a controlled pH. The solution must be slightly acidic to prevent formation of black silver(I) oxide, but too much acid prevents precipitation. Adjustment with ammonia, as outlined, may be necessary, but excess ammonia prevents precipitation by forming the diammine silver(I) ion, $[Ag(NH_3)_2]^+$.

Procedure C-15 Identification of antimony(III) ion Divide the centrifugate reserved from Procedure C-14 into two equal portions, and save *one* portion for Procedure C-16. Evaporate the *other* portion (*Hood!*) to about 0.5 ml (Note 1). Transfer to a small test tube, and add 6 *M* aqueous ammonia solution drop by drop until the system is just alkaline. Disregard any precipitate that forms (Note 2). Make acidic with 6 *M* acetic acid solution, and add 1 drop in excess. Heat to boiling, and drop into the hot solution a pinch of solid sodium thiosulfate. Allow to stand without disturbing. An orange-red precipitate of composition $\mathbf{Sb_2OS_2}$, developing at the interface between the top and bottom layers, confirms the presence of *antimony*(III) ion (Notes 3, 4).

Note 1 Removal of considerable hydrochloric acid and concentration of the solution are both effected.

Note 2 Necessary reduction in the acidity of the solution causes precipitation of antimony(III) hydroxide or basic salts. These dissolve in part in the next step, but since they are more soluble than the sulfide obtained in the final identification, their presence causes no difficulty.

Note 3 The reaction depends upon the small concentration of sulfide ion produced by the hydrolysis of the thiosulfate ion and is effective only because of the limited solubility of the antimony oxosulfide. Tin(IV) and arsenic(III) cause no interference, but a light-colored precipitate of sulfur may be noted in the absence of antimony.

Note 4 A less definitive test involves placing a few drops of the solution from Procedure C-14 on a *silver* coin and adding a small piece of tin in such a way that it is in contact both with silver and the solution. A black deposit on the coin, insoluble in sodium hypochlorite solution, confirms the presence of antimony(III) ion.

Procedure C-16 Identification of tin(IV) ion Heat the portion of solution reserved from Procedure C-15 to boiling to expel any dissolved hydrogen sulfide. Then add one small piece (0.5-cm length) of aluminum wire, and *warm gently until visible reaction ceases and the aluminum has dissolved completely* (Note 1). Filter *immediately* into 1 ml of saturated mercury(II) chloride solution (Note 2). A white precipitate of mercury(I) chloride, which may turn gray owing to further reduction to elemental mercury, confirms the presence of *tin*(IV) ion (Notes 3, 4).

Note 1 Aluminum reduces tin(IV) species to elemental tin. The latter then dissolves in the acidic solution as the tin(II) ion, but this dissolution does not occur until all the aluminum has dissolved. Antimony(III) ion and traces of arsenic(III) are also reduced to the free elements, but the resulting black deposits do not dissolve and should be ignored except in so far as their appearances may substantiate the results of Procedures C-14 and C-15.

Note 2 Immediate filtration avoids extensive reoxidation of tin(II).

Note 3 Since mercury(II) chloride is normally in excess, the usual product is white mercury(I) chloride. In some samples, there may be sufficient tin to give a gray product. The solution should be observed against a dark background to permit detection of small quantities of white precipitate.

Cation group III

The procedure to be followed with samples containing cations from the subsequent groups depends upon the absence or presence of orthophosphate ion (pages 252, 253). Regardless of the presence or absence of this anion, manganese is removed and identified by Procedure C-17. In the absence of orthophosphate ion, subsequent analyses are then made by Procedures C-18 through C-21. In its presence (Procedure C-16, page 226), Procedure C-17 is followed by Procedure C-22, which removes this anion, and then Procedures C-18 through C-21 are used. Samples containing only group

Table 11-3 Schematic outline for analysis of cation group III (orthophosphate ion absent)

<table>
<tr><td colspan="8">Centrifugate from cation group II (Procedure C-6) or solution containing only group III cations
C-17. Add 6 *M* HCl. Boil. Evaporate to dryness with 16 *M* HNO_3. Add water, 16 *M* HNO_3, solid **$KClO_3$**. Boil. Centrifuge</td></tr>
<tr><td colspan="2" rowspan="3">Precipitate: **MnO_2** (dark brown to black)
C-17. Add 6 *M* HNO_3, 0.1 *M* $NaNO_2$ Warm
Solution: Mn^{2+}, HNO_3
C-17. Add solid **$NaBiO_3$**. Warm. Centrifuge</td><td colspan="6">Centrifugate: Fe^{3+}, Al^{3+}, $Cr_2O_7^{2-}$, cations of groups IV–VI
C-18. Add CH_2O, and heat. Then add 6 *M* aq NH_3 in excess. Centrifuge</td></tr>
<tr><td colspan="5">Precipitate: **$Fe(OH)_3$** (red-brown), **$Al(OH)_3$** (white), **$Cr(OH)_3$** (gray-green)
C-19. Add 6 *M* NaOH in excess, then 3% H_2O_2. Boil. Centrifuge</td><td rowspan="5">Centrifugate: $[Co(NH_3)_6]^{2+}$, $[Ni(NH_3)_6]^{2+}$, $[Zn(NH_3)_4]^{2+}$, cations of groups V–VI (see Table 11-5)</td></tr>
<tr><td colspan="2">Residue: **$Fe(OH)_3$**
C-19. Add 6 *M* HCl. Divide in two portions</td><td colspan="3">Solution: AlO_2^- (colorless), CrO_4^{2-} (yellow)
C-20. Add 6 *M* HCl in excess, then 6 *M* aq NH_3 in excess. Centrifuge</td></tr>
<tr><td rowspan="3">Residue: **$NaBiO_3$** (discard)</td><td rowspan="3">Centrifugate: MnO_4^- (violet to purple)</td><td rowspan="3">Portion I: Fe^{3+}
C-19. Add 0.1 *M* $K_4[Fe(CN)_6]$
Precipitate: **$KFe[Fe(CN)_6]$** (dark blue)</td><td rowspan="3">Portion II: Fe^{3+}
C-19. Add *M* KNCS
Solution: $[Fe(NCS)]^{2+}$ (blood red)</td><td colspan="2">Precipitate: **$Al(OH)_3$**
C-20. Add 6 *M* HCl. Centrifuge</td><td rowspan="3">Centrifugate: CrO_4^{2-}
C-21. Add 3 *M* H_2SO_4, ether, 3% H_2O_2. Shake
Upper layer: CrO_5 (dark blue)</td></tr>
<tr><td rowspan="2">Residue: **SiO_2** (discard)</td><td>Solution: Al^{3+}
C-20. Evaporate, add aluminon and 4 *M* $NaC_2H_3O_2$–4 *M* $HC_2H_3O_2$</td></tr>
<tr><td>Precipitate: **Al compound** (red)</td></tr>
</table>

Table 11-4 Schematic outline for removal of orthophosphate ion in cation group III

Centrifugate from Procedure C-17: Fe^{3+}, Al^{3+}, $Cr_2O_7^{2-}$, cations of groups IV–VI $H_2PO_4^-$	
C-22. Add CH_2O, and heat. Add 6 *M* aq NH_3 short of precipitation, evaporate to 5 ml, add 1 ml 6 *M* HCl. Add about 0.4 *M* $ZrO(NO_3)_2$. Heat. Centrifuge	
Precipitate: $\mathbf{(ZrO)_3(PO_4)_2}$ (white) (discard)	Centrifugate: Fe^{3+}, Al^{3+}, Cr^{3+}, cations of groups IV–VI Analyze by Procedure C-18 (see Table 11-3)

III cations are treated as if orthophosphate is absent. Anions such as fluoride and oxalate ions which would behave similarly to orthophosphate ion in causing precipitation of group V cations when the solution is made alkaline, are removed in Procedure C-17. The general scheme of analysis is given in Table 11-3, and the removal of the orthophosphate ion in Table 11-4.

Procedure C-17 Removal and identification of manganese(II) ion

To the centrifugate reserved from Procedure C-6, add 0.5 ml of 6 *M* hydrochloric acid solution. Boil gently until no more hydrogen sulfide gas is evolved (Note 1). If the sample being analyzed contains only group III cations, omit this step (Note 2). Then add 1 ml of 16 *M* nitric acid solution, and evaporate just to dryness (Note 3). Dissolve the residue carefully in 1 ml of water. Add 1 ml of 16 *N* nitric acid solution and about 0.5 g of powdered potassium chlorate. Boil *cautiously* and *gently* for 1 min, and then add 0.5 ml of 16 *M* nitric acid solution and a pinch of potassium chlorate. Boil *gently*, and evaporate to about 1 ml. A dark-brown to black precipitate (presumably $\mathbf{MnO_2}$) indicates the presence of manganese(II) ion (Note 4). Dilute to 2 to 3 ml, and centrifuge. Wash the residue with 1 ml of water, and save the combined centrifugate and washings for Procedure C-18 or Procedure C-22. To the washed precipitate, add 2 to 3 ml of 6 *M* nitric acid solution and a few drops of 0.1 *M* sodium nitrite solution. Warm until a clear solution results (Note 5). Add solid sodium bismuthate from a spatula until a small quantity remains undissolved. Warm in the water bath if necessary. Centrifuge. A violet to deep-purple centrifugate, due to the permanganate ion, confirms the presence of *manganese*(II) ion (Note 6).

Note 1 Sulfide is removed to prevent its oxidation to sulfur or sulfate ion by nitric acid. The latter would effect precipitation of barium ion with manganese(IV) oxide.

Note 2 The absence of sulfide ion in unknown or known samples for this group may be assumed. To prepare a known solution, mix 2-drop volumes of 0.1 *M* manganese(II) nitrate, iron(III) chloride, aluminum nitrate, and chromium(III) chloride solutions. Use without diluting.

Note 3 Chloride ion is removed as elemental chlorine. If not removed, chloride ion would hinder oxidation to manganese(IV) oxide. Any iron(II) ion initially present is oxidized to iron(III) species.

Note 4 The reaction is specific for manganese among the cations here studied. If no precipitate results, manganese(II) ion is absent, and the solution should be analyzed directly by Procedure C-18. Chromium(III) ion is converted to orange dichromate ion in the reaction. Ignore any white crystals that form.

Note 5 A reducing agent is necessary to permit dissolution of manganese(IV) oxide in nitric acid.

Note 6 Centrifuging avoids the difficulty of observation due to suspended solids. The test can be applied directly to the manganese(IV) oxide, but the dark color of the unreacted oxide may render the color of the solution less distinct.

Orthophosphate ion absent

Procedure C-18 Precipitation of other members of group III Dilute the combined centrifugate and washings from Procedure C-17 to 10 ml. Carefully add 4 to 5 drops of formaldehyde solution, and *warm* (Note 1). Then add 6 *M* aqueous ammonia solution with vigorous stirring to just short of precipitation (Note 2). Evaporate to 5 ml, and add 6 *M* aqueous ammonia solution until the suspension is strongly alkaline (Note 3). Add 1 ml in excess, stir for 5 min (Note 4), and centrifuge. Wash the precipitate twice with 1-ml volumes of water. Reserve the centrifugate and washings for analysis for groups IV to VI (Procedure C-23) (Note 5), and analyze the precipitate by Procedure C-19.

Note 1 Formaldehyde reduces dichromate ion to chromium(III) ion. Larger quantities of formaldehyde than that recommended *may react violently* with the nitric acid present. This procedure *must be carried out very carefully* and with due protection of hands and eyes.

Note 2 Neutralization of the concentrated nitric acid from the previous step requires a considerable quantity of aqueous ammonia and gives a sizable total volume of solution. Subsequent evaporation is then necessary.

Note 3 Color changes during this neutralization reaction are important. Thus addition of aqueous ammonia gives $\mathbf{Fe(OH)_3}$ (red-brown); $\mathbf{Al(OH)_3}$ (white); $\mathbf{Cr(OH)_3}$ (gray-green); $\mathbf{Ni(OH)_2}$ (pale green), dissolving immediately to $[Ni(NH_3)_6]^{2+}$ (deep blue); $\mathbf{Co(OH)_2}$ (blue or pink), dissolving *slowly* to $[Co(NH_3)_6]^{2+}$ (tan); and $\mathbf{Zn(OH)_2}$ (white), dissolving rapidly to $[Zn(NH_3)_4]^{2+}$ (colorless). The absence of a precipitate means the absence of group III cations.

Note 4 This treatment renders dissolution of cobalt(II) hydroxide complete. Neutralization of the nitric acid provided the ammonium ion essential to preventing precipitation of magnesium hydroxide, to minimizing the dissolution of aluminum hydroxide in hydroxide ion, and to minimizing peptization of the precipitate.

Note 5 Small quantities of pink $[Cr(NH_3)_5(OH)]^{2+}$ ion may appear in the centrifugate. Potassium ion is always present in all subsequent solutions from the potassium chlorate added.

Procedure C-19 Separation and identification of iron(III) ion Suspend the residue from Procedure C-18 in 2 ml of water in a small beaker, and add 6 *M* sodium hydroxide solution with stirring until the suspension is alkaline. Then add 1 ml in excess. A red-brown residue indicates the presence of iron(III) ion. Without centrifuging, add 1 ml of 3 percent hydrogen peroxide solution (Note 1). Carefully heat the suspension to boiling, and boil for a few minutes until decomposition of the peroxide is complete (Note 2). Cool, centrifuge, and wash the precipitate with 1 ml of water. Reserve the combined centrifugate and washings for Procedure C-20. Dissolve the precipitate in 1 ml of 6 *M* hydrochloric acid solution, and divide the solution into two equal portions. To *one*, add 1 drop of 0.1 *M* potassium hexacyanoferrate(II) solution; to the *other*, add 1 drop of 0.1 *M* potassium thiocyanate solution. A dark-blue precipitate of composition $\mathbf{KFe[Fe(CN)_6]}$ and a blood-red solution of the $[Fe(NCS)]^{2+}$ ion, respectively, confirm the presence of *iron*(III) ion (Note 3).

Note 1 Oxidation of chromite ion (green) to chromate ion (yellow) is essential to avoid irreversible hydrolysis of the former to insoluble chromium(III) hydroxide and thus prevent loss of chromium and contamination of the iron(III) hydroxide precipitate.

Note 2 Unless excess peroxide is removed, acidification in Procedure C-20 will cause reduction of chromate ion to chromium(III) ion, which will then precipitate as hydroxide with the aluminum ion. A yellow centrifugate indicates the presence of chromate ion.

Note 3 The presence of iron(III) should not be reported unless both these tests are very positive. Traces of iron(III) introduced as impurity during the analysis give faint tests at this point. Nitric acid is not used as solvent here because it may yield colored products with thiocyanate ion. If iron is found, its state of oxidation in the initial sample can be established by testing portions of the solution with both hexacyanoferrate(II) and hexacyanoferrate(III) ions. Iron(II) ion gives a dark blue precipitate, $\mathbf{KFe[Fe(CN)_6]}$, with $K_3[Fe(CN)_6]$ solution and a pale blue precipitate, $\mathbf{Fe_2[Fe(CN)_6]}$, with $K_4[Fe(CN)_6]$. Iron(III) ion gives a greenish-brown solution with $K_3[Fe(CN)_6]$ solution and a dark blue precipitate, $\mathbf{KFe[Fe(CN)_6]}$, with $K_4[Fe(CN)_6]$. The two dark blue precipitates are identical as a consequence of the same distribution of the oxidation states of iron throughout the crystal lattice, i.e., $KFe^{III}[Fe^{II}(CN)_6]$.

Procedure C-20 Separation and identification of aluminum ion Evaporate the centrifugate from Procedure C-19 to 2 ml. Carefully neutralize with 6 *M* hydrochloric acid solution, and add 1 drop in excess (Note 1). Then make just alkaline with 6 *M* aqueous ammonia solution, and add no more than 2 drops in excess. Centrifuge. A white gelatinous precipitate may be aluminum hydroxide (Note 2), and a yellow centrifugate indicates chromate ion. Reserve the centrifugate for Procedure C-21. Wash the precipitate with 1 ml of water, and discard the washings. Warm the precipitate with 1 ml of 6 *M* hydrochloric acid solution, centrifuge, and save the centrifugate, but discard any undissolved residue (Note 3). Evaporate the solution to about 0.25 ml, and place 1 drop of the concentrate in a depression on a white spot plate. Add 1 drop of 0.1 percent aluminon solution and 2 drops of 4 *M* sodium acetate–4 *M* acetic acid buffer solution. A bright red precipitate, forming immediately or after a few minutes, confirms the presence of *aluminum* ion (Notes 4, 5).

Note 1 Hydrogen ion converts aluminate ion to the aluminum cation but changes chromate ion only to dichromate ion.

Note 2 Formation of a precipitate is not positive proof of the presence of aluminum ion since silica (from reaction of alkaline solutions on glass apparatus) or other metal hydroxides may form. The confirmatory test should always be run.

Note 3 The residue is probably silica.

Note 4 If the test is not conclusive, compare it with a *control* test run with 0.1 *M* aluminum nitrate solution. Aluminon, the ammonium salt of aurintricarboxylic acid, gives a red aluminum compound that can be formulated as

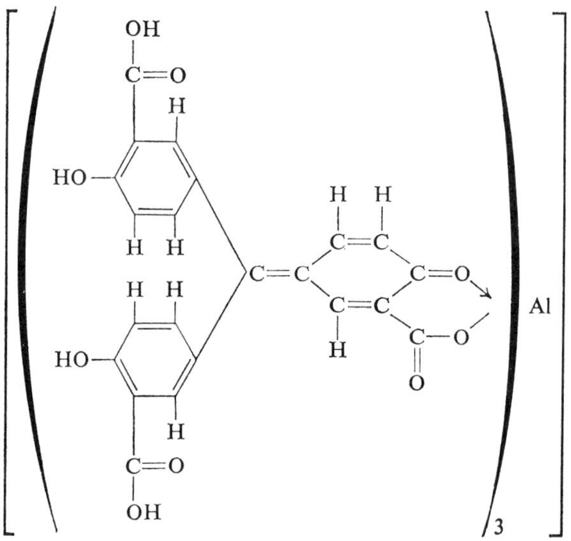

where the notation

represents the C_6 ring characteristic of the benzene molecule. The color is detectable at concentrations as low as 10^{-4} mole of Al^{3+} liter^{-1}. Iron(III), chromium(III), and other cations give colored products, many of which can be decomposed by adding ammonium carbonate solution. A pH of 4.5 is optimum for this reaction.

Note 5 Aluminum ion is often present in detectable quantities as an impurity. If even by comparison with a control test there is doubt as to its presence, consult the instructor as to a supplementary procedure that can be applied directly to the unknown sample.

Procedure C-21 Identification of chromate ion Acidify the centrifugate from Procedure C-20 with a few drops of 3 *M* sulfuric acid solution. Add 1 ml of ether and then 1 drop of 3 percent hydrogen peroxide solution (Note 1). Shake *once*, and observe. A dark-blue color, concentrating in the ether layer, proves the presence of *chromate* ion (Note 2).

Note 1 Excess acid, excess peroxide, and heat all destroy the blue peroxochromic acid very rapidly. Extraction into ether stabilizes the material. Refer to Procedure A-17 (page 226).

Note 2 Under ideal conditions, the test gives these results. Under less than ideal conditions, a fleeting blue color, giving way almost immediately to a green solution, is noted. Any conversion of yellow or orange solution to blue or through blue to green constitutes a positive test. Careful observation is required.

Orthophosphate ion present

Procedure C-22 Removal of orthophosphate ion Dilute the combined centrifugate and washings from Procedure C-17 to 10 ml. Add 4 to 5 drops of formaldehyde solution, and warm *gently* (Note 1). Then add 6 *M* aqueous ammonia solution with vigorous stirring to just short of precipitation (Note 2). Evaporate to exactly 5 ml, and add exactly 1 ml of 6 *M* hydrochloric acid solution (Note 3). Add about 0.4 *M* zirconyl nitrate solution drop by drop with stirring until precipitation appears to be complete (Note 4). Heat just to boiling, and centrifuge. Add 1 drop of the zirconyl nitrate solution to the clear liquor above the precipitate. If a precipitate results, add zirconyl nitrate solution until precipitation is complete, and recentrifuge. Wash the precipitate with 2 ml of hot water. Analyze the combined centrifugate and washings by Procedure C-18 (Note 5). Discard the precipitate (zirconyl orthophosphate).

Note 1 This reaction reduces dichromate ion to chromium(III) ion. Refer to precautions given in Procedure C-18 (page 254).

Note 2 The solution contains a large quantity of nitric acid, which must be neutralized. Addition of sufficient ammonia to cause precipitation or failure to add ammonia to just short of precipitation will cause interference in the next step.

Note 3 For optimum precipitation of zirconyl orthophosphate, an acidity of ca. 1 *N* is desirable. Under these conditions, orthophosphate ion is removed completely without precipitation of the alkaline-earth-metal salts. Insufficient acid may cause precipitation of these cations; excess acid renders precipitation of the zirconium compound incomplete.

Note 4 Excess zirconyl ion may render subsequent tests for iron(III) ion less sensitive and is to be avoided. It is better to proceed as outlined and test for completeness of precipitation than to add excess reagent initially.

Note 5 Any excess zirconyl ion will be precipitated with the hydroxides of the group III cations and carried with iron(III) in subsequent procedures. It causes no interference with the confirmatory tests for iron(III) ion, although if present in large quantity it may render these tests less sensitive owing to reactions with the reagents used.

Cation group IV

The general scheme followed in separating this group and in identifying its members is summarized in Table 11-5.

Procedure C-23 Precipitation of group IV The centrifugate reserved from Procedure C-18 is treated directly as outlined below. If the sample being analyzed contains only group IV cations, dissolve 0.5 g of solid ammonium chloride in it, make alkaline with 6 *M* aqueous ammonia solution, and add 1 ml of ammonia solution in excess (Notes 1, 2).

To the ammoniacal solution, add 10 to 15 drops of 1 *M* thioacetamide solution. Heat on the water bath for 10 min (Note 3). Centrifuge, and wash the precipitate twice with 1- to 2-ml volumes of cold water. Reserve the combined centrifugate and washings for analysis for groups V to VI (Procedure C-29) (Note 4), and analyze the precipitate by Procedure C-24 (Note 5).

Note 1 To prepare a known solution for this group, mix 2-drop volumes of 0.1 *M* cobalt(II) nitrate, 0.1 *M* nickel(II) nitrate, and 0.1 *M* zinc nitrate solutions, and dilute to 2 ml.

Note 2 Ammonium ion is added to prevent precipitation of magnesium hydroxide and to facilitate formation of the hexaamminecobalt(II) ion. The solution should be clear at this point. For the behavior of the three cations upon addition of aqueous ammonia, see Procedure C-18 (Note 3).

Note 3 Color of precipitate is again important. The compounds **CoS** and **NiS** are black, and **ZnS** is white. Traces of iron(III) remaining from group III precipitate as black $\mathbf{Fe_2S_3}$. The absence of a tangible precipitate means the absence of group IV cations.

Note 4 The solution at this point should be colorless. Occasionally, some nickel sulfide is peptized by excess sulfide ion to a dark-brown sol, or sulfide ion is partially oxidized to yellow polysulfide ion. Continued absorption of atmospheric oxygen by the alkaline solution ultimately gives sulfate ion, which then precipitates any barium ion present. It is always advisable to treat immediately all centrifugates for general unknowns and colored centrifugates for Group IV samples only with acetic acid as outlined in Procedure C-29 and return all colored precipitates so obtained to Procedure C-24.

Note 5 The precipitated sulfides undergo oxidation to sulfates upon exposure to air and thus should be carried on to Procedure C-24 as soon as possible.

Procedure C-24 Dissolution of group IV precipitate and removal of zinc ion Wash the precipitate from Procedure C-23 into a small beaker with 1 ml of water. Add 0.5 ml of 12 *M* hydrochloric acid solution, and boil gently for 2 to 3 min. If a residue remains (Note 1), add 3 drops of 15 *M* nitric acid solution, and boil gently until a clear solution results (Note 2). Dilute to 2 ml, and centrifuge to remove precipitated sulfur. Transfer the centrifugate to a small casserole (Note 3), and add 6 *M* sodium hydroxide

Table 11-5 Schematic outline for analysis of cation group IV

<table>
<tr><td colspan="6">Centrifugate from cation group III (Procedure C-18) or solution containing only group IV cations
C-23. If the latter, add solid $\mathbf{NH_4Cl}$ and excess 6 *M* aq NH_3 solution; otherwise, proceed directly
Heat with $CH_3C(S)NH_2$. Centrifuge</td></tr>
<tr><td colspan="5">Precipitate: **CoS** (black), **NiS** (black), **ZnS** (white)
C-24. Add 12 *M* HCl. Heat. Add 15 *M* HNO_3. Heat. Dilute and centrifuge</td><td rowspan="2">Centrifugate: cations of groups V-VI (see Table 11-6)</td></tr>
<tr><td>Residue: **S** (discard)</td><td colspan="4">Solution: Co^{2+}, Ni^{2+}, Zn^{2+}
C-24. Add 6 *M* NaOH, 3 percent H_2O_2. Boil. Centrifuge</td></tr>
<tr><td></td><td colspan="2">Precipitate: $\mathbf{Co(OH)_3}$ (black), $\mathbf{Ni(OH)_2}$ (pale green)
C-25. Add 12 *M* HCl. Heat. Analyze by C-25 and C-26 or by C-27</td><td colspan="2">Centrifugate: ZnO_2^{2-} (colorless)
C-28. Evaporate. Divide in two portions</td><td></td></tr>
<tr><td></td><td>C-25. Add H_2O. Divide in two portions</td><td>C-27. Add 12 *M* HCl. Adsorb on anion-exchange column. Elute with 10 *M* HCl. Detect Co^{2+} by C-25, Ni^{2+} by C-26.</td><td colspan="2"></td><td></td></tr>
<tr><td></td><td>Portion I:
Co^{2+}, Ni^{2+}
C-25. Add solid **NaF**. Add amyl alcohol—ether and solid $\mathbf{NH_4NCS}$. Shake
Alcohol layer: $[Co(NCS)_4]^{2-}$ (Blue to blue-green)</td><td>Portion II:
Co^{2+}, Ni^{2+}
C-26. Add 6 *M* aq NH_3. Add solid $\mathbf{NaC_2H_3O_2}$ and dimethylglyoxime
Precipitate: $\mathbf{[Ni(C_4H_7N_2O_2)_2]}$ (bright red)</td><td>Portion I:
ZnO_2^{2-}
C-28. Add $CH_3C(S)NH_2$
Precipitate: **ZnS** (white)</td><td>Portion II:
ZnO_2^{2-}
C-28. Add 6 *M* $HC_2H_3O_2$ and 0.1 *M* $K_4[Fe(CN)_6]$
Precipitate: $\mathbf{K_2Zn_3[Fe(CN)_6]_2}$ (white to grayish white)</td><td></td></tr>
</table>

solution with stirring until the mixture is alkaline. Add 1 ml in excess. Then add 2 to 3 drops of 3 percent hydrogen peroxide solution (Note 4). Stir vigorously, heat to boiling, and boil for 1 min. Wash the suspension into a test tube, and centrifuge. Wash the precipitate twice with 1-ml volumes of hot water. Reserve the combined centrifugate and washings for Procedure C-28, and treat the precipitate as outlined in Procedure C-25 (Note 5).

Note 1 Zinc sulfide and any contaminating iron sulfide dissolve readily in hydrochloric acid solution, but the cobalt and nickel sulfides dissolve slowly or not at all, presumably as a consequence of the transformation on standing of the more soluble α-compounds to the less soluble β-compounds.

Note 2 Oxidation of sulfide ion to elemental sulfur by nitric acid aids in the dissolution of the cobalt and nickel compounds.

Note 3 The white background renders pale-green **$Ni(OH)_2$** and blue or pink **$Co(OH)_2$** more easily detectable.

Note 4 Hydrogen peroxide gives the very insoluble and easily detectable black cobalt-(III) hydroxide but is without effect upon either nickel hydroxide or the colorless zincate ion. Blackening of the precipitate upon addition of the peroxide is a test for cobalt(II) ion.

Note 5 Any traces of iron present will carry into this precipitate.

Procedure C-25 Identification of cobalt(II) ion To the precipitate from Procedure C-24, add 0.5 ml of 12 *M* hydrochloric acid solution. Warm gently until the precipitate dissolves and liberated chlorine is removed (Notes 1, 2). Analyze *either* as follows or by Procedure C-27. Cool, dilute to 2 ml, and divide into two equal portions, saving *one* for Procedure C-26. To the *other* portion, add a pinch or two of solid sodium fluoride, and shake to saturate the solution (Note 3). Then add 0.5 to 1 ml of amyl alcohol–ether or benzyl alcohol and 1 to 2 g of solid ammonium thiocyanate (Note 4). Shake. A blue to blue-green color in the non-aqueous alcohol layer confirms the presence of *cobalt*(II) ion (Note 5).

Note 1 Cobalt(III) ion is too strongly oxidizing to exist in aqueous solution in the uncomplexed form. Dissolution of cobalt(III) hydroxide in hydrochloric acid solution occurs by reduction to cobalt(II) ion and oxidation of the chloride ion.

Note 2 If cobalt(II) ion is present, the solution will often be dark blue at this point because of the presence of the $[CoCl_4]^{2-}$ ion. Dilution converts this species into the red to pink aquo complex ion, $[Co(H_2O)_6]^{2+}$.

Note 3 Traces of iron(III) ion, which would give a red color with thiocyanate ion (Procedure C-19), are removed as the stable complex species $[FeF_6]^{3-}$. The iron(III)-thiocyanate complex is more stable than the cobalt(II) complex and thus would form preferentially.

Note 4 Solid ammonium thiocyanate avoids dilution and provides the high thiocyanate-ion concentration necessary to displace the equilibrium toward the colored complex ion,

$$\underset{\text{(reddish)}}{[Co(H_2O)_6]^{2+}} + 4NCS^- \rightleftharpoons \underset{\text{(blue-green)}}{[Co(NCS)_4]^{2-}} + 6H_2O$$

Note 5 Nickel(II) ion gives no color change with thiocyanate ion.

Procedure C-26 Identification of nickel(II) ion To the solution reserved from Procedure C-25, add 6 *M* aqueous ammonia solution drop by drop, with shaking, to just short of precipitation (Note 1). Add sufficient solid sodium acetate to saturate the solution (Note 2), and then add several drops of 1 percent dimethylglyoxime solution. A bright-red precipitate confirms the presence of *nickel*(II) ion (Note 3). If no precipitate forms and there is reason to suspect the presence of nickel(II) ion, acidify with 6 *M* acetic acid solution, and then add 6 *M* aqueous ammonia solution drop by drop with shaking. An optimum pH should be reached at which precipitation will occur.

Note 1 Excess aqueous ammonia is avoided because of the solubility of the ultimate product in strongly ammoniacal solutions.

Note 2 Accurate pH control is essential. Acetate ion removes protons without complexing the nickel(II) ion as ammonia would do.

Note 3 Cobalt(II) ion may give a brownish solution, but no precipitate results. As little as 5×10^{-5} mole of Ni^{2+} $liter^{-1}$ can be detected with dimethylglyoxime. Other common cations do not interfere. The precipitate can be formulated as

```
             O---HO
            /      \
H3C—C=N             N=C—CH3
    |    ↘        ↙   |
    |      Ni         |
    |    ↗        ↖   |
H3C—C=N             N=C—CH3
            \      /
             OH---O
```

Procedure C-27 Identification of cobalt(II) and nickel(II) ions To the solution obtained at the beginning of Procedure C-25 add 1.5 ml of 12 *M* hydrochloric acid solution. Add this solution to a column of anion-exchange resin, prepared and conditioned as outlined in Expt. II-9 (page 203). Elute as directed in Expt. II-9 and identify cobalt(II) and nickel(II) ions as directed therein, using the tests given in Procedures C-25 and C-26.

Procedure C-28 Identification of zinc ion Evaporate the solution reserved from Procedure C-24 to 2 ml, and centrifuge if not clear. Divide the clear solution into two equal portions. To *one* portion add several drops of 1 *M* thioacetamide solution and warm. A white precipitate of zinc sulfide, soluble in hydrochloric acid solution, but insoluble in acetic acid solution, confirms the presence of *zinc* ion (Notes 1, 2, 3). To the *second* portion, add 6 *M* acetic acid solution until definitely acidic. Then add several drops of 0.1 *M* potassium hexacyanoferrate(II) solution. A white to grayish-white precipitate of composition $\mathbf{K_2Zn_3[Fe(CN)_6]_2}$ confirms the presence of *zinc* ion (Note 4).

Note 1 Use of an alkaline so' 'tion renders precipitation of zinc sulfide more nearly complete.

Note 2 Sulfur may precipitate at this point due to oxidation of hydrogen sulfide. Sulfur is insoluble in either hydrochloric or acetic acid solution.

Note 3 Traces of cobalt(II) or nickel(II) ion will darken the precipitate. If the precipitate is not white, treat with 1 ml of 6 *M* hydrochloric acid solution, centrifuge, and discard the residue. Add 6 *M* aqueous ammonia solution just short of precipitation. Then add 1 drop of 0.1 *M* copper(II) nitrate solution and 2 drops of 1 *M* ammonium tetrathiocyanatomercurate(II) solution. A violet to black precipitate confirms the presence of zinc ion. Alternatively, reprecipitate *white* zinc sulfide by making the solution in hydrochloric acid alkaline and adding thioacetamide.

Note 4 Traces of cobalt(II) or nickel(II) ion color the precipitate blue to green.

Cation group V

The general scheme followed in separating this group and in identifying its members is summarized in Table 11-6.

Procedure C-29 Precipitation of group V Acidify the centrifugate from Procedure C-23 with 6 *M* acetic acid solution, and evaporate to 2 ml (Note 1). Centrifuge, and return any colored precipitate to Procedure C-24 (Note 2). Discard any white residue. Analyze the clear centrifugate as indicated below. If the sample contains only group V cations, dilute 1 ml of the solution to 2 ml, and analyze as indicated below (Note 3).

Transfer the solution to a small casserole, add 1 ml of 16 *M* nitric acid solution, and evaporate to dryness on the water bath. Heat the casserole with an open flame until no further decomposition occurs (Note 4). Cool. Add 0.5 ml of 6 *M* hydrochloric acid solution and 0.5 ml of water. Stir until all the residue is dissolved, and decant into a small beaker. Make this solution just alkaline with 6 *M* aqueous ammonia solution, ignoring any precipitate which may form (Note 5). Add an equal volume of 95 percent ethyl alcohol and 2 to 3 ml of 3 *M* ammonium carbonate solution (Note 6). Cool the mixture, and allow to stand for 30 min with frequent shaking (Note 7). A white precipitate shows the presence of group V cations. Centrifuge, and wash the precipitate with 1 ml of the ammonium carbonate solution. Reserve the combined centrifugate and washings for analysis for Group VI (Procedure C-35), and analyze the precipitate by Procedure C-30.

Note 1 The ammonium ion formed and the evaporation procedure effect flocculation of colloidal sulfides and sulfur.

Note 2 Brown to black nickel(II) sulfide and white sulfur are the substances most likely to be precipitated at this point. If the residue is light in color, it should be discarded. If it is dark in color, it should be returned to the group IV precipitate (see Procedure C-23, Note 4).

Note 3 To prepare a known solution for this group, mix 2-drop volumes of 0.1 *M* calcium, 0.1 *M* barium, and 0.1 *M* magnesium nitrate solutions, and dilute to 1 ml.

Table 11-6 Schematic outline for analysis of cation group V

<table>
<tr><td colspan="6">Centrifugate from cation group IV (Procedure C-23) or solution containing only group V cations
C-29. If the former, acidify with 6 M $HC_2H_3O_2$, and evaporate. Centrifuge. If the latter, omit, and proceed as follows:</td></tr>
<tr><td rowspan="7">Precipitate: **S** (white), **NiS** (brown to black)
Return to Procedure C-24 (Table 11-5)</td><td colspan="5">Solution: Ba^{2+}, Ca^{2+}, Mg^{2+}, NH_4^+, K^+, Na^+
C-29. Add 15 M HNO_3. Evaporate. Ignite residue. Cool, dissolve in 6 M HCl + H_2O</td></tr>
<tr><td colspan="5">Solution: Ba^{2+}, Ca^{2+}, Mg^{2+}, K^+, Na^+
C-29. Add 6 M aq NH_3, C_2H_5OH, and 3 M $(NH_4)_2CO_3$. Cool; let stand. Centrifuge</td></tr>
<tr><td colspan="4">Precipitate: **$BaCO_3$** (white), **$CaCO_3$** (white), **$MgCO_3 \cdot (NH_4)_2CO_3 \cdot 4H_2O$** (white)
C-30. Add 6 M $HC_2H_3O_2$. Heat</td><td rowspan="5">Centrifugate: cations of group VI (see Table 11-7)</td></tr>
<tr><td colspan="4">Solution: Ba^{2+}, Ca^{2+}, Mg^{2+}
C-30. Add 3 M $NH_4C_3H_3O_2$ and 0.1 M K_2CrO_4. Warm. Centrifuge</td></tr>
<tr><td rowspan="3">Precipitate: **$BaCrO_4$** (yellow)
C-30. Add 12 M HCl. Yellow-green flame test</td><td colspan="3">Centrifugate: Ca^{2+}, Mg^{2+}, $Cr_2O_7^{2-}$
C-31. Add 1 M NH_4Cl, 6 M aq NH_3, 3 M $(NH_4)_2CO_3$. Warm. Centrifuge</td></tr>
<tr><td>Precipitate: **$CaCO_3$**
C-32. Add 6 M HNO_3. Boil.</td><td colspan="2">Centrifugate: Mg^{2+}, CrO_4^{2-}
C-33. Add 6 M $HC_2H_3O_2$. Evaporate. Add 6 M aq NH_3, C_2H_5OH, 0.1 M Na_2HPO_4. Centrifuge</td></tr>
<tr><td>Add 6 M aq NH_3. Evaporate. Add 0.1 M $(NH_4)_2C_2O_4$. Centrifuge
Precipitate: **CaC_2O_4** (white)
C-32. Add 12 M HCl. Brick-red flame test</td><td>Precipitate: **$MgNH_4PO_4 \cdot 6H_2O$** (white)
C-33. Add 6 M HCl, *p*-nitrobenzeneazo-resorcinol, 6 M NaOH
Precipitate: blue **Mg lake**</td><td>Centrifugate: (discard)</td></tr>
</table>

Note 4 Excess ammonium ion, which might inhibit precipitation of the carbonates, is removed by thermal decomposition of ammonium nitrate. This evaporation may be omitted for samples known to be free from ammonium ion (e.g., group V known).

Note 5 In the absence of sufficient ammonium ion, magnesium hydroxide may precipitate. Furthermore, aqueous ammonia often contains sufficient carbonate ion to cause partial precipitation of the alkaline-earth-metal carbonates. No harm results from these precipitations.

Note 6 Alcohol so reduces the solubility of magnesium carbonate that this cation is nearly completely precipitated with this group. In the absence of alcohol, precipitation is incomplete at best.

Note 7 Precipitation is sufficiently slow so that it may be incomplete unless these conditions are observed. Heating would aid in coagulating the precipitated materials, but the magnesium compound is appreciably soluble in hot solutions. Absence of a precipitate at this point means absence of group V cations.

Procedure C-30 Dissolution of the group precipitate and identification of barium ion To the precipitate from Procedure C-29, add hot 6 *M* acetic acid solution drop by drop until dissolution is complete. Then add 5 drops of the acid solution in excess. Dilute the clear solution to 1 ml with water, and add 1 ml of 3 *M* ammonium acetate solution (Note 1). Warm the solution on the water bath, and then add 1 drop of 0.1 *M* potassium chromate solution. If a yellow precipitate forms, indicating the presence of barium ion, add the potassium chromate solution until precipitation is complete (Note 2). Centrifuge, and wash once with 1 ml of water, reserving the centrifugate and washings for Procedure C-31. To the yellow precipitate, add 2 to 4 drops of 12 *M* hydrochloric acid solution. Heat in the water bath until dissolution is complete (Note 3). Transfer the solution to a small beaker. Rinse the test tube, add the rinsings to the beaker, and evaporate to a 1- to 2-drop volume. Dip a clean platinum wire into the solution, and heat in the oxidizing flame. A green flame color confirms the presence of barium ion.

Note 1 Acetate ion acts as a buffer to reduce the acidity of the solution and ensure complete precipitation of barium chromate.

Note 2 If no precipitate forms upon the first addition of potassium chromate solution, barium ion is absent, and the addition of more chromate may be omitted. A precipitate at this point is normally confirmation of the presence of barium ion, but the confirmatory flame test should always be made.

Note 3 The large concentration of hydrogen ion in the strong acid dissolves barium chromate. Barium chloride is more volatile than the chromate and thus better suited to a flame test.

Procedure C-31 Separation of calcium from magnesium ion Add 1 ml of 1 *M* ammonium chloride or nitrate solution to the centrifugate from Procedure C-30, and add 6 *M* aqueous ammonia solution drop by drop until the solution is just alkaline. Then add 2 to 3 drops in excess. Heat almost to boiling in the water bath, and add 3 *M* ammonium carbonate solution drop by drop until precipitation is complete (Notes 1, 2). Let stand for 10 min and centrifuge. Wash with 1 ml of the ammonium carbonate solution. Reserve the centrifugate and washings for identification of the magnesium ion (Procedure C-33), and analyze the precipitate for calcium ion by Procedure C-32 (Note 3).

Note 1 The enhanced solubility of magnesium carbonate in the presence of ammonium ion and in the absence of alcohol permits the selective precipitation of calcium carbonate.

Note 2 Precipitation from hot solution gives a product which settles more rapidly. The solubility of magnesium carbonate is further increased by heating. The suspension should not be boiled because of decomposition of the reagent.

Note 3 Although the separation is quite clean, small quantities of magnesium ion may contaminate the precipitate and small quantities of calcium ion may enter the centrifugate. These cause no interference in subsequent procedures. Any unremoved barium ion will appear in the precipitate.

Procedure C-32 Identification of calcium ion Suspend the precipitate from Procedure C-31 in 1 ml of water, warm, and add 6 *M* nitric acid solution drop by drop until the precipitate has dissolved completely. Boil gently to remove dissolved carbon dioxide. Add sufficient 6 *M* aqueous ammonia solution to make the solution alkaline. Then add no more than 2 ml of 0.1 *M* ammonium oxalate solution drop by drop. A white precipitate of calcium oxalate indicates the presence of *calcium* ion (Note 1). Centrifuge, and discard the centrifugate. Moisten the residue with several drops of 12 *M* hydrochloric acid solution, and apply a flame test. A moderately persistent brick-red flame confirms the presence of *calcium* ion (Note 2).

Note 1 Anything more than a trace of precipitate is positive indication of the presence of calcium ion. Unremoved barium ion will yield white barium oxalate, but the improbability of the presence of more than a very small amount of this cation and the somewhat enhanced solubility of its oxalate never permit the formation of much precipitate.

Note 2 Inasmuch as the calcium flame color is not particularly distinctive, a control test should always be made.

Procedure C-33 Identification of magnesium ion Barely acidify the centrifugate from Procedure C-31 with 6 *M* acetic acid solution, and heat to boiling to remove liberated carbon dioxide (Note 1). Evaporate the solution to 2 ml. Add 6 *M* aqueous ammonia solution drop by drop until the solution is alkaline, and then add 1 ml in excess. Cool. Then add 2 ml of ethyl alcohol and 1 ml of 0.1 *M* disodium orthophosphate solution

(Note 2). Let stand for 5 to 10 min, and, if necessary, scratch the inside walls of the tube with a stirring rod. A white crystalline precipitate of magnesium ammonium orthophosphate indicates the presence of magnesium ion (Note 3). Centrifuge, and wash once with 1 ml of water containing several drops of 6 *M* aqueous ammonia solution (Note 4). Add 1 ml of distilled water and then 6 *M* hydrochloric acid solution (with stirring) until the precipitate dissolves. Add 1 drop of *p*-nitrobenzeneazoresorcinol solution and 6 *M* sodium hydroxide solution drop by drop until an alkaline reaction results. A brilliant-blue precipitate confirms the presence of *magnesium* ion (Notes 5, 6).

Note 1 Carbonate ion is removed to prevent subsequent precipitation of the double magnesium ammonium carbonate upon addition of alcohol. The yellow chromate ion in the solution (from Procedure C-30) causes no interference in any of the subsequent reactions.

Note 2 Magnesium ammonium orthophosphate, although appreciably soluble in water, has a sufficiently small solubility in cold alcoholic ammoniacal solutions containing excess orthophosphate ion to permit detection of small quantities of magnesium ion.

Note 3 White gelatinous precipitates result with small quantities of calcium or barium ion. The magnesium salt is definitely crystalline, although the particles may be very finely divided.

Note 4 The ammonia present represses hydrolysis in terms of the equation

$$\mathbf{MgNH_4PO_4{\cdot}6H_2O}(s) + H_2O \rightleftharpoons \mathbf{MgHPO_4}(s) + NH_4^+ + OH^- + 6H_2O$$

Magnesium hydrogen orthophosphate is white and insoluble but is gelatinous in character.

Note 5 Traces of the alkaline earth ions which may appear as phosphates do not give this test.

Note 6 *p*-Nitrobenzeneazoresorcinol, commonly called "Magneson I" or "S and O" reagent, can be formulated as

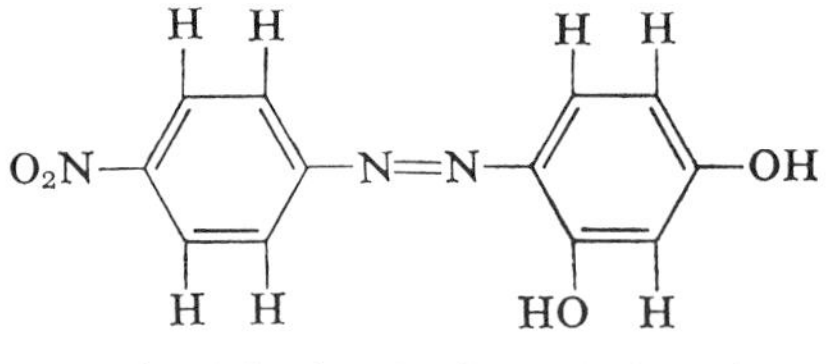

H H H H

O_2N—(benzene ring)—N=N—(benzene ring)—OH

H H HO H

This reagent is violet in alkaline solutions but colors freshly precipitated magnesium hydroxide a bright blue. The reagent is sensitive to one part of magnesium ion in 100,000 parts of solution and can detect magnesium ion even in natural waters that have been in contact with dolomitic limestone. Distilled water is thus essential throughout these procedures. Cations yielding colored hydroxides (for example, Co^{2+}, Cu^{2+}) and excess ammonium ion must be absent.

Cation group VI

The general scheme followed in identifying the members of this group is summarized in Table 11-7.

Solutions containing only group VI cations are analyzed by Procedures C-34 and C-36 through C-39. General unknown solutions are analyzed by Procedures C-34 and C-35 through C-39. The addition of ammonium ion in the precipitation of preceding cation groups and of potassium ion in Procedure C-17 requires that these ions be detected in samples of the original unknown. Contamination by sodium ion is so difficult to avoid when many reagents are used that it is best to examine the original sample for this ion. Indeed, the procedures given for the identification of the NH_4^+, Na^+, and K^+ ions are specific in the absence of ions from cation groups II to V, and can thus be applied before analyzing for these groups if a sample of the original solution is first treated by Procedure C-35. The instructor may elect either to use this alternative or to detect group VI cations in the centrifugate from cation group V. For the ultimate identification of the K^+ and Na^+ ions, alternative procedures may be used (Procedures C-38*a* or C-38*b* and Procedures C-39*a* or C-39*b*, respectively).

Procedure C-34 Identification of ammonium ion Place 0.5 ml of the *original* solution or about 0.1 g of the *original* solid (Note 1) in a 25-ml beaker or casserole. Add 2 ml of 6 *M* sodium hydroxide solution, and cover immediately with a watch glass carrying stuck to its underside a piece of moist red litmus paper. Warm the beaker gently on the water bath or with a burner, but *do not boil* its contents (Note 2). A uniform conversion of the red color of the paper to blue confirms the presence of *ammonium* ion (Notes 3, 4).

Note 1 This test is always made on an original sample because of the addition of ammonium salts in many steps of the procedure.

Note 2 Heat reduces the solubility of liberated ammonia gas, but boiling might spatter the alkaline sodium hydroxide solution onto the litmus paper.

Note 3 A uniform blue color results from hydroxide ion produced by dissolution of ammonia gas in the water on the litmus paper. Small deep-blue spots indicate spattering of the alkaline reaction mixture.

Note 4 More sensitive tests result if filter paper moistened with 0.1 *M* mercury(I) nitrate solution (page 230) or with Nessler's reagent is substituted for red litmus. However, there is often sufficient ammonia in the laboratory atmosphere to react positively with these reagents and thus to nullify the tests.

Nessler's reagent is prepared by diluting 2 drops of 0.1 *M* mercury(II) nitrate solution to 1 ml, adding 0.1 *M* potassium iodide solution until the red precipitate which first forms redissolves, and adding 6 *M* sodium hydroxide solution until the solution is definitely alkaline. Reaction with ammonia gives the yellow to brown compound Hg_2NI.

Procedure C-35 Removal of group I-V cations Dilute 0.5 ml of the *original* solution with an equal volume of water and heat in the water bath. Make just alkaline with 6 *M* aqueous ammonia solution (Note 1)

Table 11-7 Schematic outline for analysis of cation group VI

<table>
<tr>
<td>Original solid or solution
C-34. Add 6 M NaOH. Warm; do not boil</td>
<td colspan="3">Solution containing only group VI cations or portion of original solution
C-35. Dilute. Heat. Add 6 M aq NH_3 and 3 M $(NH_4)_2CO_3$. Digest. Centrifuge</td>
<td colspan="2">Centrifugate from Procedure C-29 or solution containing only group VI cations.
C-36. Add 16 M HNO_3. Evaporate to dryness. Heat residue</td>
</tr>
<tr>
<td>Gas: NH_3. Turns red litmus to blue</td>
<td>Residue: group I–V salts (discard)</td>
<td colspan="2">Centrifugate: K^+, Na^+, NH_4^+
C-36. Add 16 M HNO_3. Evaporate to dryness. Heat residue</td>
<td colspan="2"></td>
</tr>
<tr>
<td></td>
<td></td>
<td colspan="3">Residue: KNO_{2-3}, $NaNO_{2-3}$ (white)
C-36. Apply flame tests. Then divide into two equal portions</td>
<td>Gas: N_2O
H_2O</td>
</tr>
<tr>
<td></td>
<td></td>
<td>Flame test: Bulky yellow—Na^+; reddish-violet—K^+</td>
<td>Portion I (bulk): K^+, Na^+
C-38a. Add 0.1 M $Na[B(C_6H_5)_4]$
Precipitate:
$K[B(C_6H_5)_4]$ (white)
Portion I (1 drop): K^+, Na^+
C-38b. Add 0.5% $H_2[PtCl_6]$, C_2H_5OH
Precipitate:
$K_2[PtCl_6]$ (yellow)</td>
<td>Portion II (bulk): K^+, Na^+
C-39a. Add 0.25 M $HZn(UO_2)_3(CH_3COO)_9$, C_2H_5OH
Precipitate:
$NaZn(UO_2)_3(CH_3COO)_9{\cdot}H_2O$ (light yellow)
Portion II (1 drop): K^+, Na^+
C-39b. Add $HZn(UO_2)_3(CHCOO)_9$
Precipitate:
$NaZn(UO_2)_3(CH_3COO)_9{\cdot}H_2O$ (light yellow)</td>
<td></td>
</tr>
</table>

and add 3 *M* ammonium carbonate solution drop by drop until precipitation is complete (Note 2). Allow the suspension to digest in the water bath for 10 min and centrifuge while hot (Note 3). Wash the residue with 0.5 ml of hot water and add the washings to the centrifugate. Discard the residue. Treat the centrifugate and washings as outlined in Procedure C-36.

Note 1 The resulting insoluble hydroxides may make determination of alkalinity difficult. Excess aqueous ammonia solution must be avoided because of the consequent formation of soluble ammine complex ions.

Note 2 The precipitate is a mixture of hydroxides, carbonates, and basic salts. All cations, other than those of group VI, and many anions, including orthoarsenate, are removed in this fashion.

Note 3 Separation of sodium and potassium ions from the precipitate is more nearly complete in hot solutions.

Procedure C-36 Removal of ammonium ion Place the centrifugate from Procedure C-29 or Procedure C-35 or 1 ml of the solution containing only group VI cations (Note 1) in a small casserole, add 2 ml of 16 *M* nitric acid solution (Note 2), and evaporate carefully (*avoid spattering*) to dryness. Then place the casserole on a triangle (not a wire gauze), and heat strongly with the hottest flame of the burner for 5 to 10 min (Note 3). Allow the casserole to cool. If no residue remains, sodium and potassium ions are absent. If a residue remains (Note 4), add 1 ml of water and 3 to 4 drops of 6 *M* acetic acid solution. Stir, and warm gently until all the residue in the casserole dissolves. Decant the solution into a clean test tube, and reserve it for Procedure C-37.

Note 1 To prepare a known solution for this group, mix 4-drop volumes of 0.1 *M* ammonium chloride, 0.1 *M* sodium chloride, and 0.1 *M* potassium chloride solution, and dilute to 1.5 ml.

Note 2 Ammonium ion interferes with the potassium test by giving the same type of precipitate as the potassium ion. Although direct heating would decompose many ammonium salts, certain salts, e.g., the chloride, would sublime and deposit on the colder parts of the container and thus be difficult to remove completely. Conversion to the nitrate, which decomposes irreversibly to nitrogen(I) oxide and water vapor, assures complete removal.

Note 3 Heat should always be applied from below the casserole, but the flame should be directed against both the outer walls and bottom of the casserole to render decomposition complete.

Note 4 Although a residue usually indicates only sodium or potassium materials (or both), it may be due to small quantities of salts of the group V cations which were not removed in Procedure C-29. No interference will result from these cations.

Procedure C-37 Flame tests for sodium and potassium ions Apply a flame test to the solution reserved from Procedure C-36. An intense, bulky yellow, persistent flame color indicates the presence of *sodium* ion (Note 1). Repeat, but observe the flame through either a *thick* cobalt glass filter or, preferably, a didymium glass filter. A reddish-violet and not particularly persistent flame color indicates the presence of *potassium* ion (Notes 2, 3). Divide the solution into two equal portions, reserving one for Procedure C-38 and the other for Procedure C-39.

Note 1 Since there is sufficient sodium practically everywhere (on your fingers, on your towel, on the desk top) to give a yellow flame, it is best to compare the intensity of the observed color with that produced in a control test before concluding that sodium ion is present.

Note 2 In the absence of sodium ion, the potassium flame color is apparent without the use of a filter. The sodium flame color completely masks that of potassium. However, a cobalt or didymium glass filter removes the yellow light without absorbing that from the potassium flame. Again a control test on a pure potassium compound is advisable before definite conclusions can be reached.

Note 3 Inasmuch as flame tests detect sodium (especially) and potassium ions at concentrations considerably below those giving positive reactions with chemical reagents, the results of such tests must be used judiciously to avoid the reporting of trace quantities or impurities. Flame tests should be supplemented by other evidences.

Procedure C-38a Identification of potassium ion To *one* portion of the solution reserved from Procedure C-37, add two to three drops of about 0.1 *M* sodium tetraphenylborate(III) solution. A white precipitate, supplemented by a positive flame test in Procedure C-37, proves the presence of *potassium* ion (Note 1).

Note 1 Potassium tetraphenylborate(III), $\mathbf{K[B(C_6H_5)_4]}$, with a reported solubility in water of 1×10^{-6} mole liter^{-1} at 20 to 25°C, is the least soluble of the known potassium salts. In dilute acidic solutions, no other common cation except the ammonium ion gives a precipitate.

Procedure C-38b Identification of potassium ion Place one drop of the solution reserved from Procedure C-37 near one end of a clean microscope slide. Near the other end of the slide, place one drop of 0.01 *M* potassium chloride solution. Allow both solutions to evaporate to dryness. To each spot, add one drop of 0.5 percent hexachloroplatinic(IV) acid solution in 50 percent ethanol. Compare the two spots. The formation within 2 to 3 min of a yellow crystalline precipitate in the sample spot indicates the presence of *potassium* ion in concentration of at least 0.01 *M* in the solution used (Note 1). When viewed under a low-power microscope or a powerful magnifying glass, the crystals appear as yellow triangles with rounded corners.

Note 1 The NH_4^+ ion forms similar crystals. This ion must be removed as indicated in Procedure C-36.

Procedure C-39a Identification of sodium ion To the *other* portion of the solution reserved from Procedure C-37, add an equal volume of ethanol. If a precipitate forms, centrifuge, discard the precipitate, and use the centrifugate. To the alcoholic solution, add 0.5 ml of 0.25 *M* zinc uranyl acetate solution. Stir vigorously, and allow to stand for 5 to 10 min (Note 1). A light-yellow, crystalline precipitate, supplemented by a positive flame test in Procedure C-37, confirms the presence of *sodium* ion (Note 2).

Note 1 Since the sodium salt, $NaZn(UO_2)_3(CH_3COO)_9 \cdot 9H_2O$, has an appreciable solubility, optimum conditions of high concentration of reagent, adequate time for precipitation, and the presence of ethanol to minimize solubility are essential to its precipitation.

Note 2 Oxalate and orthophosphate ions give precipitates with uranyl ion and must be absent. If these anions have been detected in the sample, the above described test should be preceded by the addition of 1 *M* calcium acetate solution until precipitation is complete, centrifugation, and final evaporation of the centrifugate to a total volume of not more than 0.5 ml. Cations other than Ba^{2+} do not interfere.

Procedure C-39b Identification of sodium ion Place one drop of the solution reserved from Procedure C-37 near one end of a clean microscope slide. Near the other end, place one drop of 0.01 *M* sodium chloride solution. To each drop, add one drop of zinc uranyl acetate solution (Note 1). Compare the two tests. The immediate formation of a yellowish crystalline precipitate in the solution being tested confirms the presence of *sodium* ion in a concentration of at least 0.01 *M* (Note 2). Smaller quantities of sodium ion require longer periods for the formation of a visible precipitate. When viewed with a low-power microscope or a powerful magnifying glass, the crystals appear as regular hexagons, each containing a Star of David.

Note 1 One solution is prepared by dissolving 500 mg of $UO_2(CH_3COO)_2 \cdot 2H_2O$ in six drops of 30 percent acetic acid solution and then diluting to 5.0 ml with distilled water. A second solution is prepared by dissolving 1.5 g of $Zn(CH_3COO)_2 \cdot 2H_2O$ in three drops of 30 percent acetic acid and diluting to 5.0 ml with distilled water. The two solutions are mixed, and a tiny crystal of sodium chloride is added. The mixture is allowed to stand for 24 hr and is then filtered. The clear filtrate is used as directed.

Note 2 The NH_4^+ ion must be removed prior to the test as directed in Procedure C-36.

EXERCISES

Cation group I

11-1. Neglecting complex-ion formation, determine the quantities of lead(II), silver(I), and mercury(I) ions which can remain unprecipitated in a 0.3 *N* hydrochloric acid solution. Of what significance are your results?

11-2. How would these values compare with those which would obtain in 0.3 *N* hydroiodic acid solution? Why is hydrochloric acid chosen as precipitant for this group, rather than hydroiodic acid?

11-3. Why are lead(II) and silver(I) chlorides more soluble in 12 *M* hydrochloric acid solution than in the 6 *M* acid? Give appropriate equilibrium expressions to support your answer. Why is the solubility of mercury(I) chloride less affected by altering acid concentration?

11-4. Indicate clearly what possible modifications in analytical procedure for general unknowns would be necessitated by substitution of ammonium chloride for hydrochloric acid as a group I precipitant.

11-5. By means of ionic equations, compare the behavior of lead(II), silver(I), and mercury(I) chlorides with aqueous ammonia solution.

11-6. Although silver(I) chloride dissolves readily in 6 *M* aqueous ammonia solution, silver(I) iodide is practically insoluble in this reagent. Explain.

11-7. Using appropriate equilibrium equations and expressions, discuss the effects of pH upon the precipitation of lead(II) chromate.

11-8. Why must a solution containing the ions $[Ag(NH_3)_2]^+$ and Cl^- be acidified before silver(I) chloride will precipitate?

11-9. Give the formula for a reagent that will:

(*a*) Give precipitates with both KCl and $CaCl_2$ solutions
(*b*) Give precipitates with both NaCl and Na_2CrO_4 solutions
(*c*) Give a precipitate with Hg_2^{2+} ion but not with Hg^{2+} ion
(*d*) Give precipitates with both Hg_2^{2+} and Hg^{2+} ions
(*e*) Dissolve mercury(II) chloride but not silver(I) chloride

11-10. Indicate clearly how you could determine experimentally the solubility-product constant of lead(II) chloride at 25°C.

11-11. Is the compound $\mathbf{HgNH_2Cl}$ a complex derived from ammonia? Explain.

11-12. Why does the reaction represented by the equation

$$[Ag(NH_3)_2]^+ + 2H^+ \rightarrow Ag^+ + 2NH_4^+$$

proceed in the direction shown and not in the reverse direction?

11-13. Indicate how, by means of a single reagent in each case, one could distinguish between: (*a*) solid Hg_2Cl_2 and solid $PbCl_2$; (*b*) solid $PbSO_4$ and solid $Pb(NO_3)_2$; (*c*) solid K_2CrO_4 and solid $PbCrO_4$; (*d*) HCl and HNO_3 solutions.

11-14. Why do the following conversions occur in the sequence indicated?

$$AgNO_3 \xrightarrow{HCl} \mathbf{AgCl} \xrightarrow[(H_2O)]{NH_3} [Ag(NH_3)_2]^+ \xrightarrow{KCN} [Ag(CN)_2]^- \xrightarrow{H_2S} \mathbf{Ag_2S}$$

11-15. A white precipitate, obtained by adding dilute hydrochloric acid solution, is insoluble in both boiling water and 6 *M* aqueous ammonia solution. What cation is present? Why?

Cation group II

11-16. From available equilibrium constants, calculate the quantities of Cu^{2+}, Cd^{2+}, and Pb^{2+} ions remaining in solution when solutions of their salts in 1 *M* hydrochloric acid solution are saturated with hydrogen sulfide.

11-17. Cite some advantages of the homogeneous precipitation of sulfides by means of the decomposition of thioacetamide.

11-18. Give an ionic equation for the decomposition of thioacetamide in acidic solution. Do any of the products of this reaction suggest possible complications in selective precipitation of sulfides through pH control? Explain.

11-19. Why is mercury(II) sulfide so much less soluble in dilute nitric acid solution than in aqua regia solution?

11-20. Why is mercury(II) sulfide so much less soluble in dilute nitric acid solution than is copper(II) sulfide?

11-21. Cite three general procedures for dissolving metal sulfides, and illustrate each by means of a balanced ionic equation.

11-22. Discuss clearly the concept of amphoterism as applied to metal sulfides.

11-23. Why is copper(II) hydroxide readily soluble in 6 *M* aqueous ammonia, whereas copper(II) sulfide is completely insoluble?

11-24. Give a chemical reagent which will:

(*a*) Give precipitates with both Cu^{2+} and Cd^{2+} ions
(*b*) Dissolve tin(IV) sulfide but not tin(II) sulfide
(*c*) Give a precipitate with Cu^{2+} ion but not with Cd^{2+} ion
(*d*) Dissolve cadmium hydroxide but not bismuth(III) hydroxide
(*e*) Dissolve antimony(III) sulfide but not bismuth(III) sulfide

11-25. Illustrate each of the following conversions by means of a balanced ionic equation: (*a*) $\mathbf{BiOOH} \rightarrow \mathbf{Bi}$; (*b*) Cu(II) → Cu(I); (*c*) $H_3AsO_4 \rightarrow \mathbf{As_2S_3}$; (*d*) $\mathbf{CdS} \rightarrow Cd^{2+}$; (*e*) Hg(II) → Hg(I) → Hg(0); (*f*) **Sn(0)** → Sn(II) → Sn(IV); (*g*) $HPbO_2^- \rightarrow \mathbf{PbCrO_4}$.

11-26. Indicate the chemical basis for each of the following:

(*a*) Dissolution of arsenic(III) sulfide in aqueous ammonia solution
(*b*) Lack of precipitation of copper sulfides in the presence of cyanide ion
(*c*) Precipitation when acidified bismuth(III) chloride solution is diluted
(*d*) Failure of lead(II) sulfate to precipitate in strongly acidic solutions

11-27. Why is antimony(III) sulfide less readily soluble in sulfide ion than is arsenic(III) sulfide?

11-28. For each of the following give the formula for a single reagent that will distinguish between the materials listed: (*a*) Hg^{2+} and Cd^{2+} ions; (*b*) Hg_2^{2+} and Hg^{2+} ions; (*c*) AsO_4^{3-} and Cl^- ions; (*d*) $[AsS_2]^-$ and $[SbS_2]^-$ ions; (*e*) solid cadmium sulfide and solid tin(IV) sulfide; (*f*) sulfuric and nitric acid solutions; (*g*) Sb^{3+} and Sn^{4+} ions.

11-29. Account clearly for the observation that addition of sodium acetate solution to a solution containing the species HNO_3, H_3AsO_4, and $AgNO_3$ causes precipitation.

11-30. Account for the fact that antimony(III) hydroxide is amphoteric, whereas bismuth(III) hydroxide is not.

11-31. Account for the observation that cadmium sulfide does not dissolve in alkali-metal cyanide solutions. Give necessary equilibria and their constants.

11-32. If saturation of a solution known to contain only group II cations with hydrogen sulfide gives only a yellow precipitate, what conclusions can be drawn? Why?

11-33. Give an example of a useful identification procedure for group II cations which depends upon: (*a*) oxidation or reduction; (*b*) selective precipitation; (*c*) complex-ion formation.

11-34. List the group II cations that can function as oxidizing agents, and illustrate this characteristic of each by means of an appropriate ionic equation.

11-35. Why is lead(II) sulfate readily soluble in sodium or ammonium acetate solution but only slightly soluble in acetic acid solution?

Cation group III

11-36. By means of a balanced ionic equation, describe an oxidation-reduction reaction involving each of the following species: Mn^{2+}, Fe^{2+}, Fe^{3+}, Al^{3+}, Cr^{3+}.

11-37. Indicate how the acid-base characteristics of manganese compounds change with changes in oxidation state. Account for this trend.

11-38. For cations in group III, cite some identifying reactions that depend upon complex-ion formation.

11-39. Using appropriate equilibrium-constant data, determine the quantities of iron(II) and iron(III) ions which remain unprecipitated by 0.1 *M* aqueous ammonia plus 1 *M* ammonium nitrate solution. Account for the difference between the two.

11-40. From the chemistry of manganese, cite specific equations illustrating hydrogen peroxide as (*a*) an oxidizing agent and (*b*) a reducing agent. What factor of environment seems to be related to the oxidizing or reducing properties of hydrogen peroxide? Is this borne out by oxidation-potential data?

11-41. Compare the behavior of iron(III) ion with sulfide ion in acidic and in alkaline media. Account for the differences.

11-42. Calculate the weight of potassium chlorate needed to precipitate all the manganese from a solution containing 5 ml of 0.1 *M* manganese(II) nitrate solution. Calculate the weight of sodium nitrite needed to dissolve the product.

11-43. Identify a group III cation described by each behavior:

(*a*) No immediate precipitate with ammonium nitrate plus aqueous ammonia, but a brownish precipitate upon standing.

(*b*) Sulfide is immediately and completely decomposed by water.

(*c*) Gives a greenish hydroxide which dissolves readily in alkali-metal hydroxide solutions.

(*d*) Sulfide reacts readily with 6 *M* hydrochloric acid solution but leaves a residue of sulfur.

11-44. What effect does pH have upon the strength of iron(III) as an oxidizing agent? Cite other instances where the same effect is noted.

11-45. Which cation forms complexes the more readily, Fe^{2+} or Fe^{3+}? Support your answer by specific data from any available source.

11-46. Account for the acidic reactions shown by aqueous aluminum or chromium(III) salt solutions.

11-47. Give the formula for a chemical reagent that will: (*a*) dissolve manganese(II) sulfide but not copper(II) sulfide; (*b*) dissolve aluminum hydroxide but not iron(II) hydroxide; (*c*) dissolve iron(II) oxide but not manganese(IV) oxide; (*d*) dissolve sodium chromate but not aluminum hydroxide.

11-48. Although ammonium ion prevents precipitation of manganese(II) hydroxide, it is without appreciable effect upon the precipitation of manganese(II) sulfide. Explain, using appropriate equilibrium data.

11-49. Compare and contrast the chemical behavior of aluminum and chromium(III) ions with as many reagents as you can.

11-50. What is meant by a "lake?" Of what uses are lakes in the identification of cations?

Cation group IV

11-51. By means of a balanced ionic equation, illustrate an oxidation-reduction reaction involving each of the following species: Co^{2+}, Ni^{2+}, Zn^{2+}.

11-52. What effect does complex-ion formation have upon the ease of oxidizing cobalt(II) to cobalt(III)? Illustrate with appropriate E^0_{298} and K_{ox} data.

11-53. Cite several significant similarities and several significant differences in the behavior of Co^{2+} and Ni^{2+} ions with various reagents.

11-54. Indicate how a single reagent could be used to distinguish between the components of each of the following pairs: (*a*) Co^{2+} and Ni^{2+} ions; (*b*) solid cobalt(II) and iron(II) sulfides; (*c*) solid zinc and nickel(II) sulfides; (*d*) Fe^{2+} and Ni^{2+} ions; (*e*) Al^{3+} and Zn^{2+} ions; (*f*) Cr^{3+} and Ni^{2+} ions.

11-55. From appropriate solubility-product-constant data, calculate the solubility of cobalt(II), nickel(II), and zinc hydroxides in water. Would saturation of these aqueous suspensions with hydrogen sulfide give sulfide precipitates? Explain.

11-56. Indicate clearly why each transformation in the following scheme occurs:

$$Ni^{2+} \xrightarrow{NaOH} Ni(OH)_2 \xrightarrow{CN^-} [Ni(CN)_4]^{2-} \xrightarrow{H_2S} NiS \xrightarrow{HNO_3} Ni^{2+}$$

11-57. Cite some significant analytical similarities and differences between zinc and aluminum ions.

11-58. Both cobalt(II) and nickel(II) can be oxidized by hypobromite ion, but only cobalt(II) is oxidized by peroxide. Suggest an explanation.

11-59. Calculate the volume of hydrogen peroxide solution (3 percent H_2O_2 by weight, sp gr = 1.0) required to form 0.5 g of cobalt(III) hydroxide from the cobalt(II) compound.

11-60. Why does cobalt(III) hydroxide fail to dissolve in nitric acid unless a reducing agent is present?

11-61. What compound provides the most sensitive test for zinc ion? Cite some limitations upon this test.

11-62. Give the graphic formula for the nickel derivative of dimethylglyoxime. What is the coordination number of nickel(II) in this compound? Describe the bonding between nickel(II) and the ligand. Why is the compound soluble in both hydrochloric acid and aqueous ammonia solutions?

11-63. Account for the changes in color when a pink cobalt(II) chloride solution is first treated with 12 *M* hydrochloric acid, the solution is evaporated, and it is then diluted with water.

11-64. Is nickel(II) ion amphoteric? Explain.

11-65. Why does a cobalt(III) complex have a smaller instability constant than its cobalt(II) analogue?

Cation group V

11-66. Compare the solubility-product constants for the carbonates of the cations of group V, and draw general conclusions as to the completeness with which each carbonate can be precipitated.

11-67. From solubility and solubility-product data, summarize trends in the solubility of as many salts as you can for the series Mg^{2+}–Ca^{2+}–Sr^{2+}–Ba^{2+}. What general relationships can you discern between the characteristics of the anions involved and these trends?

11-68. Outline clearly the conditions essential to quantitative precipitation of magnesium hydroxide.

11-69. Although barium hydroxide is moderately soluble, addition of dilute aqueous ammonia or sodium hydroxide solution to a barium salt solution often gives a precipitate. Explain.

11-70. Indicate clearly the conditions under which magnesium ion can be excluded from precipitation in this group. What advantages or disadvantages might accrue from this exclusion in the analysis of a general unknown?

11-71. Indicate two significant differences between the analytical characteristics of the Ba^{2+} and Ca^{2+} ions, and indicate clearly how each might be applied with advantage to samples containing these ions.

11-72. Repeat Exercise 11-71 for the Mg^{2+} and Ca^{2+} ions.

11-73. Why is barium carbonate less soluble than magnesium carbonate in ammonium salt solutions?

11-74. Why are many alkaline-earth-metal salts less soluble in water than their alkali-metal analogues?

11-75. Indicate clearly how hydrolysis of $MgNH_4PO_4$ to $MgHPO_4$ can be avoided in the washing of this precipitate.

11-76. Outline in detail the conditions essential to use of chromate for (*a*) the separation of strontium ion from barium ion and (*b*) the identification of strontium ion. In terms of appropriate constants, account for the differences.

11-77. A solution of group V cations gives a white precipitate with 0.1 *M* ammonium oxalate solution but no precipitate with saturated calcium sulfate solution. What conclusions can be drawn?

11-78. Alkaline-earth-metal oxalates dissolve readily in dilute mineral acids, but the sulfates do not. Explain.

11-79. What group V cation is characterized by each of the following items: (1) gives a precipitate with both saturated calcium sulfate solution and 0.1 *M* ammonium sulfate solution; (2) gives no precipitate with excess ammonium sulfate solution; (3) gives no precipitate with chromate or dichromate solution under either acidic or alkaline conditions.

11-80. Account for the greater divergence in properties between the magnesium ion and the other ions of this group than between these other ions.

Cation group VI

11-81. Discuss briefly the limitations upon precipitation techniques for identifying cations of group VI. Why are alkali-metal and ammonium salts commonly soluble in water?

11-82. Indicate why certain metallic species give flame tests, whereas others do not.

11-83. Why is detection of potassium ion by means of a flame test often more reliable than detection of sodium ion by similar means?

11-84. Indicate clearly why there are greater resemblances between ammonium and potassium compounds than between ammonium and sodium compounds.

11-85. Illustrate the behavior of typical solid ammonium salts containing (*a*) oxidizing and (*b*) nonoxidizing anions upon heating. Account for fundamental differences in the nature of the products.

11-86. By means of appropriate balanced equations, illustrate some chemical characteristics (*a*) of ammonia and (*b*) of ammonium ion. Why must a clear distinction in nomenclature be maintained between these two materials?

11-87. Justify the conclusion that the ammonium ion is an ammine complex.

11-88. Of the procedures outlined for the detection of ammonia, which do you regard as best for qualitative analysis? Why?

11-89. To what extent does ammonium ion interfere with tests for sodium and potassium ions? How can these interferences be obviated?

11-90. Why is the test for the ammonium ion always made upon the original sample?

11-91. Indicate clearly how one could distinguish between the solids in each of the following pairs: NH_4Cl and KCl, K_2SO_4 and Na_2SO_4, NH_4Cl and NH_4NO_3, KCl and $NaCl$.

11-92. Fluorosilicic acid yields ammonium fluorosilicate with ammonium salts but a different precipitate with aqueous ammonia. Explain.

11-93. Why does sodium chloride give a better flame test than sodium sulfate?

11-94. To what extent can you justify conclusions that the chemistry of the cations of this group is relatively simple?

11-95. Criticize the statement "Sodium and potassium hydroxides are salts."

Cations in general

11-96. Give a concise reason for each of the following procedures in cation analysis:

(*a*) Careful adjustment of acidity is effected prior to precipitation of cation group II.

(*b*) Orthophosphate ion is removed prior to precipitation of cation group III.

(*c*) In cation group IV, cobalt(II) and nickel(II) ions are separated from zinc ions as hydroxides rather than by treating the sulfides with acid.

(*d*) Careful acidity control is effected in separating barium ion as chromate in cation group V.

(*e*) Evaporation with nitric acid and ignition of the residue is carried out in cation group VI.

11-97. Outline the simplest systematic procedure which will permit separation and identification of each cation in the following mixture: Zn^{2+}, Fe^{2+}, $Hg_2{}^{2+}$, Hg^{2+}, Ca^{2+}, $NH_4{}^+$.

11-98. Indicate clearly the chemistry involved in one or more specific cation separation procedures which depend upon each of the following: selective oxidation or reduction, selective extraction, selective precipitation, selective dissolution of a solid.

11-99. Illustrate each of the following by means of a balanced ionic equation:

(*a*) Basic character in a water-insoluble sulfide

(*b*) Acidic character in a water-insoluble sulfide

(*c*) Oxidation involving a peroxide

(*d*) Reduction involving a peroxide

(*e*) Dissolution of a precipitate as a consequence of complex-ion formation

(*f*) Formation of a precipitate at the expense of a complex ion

(*g*) A confirmatory test that involves oxidation

(*h*) A confirmatory test that involves reduction

(*i*) A confirmatory test that does not yield a compound of the ion detected

(*j*) A precipitation process that results because of hydrolysis of one reagent

11-100. Give reagents and products for each of the following *one-step* separations: Al^{3+} from Cr^{3+}; Zn^{2+} from Fe^{2+}; Ca^{2+} from Cu^{2+}; Cu^{2+} from Co^{2+}; Ag^+ from Pb^{2+}; Hg^{2+} from $Hg_2{}^{2+}$; K^+ from $NH_4{}^+$; Fe^{3+} from Ni^{2+}; Ca^{2+} from Mg^{2+}; BiO^+ from SbO^+.

11-101. Indicate what difficulties might result in systematic cation analysis if each of the following procedures were omitted:

(*a*) Oxidation with hydrogen peroxide at the beginning of group II

(*b*) Evaporation with 15 *M* nitric acid solution prior to precipitation of manganese-(IV) oxide in group III

(*c*) Evaporation of the centrifugate from group IV with nitric acid and ignition of the residue

(*d*) Inclusion of ammonium ion prior to precipitating group III

(*e*) Dilution after evaporation with sulfuric acid in the separation of lead(II) ion in group II

11-102. Cite a specific reason why the usual confirmatory test for each of the following ions cannot be applied directly to a general unknown solution: Cu^{2+}, Ca^{2+}, Pb^{2+}, SbO^+, Co^{2+}, Ba^{2+}, Al^{3+}, Cr^{3+}, $Hg_2{}^{2+}$, Na^+.

11-103. Give a cation described by each of the following items:

(*a*) Gives white precipitates with excess aqueous ammonia and sodium hydroxide solutions

(*b*) Gives a white precipitate with excess sodium hydroxide solution but no precipitate with excess aqueous ammonia solution

(*c*) Gives a white hydroxide soluble in either aqueous ammonia or sodium hydroxide solution

(*d*) Gives a yellow sulfide and a white hydroxide

(*e*) Gives a yellow chromate and a white sulfate, both insoluble in sodium hydroxide solution

(*f*) Gives a blue ammine complex ion and a green hydroxide

(*g*) Gives a white acid-soluble but base-insoluble orthophosphate

(*h*) Gives no precipitate with Cl^-, H_2S, S^{2-}, OH^-, or $CO_3{}^{2-}$

(*i*) Gives a yellow sulfide, soluble in either 12 *M* hydrochloric acid solution or sulfide ion

(*j*) Gives a sulfide which is decomposed completely by water

(*k*) Gives a white chloride, insoluble in water but readily soluble in cold 6 *M* hydrochloric acid solution

11-104. Indicate clearly how one can identify Ca^{2+} ion in the presence of Ba^{2+} ion; Pb^{2+} ion in the presence of Ba^{2+} ion; Ba^{2+} ion in the presence of Pb^{2+} ion; Ni^{2+} ion in the presence of Fe^{2+} ion; Sn^{4+} ion in the presence of Sn^{2+} ion; Al^{3+} ion in the presence of Cr^{3+} ion; Fe^{3+} ion in the presence of Fe^{2+} ion; Ni^{2+} ion in the presence of Cu^{2+} ion; BiO^+ ion in the presence of SbO^+ ion; Hg^{2+} ion in the presence of $Hg_2{}^{2+}$ ion.

11-105. Using necessary equilibrium data, determine whether or not sufficient iron(II) remains in a saturated aqueous solution of its carbonate to give a visible precipitate if this solution is saturated with hydrogen sulfide.

11-106. Give an ion-electron equation to describe each of the following oxidation-reduction couples:

(*a*) Cr(III)–Cr(VI) in acidic solution

(*b*) Mn(IV)–Mn(VII) in acidic solution

(*c*) Sn(II)–Sn(IV) in alkaline solution

(*d*) Fe(II)–Fe(III) in alkaline solution

(*e*) Cu(I)–Cu(II) in alkaline solution

11-107. A cation gives a chloride soluble in cold water and a white hydroxide soluble in aqueous ammonia or hydrochloric acid solution but insoluble in potassium hydroxide solution. What is the cation, and why?

11-108. A strongly acidic solution contains the permanganate ion. What cations cannot be present?

11-109. A strongly alkaline solution cannot contain what cations?

11-110. Suggest a set of reagents other than boiling water and aqueous ammonia solution which could be used to separate the chlorides of group I cations.

11-111. List the disadvantages of substituting thioacetamide for hydrogen sulfide.

11-112. If Ag^+ and $Hg_2{}^{2+}$ ions were not removed in group I, how might they interfere in group II?

11-113. Suggest alternative precipitants for Na^+ and K^+ ions besides those used in group VI. Cite advantages and disadvantages in their use.

11-114. Cite three instances of interference and three instances of beneficial effects offered by the ammonium ion in cation analysis.

11-115. On the basis of properties studied, construct a flow sheet for a complete system of separation and identification of the cationic species studied in this chapter without using hydrogen sulfide, sulfide ion, or any source of either.

11-116. Using whatever calculations are necessary, arrange the compounds $\mathbf{AgI}$, $\mathbf{Ag_3PO_4}$, $\mathbf{Ag_2S}$, $\mathbf{Ag_2O}$, $\mathbf{Ag_2CO_3}$, $\mathbf{AgBr}$, and $\mathbf{Ag_2CrO_4}$ in the order of decreasing silver ion concentration in their saturated aqueous solutions at room temperature.

11-117. Calculate the volume of 15 M aqueous ammonia solution required to dissolve the silver chloride obtainable from 0.51 g of silver(I) nitrate.

11-118. Calculate the volume of 3 percent hydrogen peroxide solution needed to oxidize 0.02 mole of arsenic(III) sulfide completely to AsO_4^{3-} and SO_4^{2-} ions in aqueous ammonia solution.

11-119. Determine the composition of the resulting solution when 0.54 g of elemental aluminum is added to a solution containing 0.03 mole of $[SnCl_6]^{2-}$ ion and 0.02 mole of $[SbCl_4]^-$ ion. Assume that none of the metal reacts with the hydrochloric acid that must be present.

11-120. Calculate the concentrations of the ions Ag^+, NO_3^-, H_3O^+, Cl^-, and OH^- in a solution that is prepared by adding 0.500 g of solid silver nitrate to 50 ml of 0.05 M HCl solution and allowing the system to come to complete equilibrium.

11-121. The solubility product constant of the compound $AgIO_3$ is 4.5×10^{-8} mole liter^{-1}. Calculate the quantity of this compound that can be dissolved in 500 ml of 1.0 M aqueous ammonia.

11-122. Determine the solubility of silver chloride in 0.2 M $Hg(NO_3)_2$ solution. Assume that mercury(II) nitrate is completely ionized and that the dissolution reaction is described by the equation

$$\mathbf{AgCl}(s) + Hg^{2+} \rightleftharpoons Ag^+ + [HgCl]^+$$

11-123. Arrange the following solutions in the order of decreasing equilibrium concentration of Hg^{2+} ion: saturated HgS, 0.1 M $Hg(ClO_4)_2$, 0.1 M $Na_2[HgI_4]$, and saturated $Hg(OH)_2$.

11-124. Prove the relationship that for silver bromide $\log K_{sp} = [E^\circ_{298}(\mathbf{Ag} \rightarrow Ag^+) - E^\circ_{298}(\mathbf{Ag} \rightarrow \mathbf{AgBr})]/0.059$ and use standard-potential data to calculate K_{sp} for silver bromide. ($E^\circ_{298}(\mathbf{Ag} \rightarrow \mathbf{AgBr}) = -0.073$ volt.)

11-125. Finely-divided sodium sulfate is added slowly to a solution in which Pb^{2+} and Ca^{2+} ion concentrations are initially 0.20 and 10.0 mg ml^{-1}. Calculate the concentration of Pb^{2+} ion remaining in solution when calcium sulfate just begins to precipitate. Assume that there is no volume change as the sodium sulfate is added.

12
Alloys: Analysis for Both Anions and Cations

The procedures outlined in Chap. 11 can be unified and given practicality by extension to the analysis of alloys. By the same token, a complete picture of anion and cation chemistry can be obtained by application of the procedures given in Chaps. 10 and 11 to the analysis of general unknowns consisting of synthetic mixtures of salts or even of naturally occurring minerals or *selected* commercial products such as baking powders, water softeners, or cleaning compounds.

ANALYSIS OF ALLOYS

Alloys are the products obtained by the blending together of two or more metals. Blending is usually effected by melting the components together but can in some instances be achieved by the simultaneous electrodeposition of the constituent metals, by diffusion processes, or merely by mixing the desired materials at ordinary temperatures. Alloys are remarkably diverse in their properties and compositions. In their formation, the characteristics of the component metals either disappear or change (often profoundly), and it is because of these new and often different properties that commercial alloys are made and used. Many of these properties are related to the varying abilities of metals to dissolve in each other in the solid state, to form mixtures with each other in the solid state, or to combine chemically with each other, although often in stoichiometric quantities that bear little or no relationship to the predictions of the common theories of valency.

Compositions of alloys An almost infinite variety of combinations of metals in alloys is possible. Fortunately, however, only limited numbers of these combinations have proved to be commercially interesting, and as a result the compositions of technically used alloys are limited. On the other hand, laboratory samples for analysis may have almost any composition. Technically important alloys are most commonly *binary* (two elements) in composition, less commonly *ternary* (three elements), and only rarely *quaternary* (four elements) or more complex. Conveniently, alloys are termed *ferrous*, if iron is the important component, or *nonferrous*, if they are based upon elements other than iron. The compositions of a few common alloys are given in Table 12-1. Alloys derived from mercury are termed *amalgams*. The very reactive alkali metals are uncommon in technical alloys. Calcium, strontium, and barium are used only to a limited extent. On the other hand, the majority of the other metals, the cations of which have been described, are relatively common in alloys. Ammonium is, of course, only a cation and not a metal and is therefore not a component of alloys.

Qualitative analysis of alloys Analysis is based upon the assumption that the sample used is representative of the entire specimen. Turnings, shavings, drillings, or filings, if available, are usually suitable. The size of sample selected is of critical importance because of the necessity for providing sufficient concentrations of cations to be detectable while avoiding concentrations that are too large to be handled by the recommended procedures. The student should realize that the alloy is a concentrated source of cations and should limit his sample to no more than about 0.1 g. For analysis by the procedures outlined in the preceding chapter, the alloy must be dissolved.

Oxidation-potential values (Appendix G) indicate clearly that acids that depend only upon the oxidizing strength of the hydronium ion (for example, HCl, dil H_2SO_4, dil $HClO_4$) are limited in their solvent power for metals and that they cannot act as solvents for less active metals such as copper, silver, bismuth, and mercury. On the other hand, if the anion is strongly oxidizing as well (for example, NO_3^-, SO_4^{2-}, ClO_4^- in the concentrated acid), these metals can be dissolved. The best general solvent for alloys is probably hot nitric acid solution. All common pure metals except gold and aluminum are attacked, and of these all yield soluble nitrates except

Table 12-1 Common technically important alloys

Type and name	*Major components*	*Minor components*
Ferrous alloys:		
Cast irons	Fe, C	Si
Carbon steels	Fe, C	
Alloy steels	Fe, Mn, Cr, Ni, C	V, W, Mo, Co, Si
Nonferrous alloys:		
Copper base		
Brasses	Cu, Zn	
Bronzes	Cu, Sn	
Phosphor bronze	Cu, Sn	P
Manganese bronze	Cu, Zn	Sn, Fe, Mn
Nickel silver	Cu, Zn, Ni	
Lead-tin base		
Solders	Pb, Sn	
White bearing metal	Pb, Sn	Sb, Cu
Type metal	Pb, Sb	Sn
Shot metal	Pb, As	
Battery alloys	Pb, Sb	
Zinc base		
Die-casting alloys	Zn, Al, Cu	
Nickel base		
Monel	Ni, Cu	Fe, Mn
Nichrome	Ni, Cr	
Chromel	Ni, Fe, Cr	
Inconel	Ni, Cr, Fe	Cu, Mn
Aluminum base		
Casting alloys	Al, Cu	Si
Wrought alloys	Al, Cu	Mg, Mn
Magnesium base		
Dowmetals	Mg, Al	Mn, Zn

tin and antimony, which give the white insoluble oxides $\mathbf{SnO_2}$ and $\mathbf{Sb_2O_4}$. Aluminum appears to react in alloys, but, if pure, it is rapidly coated with a resistant oxide film. Iron and related transition metals are rendered passive by 15 *M* nitric acid solution but are dissolved by the 6 *M* acid. These metals and aluminum react more readily with hydrochloric acid solution. If the alloy is resistant to attack by nitric acid, one should then try hydrochloric acid solution and finally aqua regia or hydrochloric acid and bromine. Sulfuric acid is seldom useful because of the decreased solubility of many of the sulfates.

Procedure C-40 Dissolution of alloys and preparation of a solution for analysis Place not more than 0.1 g of the alloy in a small casserole and add 2 ml of 6 *M* nitric acid solution. Heat (*Hood!*), but do not boil.

If a reaction occurs, continue heating until the reaction appears to be complete, while adding 6 *M* nitric acid solution from time to time to maintain the 2-ml volume. If a white precipitate forms, break it up with a stirring rod to allow the reagent to attack all the metallic particles (Note 1). When no more of these particles can be distinguished, evaporate just to dryness, but do not bake the residue (Note 2). Add a few drops of 16 *M* nitric acid solution, and evaporate just to dryness again (Note 3). Then add five to six drops of 6 *M* nitric acid solution and 1 ml of water. Pour into a small test tube, and wash any remaining material in the casserole into the test tube with another few drops of 6 *M* nitric acid solution plus 1 ml of water. Heat for several minutes in the water bath to complete dissolution. If the solution is clear, analyze it for cations beginning with Procedure C-2 (page 239). If a white residue is present (Note 4), centrifuge, and analyze the centrifugate by Procedure C-2. To the residue, add 0.5 ml of 6 *M* sodium hydroxide solution and 0.5 ml of 1 *M* sodium sulfide solution (Note 5). Heat on the water bath for 10 min. Dilute to 2 ml, and centrifuge. Discard any residue, and analyze the centrifugate for tin and antimony, beginning with Procedure C-13 (page 249).

If the alloy is not attacked by nitric acid after 5 to 10 min, warm a new 0.1-g sample with 2 ml of 6 *M* hydrochloric acid solution. If the reaction is slow, add 2 ml of 12 *M* hydrochloric acid solution. Heat on the water bath until reaction is complete. Evaporate to 0.5 ml, and then dilute to 2 ml. If the sample has dissolved completely, analyze the solution for cations from groups II to V beginning with Procedure C-6 (page 241) (Note 6). If any residue remains (Note 7), centrifuge, and analyze the centrifugate by Procedure C-6. Treat the residue with 6 *M* nitric acid solution as outlined above.

If the alloy is not attacked by either nitric or hydrochloric acid, heat a third 0.1-g sample with a mixture of 0.5 ml of 6 *M* nitric acid solution and 1.5 ml of 6 *M* hydrochloric acid solution as outlined for nitric acid (Note 8). When reaction is complete, evaporate to 0.5 ml, and then dilute to 1.5 ml. Centrifuge. Analyze the centrifugate by Procedure C-6 and any residue by Procedure C-3 (page 240) (Note 9).

Note 1 The compounds SnO_2 and Sb_2O_4 may coat alloy particles and protect them from the reagent.

Note 2 Excess acid is removed in this fashion, but if the residue is heated strongly, decomposition to acid-insoluble products will occur. The casserole should be removed from the flame just before the last liquid disappears, to allow the heat which it retains to complete the evaporation.

Note 3 Basic salts formed in the preceding step are thereby solubilized. The same mode of evaporation should be used.

Note 4 This residue is usually either SnO_2 or Sb_2O_4 or both, although any silicon present in the alloy will appear as SiO_2 at this point. Small black flecks are particles of carbon and may be expected with ferrous alloys. Carbon and silica may be ignored in most instances.

Note 5 The alkaline sulfide solution converts the residue of SnO_2 and Sb_2O_4 to solutions containing sulfoanions. The reaction is relatively slow. Acids attack these materials only incompletely.

Note 6 Silver and mercury would not be attacked. Lead might dissolve but would then appear in cation group II.

Note 7 The residue might be silica, but it is more likely to consist of metals not oxidized by hydrochloric acid solution.

Note 8 The dilute acids are preferred because of the necessity for removing excess acid.

Note 9 The residue will consist of chlorides of group I cations.

ANALYSIS OF SAMPLES FOR ANIONS AND CATIONS

The approach to be followed consists in detecting the anions according to the procedures outlined in Chap. 10 and the cations according to those given in Chap. 11. Necessary solutions for anion analyses are prepared according to Procedure A-1 or A-2 (page 210) and those for cation analyses according to Procedure C-1 (page 238).

The results of anion analysis interpreted in terms of the solubility rules (page 7) often give valuable information as to the cation content of the sample.

Acid-insoluble samples for cation analysis are usually ignited oxides (for example, Al_2O_3, Cr_2O_3, SnO_2) or chromates (for example, $PbCrO_4$), silver halides ($AgCl$, $AgBr$, AgI), sulfates (for example, $PbSO_4$, $CaSO_4$, $SrSO_4$, $BaSO_4$), certain silicate minerals, and calcium fluoride. These samples should be treated by Procedure C-41, an outline for which appears in Table 12-2. This procedure yields the necessary solutions for detection of the cations and permits detection of chromate ion of the anions possibly present in the acid-insoluble sample. Inasmuch as the sodium carbonate treatment (Procedure A-2, page 211) may fail to solubilize completely this anionic species, test for this anion should be made as indicated. Provision is made in Procedure A-2 for detecting anions in samples that are sufficiently difficultly soluble to resist the usual sodium carbonate treatment.

Procedure C-41 Treatment of acid-insoluble residues Mix any residue from Procedure C-1 (page 238) with about ten times its weight of potassium hydrogen sulfate (Note 1), and grind in a clean mortar. Transfer the mixture to a clean porcelain crucible, and heat carefully over an open flame until frothing subsides. Then heat strongly until a clear melt results. Allow the crucible to cool, and then place it on its side in a small beaker. Add 5 ml of water and 2 ml of 6 *M* nitric acid solution. Allow to stand, turning the crucible slowly from time to time to permit contact between its contents and the solvent. When the solid is loosened, remove the crucible. Break up lumps of any remaining solid with a stirring rod. Boil for 5 min,

Table 12-2 Schematic outline for analysis of acid-insoluble residues

<table>
<tr><td colspan="5">Residue from Procedure C-1: **Al_2O_3, Cr_2O_3, SnO_2, $PbCrO_4$, AgX, $PbSO_4$, $CaSO_4$, $SrSO_4$, $BaSO_4$, CaF_2**, silicate minerals, etc.
C-41. Add solid **$KHSO_4$**. Fuse. Cool. Add H_2O, 6 M HNO_3. Boil Centrifuge</td></tr>
<tr><td colspan="4">Residue: **$PbSO_4$, $CaSO_4$, $SrSO_4$, $BaSO_4$, SiO_2, SnO_2**, traces other materials
C-41. Add solid **Na_2CO_3, KNO_3**. Fuse. Cool. Add H_2O. Boil. Centrifuge</td><td rowspan="4">Centrifugate: cations of groups I–VI plus K^+ and $Cr_2O_7^{2-}$ (see Tables 11-1 and 10-4)</td></tr>
<tr><td colspan="3">Residue: **$PbCO_3$, $CaCO_3$, $SrCO_3$, $BaCO_3$, SnO_2**, other carbonates
C-41. Add H_2O, 6 M HNO_3, heat. Centrifuge</td><td rowspan="3">Centrifugate: SO_4^{2-}, CO_3^{2-}, NO_3^-, SiO_4^{4-}, Na^+, K^+ (discard)</td></tr>
<tr><td colspan="2">Residue: **SnO_2**
C-41. Add Na_2CO_3, S. Fuse. Cool. Add H_2O. Centrifuge</td><td rowspan="2">Centrifugate: Pb^{2+}, Ca^{2+}, Sr^{2+}, Ba^{2+}, etc. (see Table 11-2)</td></tr>
<tr><td>Residue: (discard)</td><td>Centrifugate: SnS_3^{2-} $[Sn(OH)_6]^{2-}$ [see Table 11-2 (Procedure C-13)]</td></tr>
</table>

and centrifuge. Analyze the centrifugate for cations, beginning with Procedure C-2 (page 239), and for dichromate ion, by Procedure A-17 (page 226) (Notes 2, 3). Mix the residue with 2 to 3 g of anhydrous sodium carbonate and 0.5 g of solid potassium nitrate (Note 4), and heat as before until reaction ceases. Cool and extract with 5 ml of water, as outlined before. Centrifuge, and discard the centrifugate (Note 5). Suspend the residue in 1 ml of water, warm, and add 6 M nitric acid solution drop by drop until reaction ceases. Centrifuge, and analyze the centrifugate beginning with Procedure C-6 (page 241) (Note 6). Any residue is probably tin(IV) oxide, which can be solubilized by fusing with equal parts of sodium carbonate and sulfur. Cool and leach with water. Centrifuge. Discard the residue and analyze the centrifugate for tin only, beginning with Procedure C-13 (page 249) (Note 7).

Note 1 When heated, potassium hydrogen sulfate melts and then decomposes at about 300°C to the pyrosulfate

$$2KHSO_4 \rightarrow K_2S_2O_7 + H_2O\uparrow$$

Potassium pyrosulfate at higher temperatures is a source of sulfur trioxide and is a high-temperature sulfuric acid. Thus, it yields sulfates

$$MO + K_2S_2O_7 \rightarrow MSO_4 + K_2SO_4$$

Only tin(IV) oxide resists its attack. Halides are decomposed by volatilization of hydrogen halide (HF, HCl) or elemental halogen (Br_2, I_2).

Note 2 If the quantity of residue at the beginning of this procedure is small, it may be advantageous to combine this solution with that from Procedure C-1 (page 238) before continuing with the analysis. Potassium ion is obviously present.

Note 3 The cations Pb^{2+}, Ca^{2+}, and Ba^{2+} will not be found in this solution. Dichromate ion will be present if the original sample contained an insoluble chromate. It can be detected on a small portion of the solution by Procedure A-17 (page 226).

Note 4 The oxidizing action of potassium nitrate improves the attack on materials such as Cr_2O_3. Fused sodium carbonate is a high-temperature base and reacts with acidic components.

Note 5 This solution contains sodium sulfate, carbonate, and silicate. Of these, only the silicate came from the original sample. It could be detected here.

Note 6 Any group I cations would have appeared in the preceding fusion.

Note 7 No other material is likely to be present; so analysis for tin can be made directly.

EXERCISES

12-1. By means of balanced ionic equations, formulate any reactions of each of the metals listed with each of the acids listed: metals, Zn, Fe, Al, Mn, Pb, Sn, Cu, Ag, Bi, Sb; acids, 6 *M* HCl, 6 *M* HNO_3, 3 *M* H_2SO_4, 6 *M* $HClO_4$, 15 *M* HNO_3.

12-2. Why is nitric acid the best general solvent for metals and alloys?

12-3. Discuss critically the limitations of sulfuric acid as a reagent for attacking alloys.

12-4. Does the ammonium group have any metallic properties? Give a detailed answer, consulting reference books for additional information.

12-5. What is meant by the passive state? Under what conditions can it be produced or destroyed?

12-6. Appreciable quantities of elemental copper can sometimes be dissolved upon long contact between the metal and aqueous hydrochloric acid solutions. Explain. Why is evolution of hydrogen gas not a reasonable explanation?

12-7. What limitations exist upon attack of alloys by strongly alkaline solutions? Illustrate any specific cases of dissolution by means of balanced ionic equations.

12-8. Suggest some possible alloys that might be attacked by fusion with potassium hydrogen sulfate, and describe the behavior of component metals by appropriate equations.

12-9. For each of the following alloys, outline in flow-sheet form the simplest possible complete system of separation and identification of the components: Fe, Cr, Ni, C; Sn, Sb, Pb; Al, Cu, Mn, Mg; Cu, Zn, Ni.

12-10. An alloy, which was unreactive toward water, gave with aqua regia a white curdy residue and a colorless solution and with nitric acid a colorless solution and a finely divided white residue. What conclusions can be drawn with regard to the composition of this alloy?

12-11. An alloy was found to dissolve completely in 6 *M* nitric acid solution, in 6 *M* hydrochloric acid solution, and in 3 *M* sulfuric acid solution. In each case, the resulting solution was colorless. What conclusions can be drawn with regard to the composition of the alloy?

12-12. Feldspars are commonly treated by fusion with an alkali-metal carbonate. How, then, could one test for sodium and potassium ions in these minerals?

12-13. For each of the following give the simplest and best solvent, and illustrate its behavior with a balanced equation: **$AgCl$, CuS, $CuSO_4$, $CaCrO_4$, $FeCO_3$, KH_2AsO_4, $BaSO_4$, $Pb(ClO_4)_2$, Ag_2S.**

12-14. A metal ion gives a water-soluble chloride and a white water-insoluble hydroxide, which dissolves in either hydrochloric acid solution or sodium hydroxide solution but not in aqueous ammonia solution. What is the metal ion?

12-15. An unknown sample dissolves readily in water and contains only carbonate and nitrate as anions. What cations cannot be present?

12-16. An unknown sample dissolves readily in water and contains nickel(II) and barium as the only cations. What anions can be present?

12-17. Indicate clearly why fusion of a water-insoluble sulfate with sodium carbonate is more likely to cause its decomposition than mere digestion with sodium carbonate solution.

12-18. A greenish powder, known to contain but a single cation, dissolves readily in water to give a green solution which turns dark blue upon addition of excess aqueous ammonia solution. What anions can be present in the initial powder?

12-19. A cation gives a water-insoluble sulfide that dissolves in 6 *M* hydrochloric acid solution, and a water-insoluble hydroxide that dissolves in both 6 *M* sodium hydroxide and aqueous ammonia solutions. Identify the cation.

12-20. Indicate clearly how one could identify: Cr^{3+} ion in a sample containing CrO_4^{2-} ion; MnO_4^- ion in a sample containing Mn^{2+} ion.

12-21. What is a flux? Cite some examples of acid and of base fluxes (besides those employed in Procedure C-40), and illustrate the chemical behavior of each with an appropriate equation.

12-22. Why do the metals platinum and gold dissolve in aqua regia, whereas they are insoluble in nitric acid solution?

12-23. Compare the behavior of elemental arsenic, antimony, and bismuth with 15 *M* nitric acid solution. Account for any trend you distinguish.

12-24. Why does a metal (e.g., iron) dissolve more readily in acids when in contact with a more noble metal (e.g., copper) than when not in contact with it?

12-25. Why do less noble metals protect more noble ones from oxidation by hydronium ion?

12-26. Determine the volume of 0.5 *M* HNO_3 solution needed to dissolve completely 0.5 g of an alloy containing 10 percent Al, 70 percent Zn, and 20 percent Cu, by weight.

12-27. Five grams of an antimonial lead battery plate containing no tin yields 0.75 g of a white, insoluble residue after heating with 6 *M* HNO_3 solution, washing with water, and drying. Calculate the percentage of antimony in this alloy.

12-28. A sample of ore is 5 percent chromite ($FeCr_2O_4$). This material is decomposed by fusion with potassium hydrogen sulfate and the chromium values ultimately oxidized by peroxide in alkaline medium. Calculate the volume of 0.01 *M* $FeSO_4$ solution needed to reduce completely under acidic conditions the chromium(VI) produced from a 2.000-g sample of the ore.

12-29. A solution is 0.02 *M* in Na_2SO_3 and 0.04 *M* in $Na_2C_2O_4$. Calculate the volume of 0.01 *M* potassium permanganate required to react completely with 50.0 ml of this solution under acidic conditions.

12-30. A solution is 0.02 *M* in each of the following cations: Ni^{2+}, Cu^{2+}, Ag^+, and Fe^{3+}. Calculate the volume of 0.05 *M* aqueous ammonia required to react completely with 25.0 ml of this solution.

Appendixes

Appendix A: Chemical Arithmetic

The solving of appropriate numerical problems is essential to a complete understanding of the principles of solution chemistry. The problems needed to illustrate these principles are designed not to be exercises in mathematical proficiency but rather to aid in developing logical chemical thought and to emphasize the quantitative aspect of chemical reactions. As such, these problems are not difficult and require for their solution no more than a clear understanding of arithmetic and the rudiments of algebra. The student has performed the majority of the necessary operations many times in his courses in arithmetic and algebra. There is need, therefore, only to refresh his memory and to direct his attention to a few notations and applications.

UNITS OF MEASUREMENT

A measured quantity recorded without the unit of measurement is meaningless. The correct solution of every problem requires, therefore, that all numerical quantities be accompanied by their units (e.g., milliliters, grams, moles liter^{-1}, etc.). These units should be retained through each mathematical operation. Units are handled according to the rules of algebra. Thus, if

x and y represent any simple or complex units of measurement, these rules can be illustrated as

Addition

$$3x + 4x = 7x$$
$$4x^3 + x^3 + 0.5x^3 = 5.5x^3$$

Subtraction

$$8y - 2y = 6y$$
$$10y^2 - 3y^2 - 5y^2 = 2y^2$$

Multiplication

$$(4x)(8y) = 32xy$$
$$\left(12\frac{x}{y}\right)\left(4\frac{x}{y}\right) = 48\frac{x^2}{y^2}$$

Division

$$35y \div 5x = 7\frac{y}{x}$$
$$15\frac{x^2}{y^2} \div 3\frac{x}{y} = 5\frac{x}{y}$$

If a problem is solved correctly, the answer will be expressed logically in the correct units. Use of units is important to the dimensional approach outlined below for the solution of problems.

SIGNIFICANT FIGURES

Since no measuring instrument or mechanism is completely without error, every measured quantity reflects some degree of uncertainty. If several instruments or mechanisms are employed, the accuracy of the final result is limited by the least reliable measurement. In any computation, therefore, the numerical result must not suggest greater accuracy than the experimental data warrant. Significant figures are important, therefore, both in the initial data and in the final answer. The all too common practice of carrying all figures in series of arithmetical operations reflects lack of understanding of the importance of significant figures.

If a volume is recorded as 25.6 ml, for example, the implication is that the figures 2 and 5 are known with certainty and that the figure 6 is doubtful. Both the 2 and the 5 are significant parts of the measurement. In the sense that the 6 represents a best approximation, it is also of significance.

The total number of significant figures is then *three*, i.e., the number known with certainty and the estimated last figure. The volume 25.6 ml implies that the exact volume lies between 25.5 ml and 25.7 ml (an uncertainty of 1 part in 256). A more refined measurement of 25.65 ml would give a result expressed to *four* significant figures and would indicate an exact value between 25.64 and 25.66 ml (an uncertainty of 1 part in 2,565).

The zero is used both to represent a quantity and to fix a decimal point. When used for the latter purpose, the zero is not considered as a significant figure. A zero at the end of a number, however, may or may not be significant. In the volume 1,800 ml, both zeros are significant if they are used to indicate measured figures (i.e., the exact volume between 1,795 ml and 1,805 ml), but not if they are used only to indicate the decimal point. This ambiguity disappears if the quantity is expressed exponentially. If the measurement were reliable to two significant figures, it would then be recorded as 1.8×10^3 ml; but if four figures were significant, it would be expressed as 1.800×10^3 ml. Here all zeros are significant. A zero appearing between other digits in a number is always significant. An exact number, e.g., exactly 100 cm in 1 m, is considered to have an unlimited number of significant figures.

The following rules governing the handling of significant figures are useful:

Addition and subtraction The last significant figure in the result is determined by the column containing the first doubtful figure. For example, in the addition

$$\text{Measured volumes} = \begin{cases} 37.25 \text{ ml} & \text{(4 significant figures)} \\ 3.5 \text{ ml} & \text{(2 significant figures)} \\ 0.92 \text{ ml} & \text{(2 significant figures)} \end{cases}$$

$$\text{Calculated volume} = 41.7 \text{ ml} \quad \text{(3 significant figures)}$$

the direct result of addition, 41.67 ml, is corrected by dropping the last (and uncertain) column. The last figure retained in the recorded answer is increased by one if the dropped figure is equal to or greater than half a unit.

Multiplication and division The number of significant figures in the answer is determined by the quantity that contains the least number of significant figures. Thus, in the multiplication

$$\text{Measured values} = \begin{cases} 2.024 \text{ moles liter}^{-1} & \text{(4 significant figures)} \\ \times\ 2.50 \text{ liters} & \text{(3 significant figures)} \end{cases}$$

$$\text{Calculated value} = 5.06 \text{ moles} \quad \text{(3 significant figures)}$$

the zero in 2.50 liters is significant and does not determine the location of the decimal point. As a general rule, the accuracy of the result should always reflect the accuracy of the measurement.

DIMENSIONAL ANALYSIS

Numerical problems essential to the elucidation of chemical principles differ from problems encountered in everyday economic life only in the units used. The objective in solving a numerical problem in chemistry is to increase understanding of the quantitative aspects of the science. Therefore, rote approaches involving the liberal use of either proportions or mathematical formulas are discouraged. For a given problem, a logical combination of numerical values leads to the desired answer. In nearly all instances, this combination can be achieved by the multiplication of a series of factors so arranged that the initial units are converted successively through various combinations of units to the desired units of the answer. The inclusion of the appropriate units in each factor and the correct cancellation of units in each multiplication are essential both to the success of the calculation and to checking the correctness of the method of calculation. Any deviation from logical units in the answer is an immediate indication of an error in calculation. The factor-unit approach, or dimensional analysis, is sufficiently versatile that the student will wish to apply it to many related calculations.

This approach is utilized in the illustrative problems that appear in this book. Two specific examples emphasize the details of the method.

Example A-1 Calculate the weight of arsenic(III) sulfide (in g) that must be dissolved as

$$As_2S_3(s) + 3S_x^{2-} \rightarrow 2[AsS_4]^{3-} + (3x - 5)S$$

to give 500 ml of a solution that is 0.025 M in tetrasulfoarsenate(V) ion.

Solution Using the relationship that 2 moles of $[AsS_4]^{3-}$ ion result from 1 mole of $As_2S_3(s)$ and realizing that a 0.025 M solution of the former contains 0.025 mole liter^{-1} of the ion, we can write

$$\begin{aligned}\text{Wt. of } As_2S_3(s) \text{ (in g)} &= \frac{0.025 \cancel{\text{ mole } [AsS_4]^{3-}}}{1 \cancel{\text{ liter}}} \times 0.500 \cancel{\text{ liter}} \\ &\quad \times \frac{2 \cancel{\text{ moles } As_2S_3}}{1 \cancel{\text{ mole } [AsS_4]^{3-}}} \times \frac{246.02 \text{ g } As_2S_3}{1 \cancel{\text{ mole } As_2S_3}} \\ &= 6.15 \text{ g}\end{aligned}$$

Example A-2 Calculate the weight of aluminum (in g) necessary to react with excess nitrate ion, as

$$8Al(s) + 5OH^- + 18H_2O + 3NO_3^- \rightarrow 3NH_3 + 8[Al(OH)_4]^-$$

to form sufficient ammonia that if it were all dissolved in water 1500 ml of 0.15 M solution would result.

Solution Using the ratio of 3 moles of ammonia released to 8 moles of aluminum reacting, as established by the equation, and recognizing that 1500 ml of a 0.15 *M* solution of NH_3 contains

$$\frac{0.15 \text{ mole } NH_3}{1 \text{ liter}} \times 1500 \text{ ml} \times \frac{1 \text{ liter}}{1000 \text{ ml}}$$

or 0.225 mole of NH_3, we calculate

$$\text{Wt. of Al (in g)} = 0.225 \text{ mole } NH_3 \times \frac{8 \text{ moles Al}}{3 \text{ moles } NH_3} \times \frac{26.98 \text{ g Al}}{1 \text{ mole Al}} = 16.2 \text{ g}$$

Of course, the calculation can be performed in a single step,

$$\text{Wt. of Al (in g)} = \frac{0.15 \text{ mole } NH_3}{1 \text{ liter}} \times 1500 \text{ ml} \times \frac{1 \text{ liter}}{1000 \text{ ml}} \times \frac{8 \text{ moles Al}}{3 \text{ moles } NH_3} \times \frac{26.98 \text{ g Al}}{1 \text{ mole Al}} = 16.2 \text{ g}$$

Success in the solving of numerical problems by this means depends upon:

1. A complete understanding of what data are given and of what results are required.
2. A systematic conversion of the units that are given to the units in which the answer is to be expressed, using necessary conversion factors, retaining units in all factors, and cancelling units appropriately.
3. A final examination of the reasonableness of the answer in relationship to the data given and the statement of the problem.

The table of four-place logarithms given as Appendix B will be useful to the student in many of the problems that are assigned. The student can do a large number of the calculations with a good slide rule, however.

Appendix B: Table of Logarithms

No.	0	1	2	3	4	5	6	7	8	9	1	2	3	4	5	6	7	8	9
10	0000	0043	0086	0128	0170	0212	0253	0294	0334	0374	4	8	12	17	21	25	29	33	37
11	0414	0453	0492	0531	0569	0607	0645	0682	0719	0755	4	8	11	15	19	23	26	30	34
12	0792	0828	0864	0899	0934	0969	1004	1038	1072	1106	3	7	10	14	17	21	24	28	31
13	1139	1173	1206	1239	1271	1303	1335	1367	1399	1430	3	6	10	13	16	19	23	26	29
14	1461	1492	1523	1553	1584	1614	1644	1673	1703	1723	3	6	9	12	15	18	21	24	27
15	1761	1790	1818	1847	1875	1903	1931	1959	1987	2014	3	6	8	11	14	17	20	22	25
16	2041	2068	2095	2122	2148	2175	2201	2227	2253	2279	3	5	8	11	13	16	18	21	24
17	2304	2330	2355	2380	2405	2430	2455	2480	2504	2529	2	5	7	10	12	15	17	20	22
18	2553	2577	2601	2625	2648	2672	2695	2718	2742	2765	2	5	7	9	12	14	16	19	21
19	2788	2810	2833	2856	2878	2900	2923	2945	2967	2989	2	4	7	9	11	13	16	18	20
20	3010	3032	3054	3075	3096	3118	3139	3160	3181	3201	2	4	6	8	11	13	15	17	19
21	3222	3243	3263	3284	3304	3324	3345	3365	3385	3404	2	4	6	8	10	12	14	16	18
22	3424	3444	3464	3483	3502	3522	3541	3560	3579	3598	2	4	6	8	10	12	14	15	17
23	3617	3636	3655	3674	3692	3711	3729	3747	3766	3784	2	4	6	7	9	11	13	15	17
24	3802	3820	3838	3856	3874	3892	3909	3927	3945	3962	2	4	5	7	9	11	12	14	16
25	3979	3997	4014	4031	4048	4065	4082	4099	4116	4133	2	3	5	7	9	10	12	14	15
26	4150	4166	4183	4200	4216	4232	4249	4265	4281	4298	2	3	5	7	8	10	11	13	15
27	4314	4330	4346	4362	4378	4393	4409	4425	4440	4456	2	3	5	6	8	9	11	13	14
28	4472	4487	4502	4518	4533	4548	4564	4579	4594	4609	2	3	5	6	8	9	11	12	14
29	4624	4639	4654	4669	4683	4698	4713	4728	4742	4757	1	3	4	6	7	9	10	12	13
30	4771	4786	4800	4814	4829	4843	4857	4871	4886	4900	1	3	4	6	7	9	10	11	13
31	4914	4928	4942	4955	4969	4983	4997	5011	5024	5038	1	3	4	6	7	8	10	11	12
32	5051	5065	5079	5092	5105	5119	5132	5145	5159	5172	1	3	4	5	7	8	9	11	12
33	5185	5198	5211	5224	5237	5250	5263	5276	5289	5302	1	3	4	5	6	8	9	10	12
34	5315	5328	5340	5353	5366	5378	5391	5403	5416	5428	1	3	4	5	6	8	9	10	11
35	5441	5453	5465	5478	5490	5502	5514	5527	5539	5551	1	2	4	5	6	7	9	10	11
36	5563	5575	5587	5599	5611	5623	5635	5647	5658	5670	1	2	4	5	6	7	8	10	11
37	5682	5694	5705	5717	5729	5740	5752	5763	5775	5786	1	2	3	5	6	7	8	9	10
38	5789	5809	5821	5832	5843	5855	5866	5877	5888	5899	1	2	3	5	6	7	8	9	10
39	5911	5922	5933	5944	5955	5966	5977	5988	5999	6010	1	2	3	4	5	7	8	9	10
40	6021	6031	6042	6053	6064	6075	6085	6096	6107	6117	1	2	3	4	5	6	8	9	10
41	6128	6138	6149	6160	6170	6180	6191	6201	6212	6222	1	2	3	4	5	6	7	8	9
42	6232	6243	6253	6263	6274	6284	6294	6304	6314	6325	1	2	3	4	5	6	7	8	9
43	6335	6345	6355	6365	6375	6386	6395	6405	6415	6425	1	2	3	4	5	6	7	8	9
44	6435	6444	6454	6464	6474	6484	6493	6503	6513	6522	1	2	3	4	5	6	7	8	9
45	6532	6542	6551	6561	6571	6580	6590	6599	6609	6618	1	2	3	4	5	6	7	8	9
46	6628	6637	6646	6656	6665	6675	6684	6693	6702	6712	1	2	3	4	5	6	7	7	8
47	6721	6730	6739	6749	6758	6767	6776	6785	6794	6803	1	2	3	4	5	5	6	7	8
48	6812	6821	6830	6839	6848	6857	6866	6875	6884	6893	1	2	3	4	4	5	6	7	8
49	6902	6911	6920	6928	6937	6946	6955	6964	6972	6981	1	2	3	4	4	5	6	7	8
50	6990	6998	7007	7016	7024	7033	7042	7050	7059	7067	1	2	3	3	4	5	6	7	8
51	7076	7084	7093	7101	7110	7118	7126	7135	7143	7152	1	2	3	3	4	5	6	7	8
52	7160	7168	7177	7185	7193	7202	7210	7218	7226	7235	1	2	2	3	4	5	6	7	7
53	7243	7251	7259	7267	7275	7284	7292	7300	7308	7316	1	2	2	3	4	5	6	6	7
54	7324	7332	7340	7348	7356	7364	7372	7380	7388	7396	1	2	2	3	4	5	6	6	7

No.	0	1	2	3	4	5	6	7	8	9	1	2	3	4	5	6	7	8	9
55	7404	7412	7419	7427	7435	7443	7451	7459	7466	7474	1	2	2	3	4	5	5	6	7
56	7482	7490	7497	7505	7513	7520	7528	7536	7543	7551	1	2	2	3	4	5	5	6	7
57	7559	7566	7574	7582	7589	7597	7604	7612	7619	7627	1	2	2	3	4	5	5	6	7
58	7634	7642	7649	7657	7664	7672	7679	7686	7694	7701	1	1	2	3	4	4	5	6	7
59	7709	7716	7723	7731	7738	7745	7752	7760	7767	7774	1	1	2	3	4	4	5	6	7
60	7782	7789	7796	7803	7810	7818	7825	7832	7839	7846	1	1	2	3	4	4	5	6	6
61	7853	7860	7868	7875	7882	7889	7896	7903	7910	7917	1	1	2	3	4	4	5	6	6
62	7924	7931	7938	7945	7952	7959	7966	7973	7980	7987	1	1	2	3	3	4	5	6	6
63	7992	8000	8007	8014	8021	8028	8035	8041	8048	8055	1	1	2	3	3	4	5	5	6
64	8062	8069	8075	8082	8089	8096	8102	8109	8116	8122	1	1	2	3	3	4	5	5	6
65	8129	8136	8142	8149	8156	8162	8169	8176	8182	8189	1	1	2	3	3	4	5	5	6
66	8195	8202	8209	8215	8222	8228	8235	8241	8248	8254	1	1	2	3	3	4	5	5	6
67	8261	8267	8274	8280	8287	8293	8299	8306	8312	8319	1	1	2	3	3	4	5	5	6
68	8325	8331	8338	8344	8351	8357	8363	8370	8376	8382	1	1	2	3	3	4	4	5	6
69	8388	8395	8401	8407	8414	8420	8426	8432	8439	8445	1	1	2	2	3	4	4	5	6
70	8451	8457	8463	8470	8476	8482	8488	8494	8500	8506	1	1	2	2	3	4	4	5	6
71	8513	8519	8525	8531	8537	8543	8549	8555	8561	8567	1	1	2	2	3	4	4	5	5
72	8573	8579	8585	8591	8597	8603	8609	8615	8621	8627	1	1	2	2	3	4	4	5	5
73	8633	8639	8645	8651	8657	8663	8669	8675	8681	8686	1	1	2	2	3	4	4	5	5
74	8692	8698	8704	8710	8716	8722	8727	8733	8739	8745	1	1	2	2	3	4	4	5	5
75	8751	8756	8762	8768	8774	8779	8785	8791	8797	8802	1	1	2	2	3	3	4	5	5
76	8808	8814	8820	8825	8831	8837	8842	8848	8854	8859	1	1	2	2	3	3	4	5	5
77	8865	8871	8876	8882	8887	8893	8899	8904	8910	8915	1	1	2	2	3	3	4	4	5
78	8921	8927	8932	8938	8943	8949	8954	8960	8965	8971	1	1	2	2	3	3	4	4	5
79	8976	8982	8987	8993	8998	9004	9009	9015	9020	9025	1	1	2	2	3	3	4	4	5
80	9031	9036	9042	9047	9053	9058	9063	9069	9074	9079	1	1	2	2	3	3	4	4	5
81	9085	9090	9096	9101	9106	9112	9117	9122	9128	9133	1	1	2	2	3	3	4	4	5
82	9138	9143	9149	9154	9159	9165	9170	9175	9180	9186	1	1	2	2	3	3	4	4	5
83	9191	9196	9201	9206	9212	9217	9222	9227	9232	9238	1	1	2	2	3	3	4	4	5
84	9243	9248	9253	9258	9263	9269	9274	9279	9284	9289	1	1	2	2	3	3	4	4	5
85	9294	9299	9304	9309	9315	9320	9325	9330	9335	9340	1	1	2	2	3	3	4	4	5
86	9345	9350	9355	9360	9365	9370	9375	9380	9385	9390	1	1	2	2	3	3	4	4	5
87	9395	9400	9405	9410	9415	9420	9425	9430	9435	9440	0	1	1	2	2	3	3	4	4
88	9445	9450	9455	9460	9465	9469	9474	9479	9484	9489	0	1	1	2	2	3	3	4	4
89	9494	9499	9504	9509	9513	9518	9523	9528	9533	9538	0	1	1	2	2	3	3	4	4
90	9542	9547	9552	9557	9562	9566	9571	9576	9581	9586	0	1	1	2	2	3	3	4	4
91	9590	9595	9600	9605	9609	9614	9619	9624	9628	9633	0	1	1	2	2	3	3	4	4
92	9638	9643	9647	9652	9657	9661	9666	9671	9675	9680	0	1	1	2	2	3	3	4	4
93	9685	9689	9594	9699	9703	9708	9713	9717	9722	9727	0	1	1	2	2	3	3	4	4
94	9731	9736	9741	9745	9750	9754	9759	9763	9768	9773	0	1	1	2	2	3	3	4	4
95	9777	9782	9786	9791	9795	9800	9805	9809	9814	9818	0	1	1	2	2	3	3	4	4
96	9823	9827	9832	9836	9841	9845	9850	9854	9859	9863	0	1	1	2	2	3	3	4	4
97	9868	9872	9877	9881	9886	9890	9894	9899	9903	9908	0	1	1	2	2	3	3	4	4
98	9912	9917	9921	9926	9930	9934	9939	9943	9948	9952	0	1	1	2	2	3	3	4	4
99	9956	9961	9965	9969	9974	9978	9983	9987	9991	9996	0	1	1	2	2	3	3	3	4

Appendix C: Ionization Constants of Acids*

Acid	*Equilibrium equation*	K†	pK
Acetic	$CH_3CO_2H + H_2O \rightleftharpoons H_3O^+ + CH_3CO_2^-$	1.80×10^{-5} (K_a)	4.7
Aluminum hydroxide	$\mathbf{Al(OH)_3} \rightleftharpoons H_3O^+ + AlO_2^-$	4×10^{-13} (K_a)	12.4
Aluminum ion	$Al^{3+} + 2H_2O \rightleftharpoons H_3O^+ + AlOH^{2+}$	1.4×10^{-5} (K_{h_1})	4.9
Ammonium ion	$NH_4^+ + H_2O \rightleftharpoons H_3O^+ + NH_3$	5.6×10^{-10} (K_h)	9.3
Antimony(III) hydroxide	$\mathbf{SbOOH} + H_2O \rightleftharpoons H_3O^+ + SbO_2^-$	1×10^{-11} (K_a)	11.0
Arsenic (ortho)	$H_3AsO_4 + H_2O \rightleftharpoons H_3O^+ + H_2AsO_4^-$	5.0×10^{-3} (K_{a_1})	2.3
	$H_2AsO_4^- + H_2O \rightleftharpoons H_3O^+ + HAsO_4^{2-}$	1.6×10^{-7} (K_{a_2})	6.8
	$HAsO_4^{2-} + H_2O \rightleftharpoons H_3O^+ + AsO_4^{3-}$	2.5×10^{-12} (K_{a_3})	11.6
Arsenous (meta)	$HAsO_2 + H_2O \rightleftharpoons H_3O^+ + AsO_2^-$	6×10^{-10} (K_a)	9.2
Benzoic	$\mathbf{C_6H_5CO_2H} + H_2O \rightleftharpoons H_3O + C_6H_5CO_2^-$	6.6×10^{-5} (K_a)	4.2
Bismuth(III) ion	$Bi^{3+} + 2H_2O \rightleftharpoons H_3O^+ + BiOH^{2+}$	1×10^{-2} (K_{h_1})	2.0
Boric (ortho)	$H_3BO_3 + H_2O \rightleftharpoons H_3O^+ + H_2BO_3^-$	6.0×10^{-10} (K_{a_1})	9.2
Carbonic	$H_2CO_3 + H_2O \rightleftharpoons H_3O^+ + HCO_3^-$	4.2×10^{-7} (K_{a_1})	6.4
	$HCO_3^- + H_2O \rightleftharpoons H_3O^+ + CO_3^{2-}$	4.8×10^{-11} (K_{a_2})	10.3
Chromic	$H_2CrO_4 + H_2O \rightleftharpoons H_3O^+ + HCrO_4^-$	1.8×10^{-1} (K_{a_1})	0.74
	$HCrO_4^- + H_2O \rightleftharpoons H_3O^+ + CrO_4^{2-}$	3.2×10^{-7} (K_{a_2})	6.5
	$2HCrO_4^- \rightleftharpoons Cr_2O_7^{2-} + H_2O$	$4.3 \times 10^{+1}$ (K) (approx)	-0.4 (approx)
Chromium(III) hydroxide	$\mathbf{Cr(OH)_3} \rightleftharpoons H_3O^+ + CrO_2^-$	9×10^{-17} (K_a)	16.1
Chromium(III) ion	$Cr^{3+} + 2H_2O \rightleftharpoons H_3O^+ + CrOH^{2+}$	1×10^{-4} (K_{h_1})	4.0
Copper(II) hydroxide	$\mathbf{Cu(OH)_2} + H_2O \rightleftharpoons H_3O^+ + HCuO_2^-$	1×10^{-19} (K_{a_1})	19.0
	$HCuO_2^- + H_2O \rightleftharpoons H_3O^+ + CuO_2^{2-}$	7.9×10^{-14} (K_{a_2})	13.1
Copper(II) ion	$Cu^{2+} + 2H_2O \rightleftharpoons H_3O^+ + CuOH^+$	1×10^{-8} (K_{h_1})	8.0
Formic	$HCO_2H + H_2O \rightleftharpoons H_3O^+ + HCO_2^-$	2.1×10^{-4} (K_a)	3.7
Hydriodic	$HI + H_2O \rightleftharpoons H_3O^+ + I^-$	3.2×10^{9} (K_a)	-9.5
Hydrobromic	$HBr + H_2O \rightleftharpoons H_3O^+ + Br^-$	1×10^{9} (K_a)	-9 (approx)

Hydrochloric	$HCl + H_2O \rightleftharpoons H_3O^+ + Cl^-$	1×10^6 (K_a)	−6 (approx)
Hydrocyanic	$HCN + H_2O \rightleftharpoons H_3O^+ + CN^-$	4.8×10^{-10} (K_a)	9.3
Hydrofluoric	$HF + H_2O \rightleftharpoons H_3O^+ + F^-$	6.9×10^{-4} (K_a)	3.2
Hydrogen peroxide	$H_2O_2 + H_2O \rightleftharpoons H_3O^+ + HO_2^-$	2.4×10^{-12} (K_{a1})	11.6
Hydrosulfuric	$H_2S + H_2O \rightleftharpoons H_3O^+ + HS^-$	1×10^{-7} (K_{a1})	7.0
	$HS^- + H_2O \rightleftharpoons H_3O^+ + S^{2-}$	1.3×10^{-13} (K_{a2})	12.9
Hypochlorous	$HClO + H_2O \rightleftharpoons H_3O^+ + ClO^-$	3.2×10^{-8} (K_a)	7.5
Iron(III) ion	$Fe^{3+} + 2H_2O \rightleftharpoons H_3O^+ + FeOH^{2+}$	4.0×10^{-3} (K_{h1})	2.4
Iron(II) ion	$Fe^{2+} + 2H_2O \rightleftharpoons H_3O^+ + FeOH^+$	1.2×10^{-6} (K_{h1})	5.9
Lead(II) hydroxide	$\mathbf{Pb(OH)_2} + H_2O \rightleftharpoons H_3O^+ + HPbO_2^-$	4.6×10^{-16} (K_{a1})	15.3
Magnesium ion	$Mg^{2+} + 2H_2O \rightleftharpoons H_3O^+ + MgOH^+$	2×10^{-12} (K_{h1})	11.7
Mercury(II) ion	$Hg^{2+} + 2H_2O \rightleftharpoons H_3O^+ + HgOH^+$	2×10^{-3} (K_{h1})	2.7
Nitric	$HNO_3 + H_2O \rightleftharpoons H_3O^+ + NO_3^-$	2.3×10 (K_a)	−1.37
Nitrous	$HNO_2 + H_2O \rightleftharpoons H_3O^+ + NO_2^-$	4.5×10^{-4} (K_a)	3.4
Oxalic	$H_2C_2O_4 + H_2O \rightleftharpoons H_3O^+ + HC_2O_4^-$	6.3×10^{-2} (K_{a1})	1.2
	$HC_2O_4^- + H_2O \rightleftharpoons H_3O^+ + C_2O_4^{2-}$	6.3×10^{-5} (K_{a2})	4.2
Perchloric	$HClO_4 + H_2O \rightleftharpoons H_3O^+ + ClO_4^-$	2.0×10^7 (K_a)	−7.3
Permanganic	$HMnO_4 + H_2O \rightleftharpoons H_3O^+ + MnO_4^-$	2.0×10^3 (K_a)	−2.3
Phosphoric (ortho)	$H_3PO_4 + H_2O \rightleftharpoons H_3O^+ + H_2PO_4^-$	7.5×10^{-3} (K_{a1})	2.1
	$H_2PO_4^- + H_2O \rightleftharpoons H_3O^+ + HPO_4^{2-}$	6.2×10^{-8} (K_{a2})	7.2
	$HPO_4^{2-} + H_2O \rightleftharpoons H_3O^+ + PO_4^{3-}$	2.0×10^{-13} (K_{a3})	12.7
Propionic	$CH_3CH_2CO_2H + H_2O \rightleftharpoons H_3O^+ + \mathbf{CH_3CH_2CO_2^-}$	1.4×10^{-5} (K_a)	4.9
Silicic (meta)	$H_2SiO_3 + H_2O \rightleftharpoons H_3O^+ + HSiO_3^-$	3.2×10^{-10} (K_{a1})	9.5
	$HSiO_3^- + H_2O \rightleftharpoons H_3O^+ + SiO_3^{2-}$	6.3×10^{-12} (K_{a2})	11.8
Sulfuric	$H_2SO_4 + H_2O \rightleftharpoons H_3O^+ + HSO_4^-$	Large (K_{a1})	Neg
	$HSO_4^- + H_2O \rightleftharpoons H_3O^+ + SO_4^{2-}$	1.26×10^{-2} (K_{a2})	1.9
Sulfurous	$H_2SO_3 + H_2O \rightleftharpoons H_3O^+ + HSO_3^-$	1.6×10^{-2} (K_{a1})	1.8
	$HSO_3^- + H_2O \rightleftharpoons H_3O^+ + SO_3^{2-}$	1.3×10^{-7} (K_{a2})	6.9
Thiocyanic	$HNCS + H_2O \rightleftharpoons H_3O^+ + NCS^-$	Large (K_a)	Neg
Thiosulfuric	$H_2S_2O_3 + H_2O \rightleftharpoons H_3O^+ + HS_2O_3^-$	2.0×10^{-2} (K_{a1})	1.7
	$HS_2O_3^- + H_2O \rightleftharpoons H_3O^+ + S_2O_3^{2-}$	3.2×10^{-3} (K_{a2})	2.5

Acid	*Equilibrium equation*	K†	pK
Tin(IV) hydroxide	$\mathbf{Sn(OH)_4} + 4H_2O \rightleftharpoons 2H_3O^+ + [Sn(OH)_6]^{2-}$	10^{-32} (K_a) (approx)	32 (approx)
Tin(II) hydroxide	$\mathbf{Sn(OH)_2} + H_2O \rightleftharpoons H_3O^+ + HSnO_2^-$	3.8×10^{-15} (K_{a_1})	14.4
Zinc hydroxide	$\mathbf{Zn(OH)_2} + 2H_2O \rightleftharpoons 2H_3O^+ + ZnO_2^{2-}$	1.0×10^{-29} (K_a)	29.0
Zinc ion	$Zn^{2+} + 2H_2O \rightleftharpoons H_3O^+ + ZnOH^+$	2.5×10^{-10} (K_{h_1})	9.6

*Constants for anionic and molecular acids are indicated as K_a; those for cationic acids as K_h. The K_h notation emphasizes the common use of the term hydrolysis to describe the general process

$$M^{n+} + 2H_2O \rightleftharpoons H_3O^+ + M(OH)^{(n-1)+}$$

Distinction between the two is a matter of convention only (page 75). All numerical values in this and subsequent tables are for room temperature (about 20 to 25°C).

Appendix D: Ionization Constants of Bases*

Base	*Equilibrium equation*	*K*	p*K*
Acetate ion	$CH_3CO_2^- + H_2O \rightleftharpoons CH_3CO_2H + OH^-$	5.6×10^{-10} (K_h)	9.3
Ammonia	$NH_3 + H_2O \rightleftharpoons NH_4^+ + OH^-$	1.80×10^{-5} (K_b)	4.7
Aniline	$C_6H_5NH_2 + H_2O \rightleftharpoons C_6H_5NH_3^+ + OH^-$	3.8×10^{-10} (K_b)	9.4
Arsenate ion (ortho)	$AsO_4^{3-} + H_2O \rightleftharpoons HAsO_4^{2-} + OH^-$	1.7×10^{-3} (K_{h1})	2.4
	$HAsO_4^{2-} + H_2O \rightleftharpoons H_2AsO_4^- + OH^-$	6.3×10^{-8} (K_{h2})	7.2
	$H_2AsO_4^- + H_2O \rightleftharpoons H_3AsO_4 + OH^-$	2.0×10^{-12} (K_{h3})	11.7
Arsenate ion (meta)	$AsO_2^- + H_2O \rightleftharpoons HAsO_2 + OH^-$	1.6×10^{-5} (K_h)	4.8
Borate ions	$H_2BO_3^- + H_2O \rightleftharpoons H_3BO_3 + OH^-$	1.6×10^{-5} (K_{h1})	4.8
	$B_4O_7^{2-} + 5H_2O \rightleftharpoons 2H_2BO_3^- + 2H_3BO_3$	10^{-3} (K) (approx)	3 (approx)
Bromide ion	$Br^- + H_2O \rightleftharpoons HBr + OH^-$	1.0×10^{-23} (K_h)	23
Carbonate ion	$CO_3^{2-} + H_2O \rightleftharpoons HCO_3^- + OH^-$	2.1×10^{-4} (K_{h1})	3.7
	$HCO_3^- + H_2O \rightleftharpoons H_2CO_3 + OH^-$	2.4×10^{-8} (K_{h2})	7.6
Chloride ion	$Cl^- + H_2O \rightleftharpoons HCl + OH^-$	1.0×10^{-20} (K_h)	20
Chromate ion	$CrO_4^{2-} + H_2O \rightleftharpoons HCrO_4^- + OH^-$	3×10^{-8} (K_{h1})	7.5
Cyanide ion	$CN^- + H_2O \rightleftharpoons HCN + OH^-$	2.1×10^{-5} (K_h)	4.7
Diethylamine	$(C_2H_5)_2NH + H_2O \rightleftharpoons (C_2H_5)_2NH_2^+ + OH^-$	9.6×10^{-4}(K_b)	3.0
Dimethylamine	$(CH_3)_2NH + H_2O \rightleftharpoons (CH_3)_2NH_2^+ + OH^-$	5.1×10^{-4} (K_b)	3.3
Ethylamine	$C_2H_5NH_2 + H_2O \rightleftharpoons C_2H_5NH_3^+ + OH^-$	5.6×10^{-4} (K_b)	3.3
Fluoride ion	$F^- + H_2O \rightleftharpoons HF + OH^-$	1.5×10^{-11} (K_h)	10.8
Iodide ion	$I^- + H_2O \rightleftharpoons HI + OH^-$	3.2×10^{-24} (K_h)	23.5
Methylamine	$CH_3NH_2 + H_2O \rightleftharpoons CH_3NH_3^+ + OH^-$	5.0×10^{-4} (K_b)	3.3
Nitrate ion	$NO_3^- + H_2O \rightleftharpoons HNO_3 + OH^-$	4.0×10^{-16} (K_h)	15.4
Nitrite ion	$NO_2^- + H_2O \rightleftharpoons HNO_2 + OH^-$	2.2×10^{-11} (K_h)	10.7
Oxalate ion	$C_2O_4^{2-} + H_2O \rightleftharpoons HC_2O_4^- + OH^-$	1.6×10^{-10} (K_h)	9.8
Permanganate ion	$MnO_4^- + H_2O \rightleftharpoons HMnO_4 + OH^-$	5.0×10^{-17} (K_h)	16.3
Phosphate ion (ortho)	$PO_4^{3-} + H_2O \rightleftharpoons HPO_4^{2-} + OH^-$	5.0×10^{-2} (K_{h1})	1.3
	$HPO_4^{2-} + H_2O \rightleftharpoons H_2PO_4^- + OH^-$	1.6×10^{-7} (K_{h2})	6.8
	$H_2PO_4^- + H_2O \rightleftharpoons H_3PO_4 + OH^-$	1.3×10^{-12} (K_{h3})	11.9

Silicate ion (meta)	$SiO_3^{2-} + H_2O \rightleftharpoons HSiO_3^- + OH^-$	1.6×10^{-3} (K_{h1})	2.8
	$HSiO_3^- + H_2O \rightleftharpoons H_2SiO_3 + OH^-$	3.1×10^{-5} (K_{h2})	4.5
Sulfate ion	$SO_4^{2-} + H_2O \rightleftharpoons HSO_4^- + OH^-$	8.0×10^{-13} (K_{h1})	12.1
Sulfite ion	$SO_3^{2-} + H_2O \rightleftharpoons HSO_3^- + OH^-$	1.3×10^{-8} (K_{h1})	7.1
	$HSO_3^- + H_2O \rightleftharpoons H_2SO_3 + OH^-$	6.3×10^{-13} (K_{h2})	12.2
Sulfide ion	$S^{2-} + H_2O \rightleftharpoons HS^- + OH^-$	7.7×10^{-2} (K_{h1})	1.1
	$HS^- + H_2O \rightleftharpoons H_2S + OH^-$	1×10^{-7} (K_{h2})	7.0
Thiocyanate ion	$NCS^- + H_2O \rightleftharpoons HNCS + OH^-$	Very small (K_h)	Large
Thiosulfate ion	$S_2O_3^{2-} + H_2O \rightleftharpoons HS_2O_3^- + OH^-$	3.1×10^{-12} (K_{h1})	11.5
Triethylamine	$(C_2H_5)_3N + H_2O \rightleftharpoons (C_2H_5)_3NH^+ + OH^-$	5.8×10^{-4}(K_b)	3.2
Trimethylamine	$(CH_3)_3N + H_2O \rightleftharpoons (CH_3)_3NH^+ + OH^-$	5.3×10^{-5} (K_b)	4.3

* Constants for molecular bases are indicated as K_b; those for anionic bases as K_h. Again the K_h notation is a conventional emphasis upon hydrolysis as a name for a reaction described by the type equation

$$A^{n-} + H_2O \rightleftharpoons HA^{(n-1)-} + OH^-$$

The solubility-product-constant data for metal hydroxides (Appendix E) are pertinent also.

Appendix E: Solubility-product Constants*

Anion	*Equilibrium equation*	K_{sp}	pK_{sp}
Acetate	$\mathbf{AgCH_3CO_2} \rightleftharpoons Ag^+ + CH_3CO_2^-$	4.0×10^{-3}	2.4
	$\mathbf{Hg_2(CH_3CO_2)_2} \rightleftharpoons Hg_2^{2+} + 2CH_3CO_2^-$	3.6×10^{-10}	9.4
Arsenate (ortho)	$\mathbf{Ag_3AsO_4} \rightleftharpoons 3Ag^+ + AsO_4^{3-}$	1.0×10^{-22}	22.0
Arsenite (meta)†			
Bromide	$\mathbf{PbBr_2} \rightleftharpoons Pb^{2+} + 2\ Br^-$	1.0×10^{-6}	5.0
	$\mathbf{CuBr} \rightleftharpoons Cu^+ + Br^-$	5.0×10^{-9}	8.3
	$\mathbf{AgBr} \rightleftharpoons Ag^+ + Br^-$	5.0×10^{-13}	12.3
	$\mathbf{Hg_2Br_2} \rightleftharpoons Hg_2^{2+} + 2Br^-$	5.0×10^{-23}	22.3
Carbonate	$\mathbf{MgCO_3} \rightleftharpoons Mg^{2+} + CO_3^{2-}$	4.0×10^{-5}	4.4
	$\mathbf{NiCO_3} \rightleftharpoons Ni^{2+} + CO_3^{2-}$	1.4×10^{-7}	6.9
	$\mathbf{CaCO_3} \rightleftharpoons Ca^{2+} + CO_3^{2-}$	4.7×10^{-9}	8.3
	$\mathbf{BaCO_3} \rightleftharpoons Ba^{2+} + CO_3^{2-}$	1.6×10^{-9}	8.8
	$\mathbf{SrCO_3} \rightleftharpoons Sr^{2+} + CO_3^{2-}$	7.0×10^{-10}	9.2
	$\mathbf{MnCO_3} \rightleftharpoons Mn^{2+} + CO_3^{2-}$	4.0×10^{-10}	9.4
	$\mathbf{CuCO_3} \rightleftharpoons Cu^{2+} + CO_3^{2-}$	2.5×10^{-10}	9.6
	$\mathbf{FeCO_3} \rightleftharpoons Fe^{2+} + CO_3^{2-}$	2.0×10^{-11}	10.7
	$\mathbf{ZnCO_3} \rightleftharpoons Zn^{2+} + CO_3^{2-}$	3.0×10^{-11}	10.8
	$\mathbf{Ag_2CO_3} \rightleftharpoons 2Ag^+ + CO_3^{2-}$	8.2×10^{-12}	11.1
	$\mathbf{CdCO_3} \rightleftharpoons Cd^{2+} + CO_3^{2-}$	5.2×10^{-12}	11.3
	$\mathbf{CoCO_3} \rightleftharpoons Co^{2+} + CO_3^{2-}$	1.6×10^{-13}	12.8
	$\mathbf{PbCO_3} \rightleftharpoons Pb^{2+} + CO_3^{2-}$	1.5×10^{-13}	12.8
Chloride	$\mathbf{PbCl_2} \rightleftharpoons Pb^{2+} + 2Cl^-$	1.6×10^{-5}	4.8
	$\mathbf{CuCl} \rightleftharpoons Cu^+ + Cl^-$	2.0×10^{-7}	6.7
	$\mathbf{AgCl} \rightleftharpoons Ag^+ + Cl^-$	2.8×10^{-10}	9.6
	$\mathbf{Hg_2Cl_2} \rightleftharpoons Hg_2^{2+} + 2Cl^-$	1.1×10^{-18}	17.9
Chromate	$\mathbf{CaCrO_4} \rightleftharpoons Ca^{2+} + CrO_4^{2-}$	7.1×10^{-4}	3.2
	$\mathbf{SrCrO_4} \rightleftharpoons Sr^{2+} + CrO_4^{2-}$	5.7×10^{-5}	4.2
	$\mathbf{Hg_2CrO_4} \rightleftharpoons Hg_2^{2+} + CrO_4^{2-}$	2.0×10^{-9}	8.7

	$BaCrO_4 \rightleftharpoons Ba^{2+} + CrO_4^{2-}$	8.5×10^{-11}	10.1
	$Ag_2CrO_4 \rightleftharpoons 2Ag^+ + CrO_4^{2-}$	1.9×10^{-12}	11.7
	$PbCrO_4 \rightleftharpoons Pb^{2+} + CrO_4^{2-}$	1.8×10^{-14}	13.7
Cyanide	$AgCN \rightleftharpoons Ag^+ + CN^-$	1.6×10^{-14}	13.8
Fluoride	$BaF_2 \rightleftharpoons Ba^{2+} + 2F^-$	1.0×10^{-7}	6.0
	$MgF_2 \rightleftharpoons Mg^{2+} + 2F^-$	8.0×10^{-8}	7.9
	$SrF_2 \rightleftharpoons Sr^{2+} + 2F^-$	2.5×10^{-9}	8.6
	$CaF_2 \rightleftharpoons Ca^{2+} + 2F^-$	1.7×10^{-10}	9.3
	$ThF_4 \rightleftharpoons Th^{4+} + 4F^-$	4.0×10^{-28}	27.4
Hexacyanoferrate(II)	$KFe[Fe(CN)_6] \rightleftharpoons K^+ + Fe^{3+} + [Fe(CN)_6]^{4-}$	3.2×10^{-41}	40.5
	$Ag_4[Fe(CN)_6] \rightleftharpoons 4Ag^+ + [Fe(CN)_6]^{4-}$	1.6×10^{-41}	40.8
	$K_2Zn_3[Fe(CN)_6]_2 \rightleftharpoons 2K^+ + 3Zn^{2+} + 2[Fe(CN)_6]^{4-}$	1.0×10^{-95}	95.0
Hydroxide	$Ag_2O + H_2O \rightleftharpoons 2Ag^+ + 2OH^-$	2.0×10^{-8}	7.7
	$Mg(OH)_2 \rightleftharpoons Mg^{2+} + 2OH^-$	8.9×10^{-12}	11.1
	$BiOOH \rightleftharpoons BiO^+ + OH^-$	1.0×10^{-12}	12.0
	$Mn(OH)_2 \rightleftharpoons Mn^{2+} + 2OH^-$	2.0×10^{-13}	12.7
	$Cd(OH)_2 \rightleftharpoons Cd^{2+} + 2OH^-$	2.0×10^{-14}	13.7
	$Pb(OH)_2 \rightleftharpoons Pb^{2+} + 2OH^-$	4.2×10^{-15}	14.4
	$Fe(OH)_2 \rightleftharpoons Fe^{2+} + 2OH$	1.8×10^{-15}	14.7
	$Ni(OH_2) \rightleftharpoons Ni^{2+} + 2OH^-$	1.6×10^{-16}	15.8
	$Co(OH)_2 \rightleftharpoons Co^{2+} + 2OH^-$	2.0×10^{-16}	15.7
	$Zn(OH)_2 \rightleftharpoons Zn^{2+} + 2OH^-$	4.5×10^{-17}	16.3
	$SbOOH \rightleftharpoons SbO^+ + OH^-$	1.0×10^{-17}	17.0
	$Cu(OH)_2 \rightleftharpoons Cu^{2+} + 2OH^-$	1.6×10^{-19}	18.8
	$Hg(OH)_2 \rightleftharpoons Hg^{2+} + 2OH^-$	3.2×10^{-26}	25.5
	$Sn(OH)_2 \rightleftharpoons Sn^{2+} + 2OH^-$	6.3×10^{-27}	26.2
	$Cr(OH)_3 \rightleftharpoons Cr^{3+} + 3OH^-$	6.7×10^{-31}	30.2
	$Al(OH)_3 \rightleftharpoons Al^{3+} + 3OH^-$	5.0×10^{-33}	32.3
	$Fe(OH)_3 \rightleftharpoons Fe^{3+} + 3OH^-$	6.0×10^{-38}	37.2
	$Sn(OH)_4 \rightleftharpoons Sn^{4+} + 4OH^-$	10^{-57} (approx)	57 (approx)

Anion	Equilibrium equation	K_{sp}	pK_{sp}
Iodide	$PbI_2 \rightleftharpoons Pb^{2+} + 2I^-$	8.3×10^{-9}	8.1
	$CuI \rightleftharpoons Cu^+ + I^-$	1.0×10^{-12}	12.0
	$AgI \rightleftharpoons Ag^+ + I^-$	8.5×10^{-17}	16.1
	$HgI_2 \rightleftharpoons Hg^{2+} + 2I^-$	2.5×10^{-26}	25.6
	$Hg_2I_2 \rightleftharpoons Hg_2^{2+} + 2I^-$	4.5×10^{-29}	28.3
Nitrate	$BiONO_3 + 2H^+ \rightleftharpoons Bi^{3+} + NO_3^- + H_2O$	2.5×10^{-3}	2.6
Nitrite	$AgNO_2 \rightleftharpoons Ag^+ + NO_2^-$	1.2×10^{-4}	3.9
Oxalate	$MgC_2O_4 \rightleftharpoons Mg^{2+} + C_2O_4^{2-}$	8.6×10^{-5}	4.1
	$CoC_2O_4 \rightleftharpoons Co^{2+} + C_2O_4^{2-}$	4.0×10^{-6}	5.4
	$FeC_2O_4 \rightleftharpoons Fe^{2+} + C_2O_4^{2-}$	2.0×10^{-7}	6.7
	$NiC_2O_4 \rightleftharpoons Ni^{2+} + C_2O_4^{2-}$	1.0×10^{-7}	7.0
	$SrC_2O_4 \rightleftharpoons Sr^{2+} + C_2O_4^{2-}$	5.6×10^{-8}	7.3
	$CuC_2O_4 \rightleftharpoons Cu^{2+} + C_2O_4^{2-}$	3.0×10^{-8}	7.5
	$BaC_2O_4 \rightleftharpoons Ba^{2+} + C_2O_4^{2-}$	1.5×10^{-8}	7.8
	$CdC_2O_4 \rightleftharpoons Cd^{2+} + C_2O_4^{2-}$	1.5×10^{-8}	7.8
	$ZnC_2O_4 \rightleftharpoons Zn^{2+} + C_2O_2^{2-}$	1.5×10^{-9}	8.8
	$CaC_2O_4 \rightleftharpoons Ca^{2+} + C_2O_4^{2-}$	1.3×10^{-9}	8.9
	$Ag_2C_2O_4 \rightleftharpoons 2Ag^+ + C_2O_4^{2-}$	1.0×10^{-11}	11.0
	$PbC_2O_4 \rightleftharpoons Pb^{2+} + C_2O_4^{2-}$	8.3×10^{-12}	11.1
	$Hg_2C_2O_4 \rightleftharpoons Hg_2^{2+} + C_2O_4^{2-}$	1.0×10^{-13}	13.0
	$MnC_2O_4 \rightleftharpoons Mn^{2+} + C_2O_4^{2-}$	1.1×10^{-15}	15.0
	$La_2(C_2O_4)_3 \rightleftharpoons 2La^{3+} + 3C_2O_4^{2-}$	2.0×10^{-28}	27.7
Permanganate†			
Phosphate (ortho)	$Li_3PO_4 \rightleftharpoons 3Li^+ + PO_4^{3-}$	3.2×10^{-13}	12.5
	$MgNH_4PO_4 \rightleftharpoons Mg^{2+} + NH_4^+ + PO_4^{3-}$	2.5×10^{-13}	12.6
	$Ag_3PO_4 \rightleftharpoons 3Ag^+ + PO_4^{3-}$	1.6×10^{-18}	17.8
	$AlPO_4 \rightleftharpoons Al^{3+} + PO_4^{3-}$	6.3×10^{-19}	18.2
	$Mn_3(PO_4)_2 \rightleftharpoons 3Mn^{2+} + 2PO_4^{3-}$	1.0×10^{-22}	22.0
	$Ba_3(PO_4)_2 \rightleftharpoons 3Ba^{2+} + 2PO_4^{3-}$	3.2×10^{-23}	22.5

	$BiPO_4 \rightleftharpoons Bi^{3+} + PO_4^{3-}$	3.2×10^{-24}	23.5
	$Sr_3(PO_4)_2 \rightleftharpoons 3Sr^{2+} + 2PO_4^{3-}$	4.0×10^{-28}	27.4
	$Ca_3(PO_4)_2 \rightleftharpoons 3Ca^{2+} + 2PO_4^{3-}$	1.3×10^{-32}	31.9
	$Mg_3(PO_4)_2 \rightleftharpoons 3Mg^{2+} + 2PO_4^{3-}$	10^{-32} (approx)	32 (approx)
	$Pb_3(PO_4)_2 \rightleftharpoons 3Pb^{2+} + 2PO_4^{3-}$	1.0×10^{-32}	32.0
Sulfate	$Ag_2SO_4 \rightleftharpoons 2Ag^{+} + SO_4^{2-}$	6.4×10^{-5}	4.2
	$CaSO_4 \rightleftharpoons Ca^{2+} + SO_4^{2-}$	2.5×10^{-5}	4.8
	$Hg_2SO_4 \rightleftharpoons Hg_2^{2+} + SO_4^{2-}$	5.0×10^{-7}	6.3
	$SrSO_4 \rightleftharpoons Sr^{2+} + SO_4^{2-}$	3.2×10^{-7}	6.5
	$PbSO_4 \rightleftharpoons Pb^{2+} + SO_4^{2-}$	1.3×10^{-8}	7.9
	$BaSO_4 \rightleftharpoons Ba^{2+} + SO_4^{2-}$	7.9×10^{-11}	10.1
Sulfide	$MnS(pink) \rightleftharpoons Mn^{2+} + S^{2-}$	7.0×10^{-16}	15.2
	$FeS \rightleftharpoons Fe^{2+} + S^{2-}$	4.0×10^{-19}	18.4
	$NiS(\alpha) \rightleftharpoons Ni^{2+} + S^{2-}$	3.0×10^{-21}	20.5
	$CoS(\alpha) \rightleftharpoons Co^{2+} + S^{2-}$	5.0×10^{-22}	21.3
	$ZnS \rightleftharpoons Zn^{2+} + S^{2-}$	1.6×10^{-23}	22.8
	$SnS \rightleftharpoons Sn^{2+} + S^{2-}$	1.0×10^{-26}	26.0
	$CdS \rightleftharpoons Cd^{2+} + S^{2-}$	1.0×10^{-28}	28.0
	$PbS \rightleftharpoons Pb^{2+} + S^{2-}$	7.0×10^{-29}	28.2
	$CuS \rightleftharpoons Cu^{2+} + S^{2-}$	8.0×10^{-37}	36.1
	$Cu_2S \rightleftharpoons 2Cu^{+} + S^{2-}$	1.2×10^{-49}	48.9
	$Ag_2S \rightleftharpoons 2Ag^{+} + S^{2-}$	5.5×10^{-51}	50.3
	$HgS \rightleftharpoons Hg^{2+} + S^{2-}$	1.6×10^{-54}	53.8
	$Fe_2S_3 \rightleftharpoons 2Fe^{3+} + 3S^{2-}$	1×10^{-88}	88.0

Anion	*Equilibrium equation*	K_{sp}	pK_{sp}
Sulfite	$\mathbf{CaSO_3} \rightleftharpoons Ca^{2+} + SO_3^{2-}$	1.6×10^{-5}	4.8
	$\mathbf{BaSO_3} \rightleftharpoons Ba^{2+} + SO_3^{2-}$	1.0×10^{-8}	8.0
Thiocyanate	$\mathbf{AgNCS} \rightleftharpoons Ag^{+} + NCS^{-}$	1×10^{-12}	12.0
	$\mathbf{CuNCS} \rightleftharpoons Cu^{+} + NCS^{-}$	4×10^{-14}	13.4
	$\mathbf{Hg_2(NCS)_2} \rightleftharpoons Hg_2^{2+} + 2NCS^{-}$	3×10^{-20}	19.5
Thiosulfate	$\mathbf{BaS_2O_3} \rightleftharpoons Ba^{2+} + S_2O_3^{2-}$	1.6×10^{-5}	4.8

* Data taken largely from L. G. Sillén and A. E. Martell, *Stability Constants of Metal-Ion Complexes*, Special Publication No. 17, The Chemical Society, London (1964).

† Reliable data not available.

Appendix F: Formation Constants of Complexes*†

Donor group	Equilibrium equation	Stepwise		Overall	
		K_f	$\log K_f$	$\beta_n(n)$	$\log \beta_n(n)$
CH_3COO^-	$Pb^{2+} + 4CH_3COO^- \rightleftharpoons [Pb(CH_3COO)_4]^{2-}$			1×10^2 (4) (approx)	2 (4) (approx)
NH_3	$Co^{2+} + 6NH_3 \rightleftharpoons [Co(NH_3)_6]^{2+}$			5.0×10^4 (6)	4.90 (6)
	$Ag^+ + NH_3 \rightleftharpoons [Ag(NH_3)]^+$	2.0×10^3	3.31		
	$[Ag(NH_3)]^+ + NH_3 \rightleftharpoons [Ag(NH_3)_2]^+$	7.9×10^3	3.90		
	$Ag^+ + 2NH_3 \rightleftharpoons [Ag(NH_3)_2]^+$			1.6×10^7 (2)	7.21 (2)
	$Cd^{2+} + 4NH_3 \rightleftharpoons [Cd(NH_3)_4]^{2+}$			1.4×10^6 (4)	6.14 (4)
	$Ni^{2+} + 6NH_3 \rightleftharpoons [Ni(NH_3)_6]^{2+}$			2.0×10^8 (6)	8.30 (6)
	$Zn^{2+} + 4NH_3 \rightleftharpoons [Zn(NH_3)_4]^{2+}$			2.8×10^9 (4)	9.45 (4)
	$Cu^{2+} + NH_3 \rightleftharpoons [Cu(NH_3)]^{2+}$	1.9×10^4	4.27		
	$[Cu(NH_3)]^{2+} + NH_3 \rightleftharpoons [Cu(NH_3)_2]^{2+}$	3.9×10^3	3.59		
	$[Cu(NH_3)_2]^{2+} + NH_3 \rightleftharpoons [Cu(NH_3)_3]^{2+}$	1.0×10^3	3.00		
	$[Cu(NH_3)_3]^{2+} + NH_3 \rightleftharpoons [Cu(NH_3)_4]^{2+}$	1.5×10^2	2.19		
	$Cu^{2+} + 4NH_3 \rightleftharpoons [Cu(NH_3)_4]^{2+}$			1.1×10^{13} (4)	13.05 (4)
	$Hg^{2+} + 4NH_3 \rightleftharpoons [Hg(NH_3)_4]^{2+}$			1.8×10^{19} (4)	19.26 (4)
	$Co^{3+} + 6NH_3 \rightleftharpoons [Co(NH_3)_6]^{3+}$			4.6×10^{33} (6)	33.66 (6)
Br^-	$Cd^{2+} + 4Br^- \rightleftharpoons [CdBr_4]^{2-}$			8.5×10^2 (4)	2.93 (4)
	$Hg^{2+} + 4Br^- \rightleftharpoons [HgBr_4]^{2-}$			1.0×10^{21} (4)	21.00 (4)
Cl^-	$Sn^{4+} + 6Cl^- \rightleftharpoons [SnCl_6]^{2-}$			6.6 (6)	0.82 (6)
	$Pb^{2+} + 3Cl^- \rightleftharpoons [PbCl_3]^-$			2.5×10^1 (3)	1.4 (3)
	$Sn^{2+} + 4Cl^- \rightleftharpoons [SnCl_4]^{2-}$			1.8×10^1 (4)	1.25 (4)
	$Cd^{2+} + 4Cl^- \rightleftharpoons [CdCl_4]^{2-}$			4.0×10^1 (4)	1.6 (4)
	$Hg^{2+} + Cl^- \rightleftharpoons [HgCl]^+$	5.5×10^6	6.74		
	$[HgCl]^+ + Cl^- \rightleftharpoons [HgCl_2]$	3.0×10^6	6.48		
	$[HgCl_2] + Cl^- \rightleftharpoons [HgCl_3]^-$	7.1	0.85		
	$[HgCl_3]^- + Cl^- \rightleftharpoons [HgCl_4]^{2-}$	1.0×10^1	1.00		
	$Hg^{2+} + Cl^- \rightleftharpoons [HgCl_4]^{2-}$			5.0×10^{15} (4)	15.07 (4)

CN^-	$Cd^{2+} + 4CN^- \rightleftharpoons [Cd(CN)_4]^{2-}$			5.9×10^{18} (4)	18.77 (4)
	$Cu^+ + 4CN^- \rightleftharpoons [Cu(CN)_4]^{4-}$			3.6×10^{27} (4)	27.56 (4)
	$Fe^{2+} + 6CN^- \rightleftharpoons [Fe(CN)_6]^{4-}$			1.0×10^{24} (6)	24.0 (6)
	$Fe^{3+} + 6CN^- \rightleftharpoons [Fe(CN)_6]^{3-}$			1.0×10^{31} (6)	31.0 (6)
F^-	$Fe^{3+} + F^- \rightleftharpoons [FeF]^{2+}$	1.4×10^5	5.17		
	$[FeF]^{2+} + F^- \rightleftharpoons [FeF_2]^+$	8.3×10^3	3.92		
	$[FeF_2]^+ + F^- \rightleftharpoons [FeF_3]$	8.1×10^2	2.91		
	$Fe^{3+} + 3F \rightleftharpoons [FeF_3]$			1.0×10^{12} (3)	12.00 (3)
	$Sn^{4+} + 6F^- \rightleftharpoons [SnF_6]^{2-}$			1×10^{25} (6) (approx)	25 (6) (approx)
	$Al^{3+} + 6F^- \rightleftharpoons [AlF_6]^{3-}$			1×10^{20} (6) (approx)	20 (6) (approx)
I^-	$Pb^{2+} + 4I^- \rightleftharpoons [PbI_4]^{2-}$			1.6×10^6 (4)	6.20 (4)
	$Cd^{2+} + 4I^- \rightleftharpoons [CdI_4]^{2-}$			2.5×10^6 (4)	6.40 (4)
	$Hg^{2+} + 4I^- \rightleftharpoons [HgI_4]^{2-}$			1.9×10^{30} (4)	30.28 (4)
$C_2O_4{}^{2-}$	$Al^3 + 3C_2O_4{}^{2-} \rightleftharpoons [Al(C_2O_4)_3]^{3-}$			2.0×10^{16} (3)	16.3 (3)
	$Fe^{3+} + 3C_2O_4{}^{2-} \rightleftharpoons [Fe(C_2O_4)_3]^{3-}$			4.0×10^{21} (3)	21.6 (3)
S^{2-}	$Hg^{2+} + 2S^{2-} \rightleftharpoons [HgS_2]^{2-}$			3.5×10^{54} (2)	54.7 (2)
$SO_3{}^{2-}$	$Ag^+ + 2SO_3{}^{2-} \rightleftharpoons [Ag(SO_3)_2]^{3-}$			2.5×10^8 (2)	8.40 (2)
NCS^-	$Fe^{3+} + NCS^- \rightleftharpoons [Fe(NCS)]^{2+}$			1.1×10^3 (1)	3.03 (1)
	$Hg^{2+} + 4NCS^- \rightleftharpoons [Hg(NCS)_4]^{2-}$			7.8×10^{21} (4)	21.89 (4)
$S_2O_3{}^{2-}$	$Ag^+ + 2S_2O_3{}^{2-} \rightleftharpoons [Ag(S_2O_3)_2]^{3-}$			2.9×10^{13} (2)	13.46 (2)

* Data taken largely from L. G. Sillén and A. E. Martell, *Stability Constants of Metal-Ion Complexes*, Special Publication No. 17, The Chemical Society, London (1964).

† Coordinated water molecules not shown, for example, $[Cu(NH_3)_4]^{2+}$ better represented as $[Cu(NH_3)_4(H_2O)_2]^{2+}$ (Chap. 6).

Appendix G: Oxidation Potentials

Couple	*Equilibrium equation*	E^0_{298},* *volts*
	I. *Acidic Solutions*	
K(0)–K(I)	**K** $\rightleftharpoons K^+ + e^-$	+2.925
Ba(0)–Ba(II)	**Ba** $\rightleftharpoons Ba^{2+} + 2e^-$	2.90
Sr(0)–Sr(II)	**Sr** $\rightleftharpoons Sr^{2+} + 2e^-$	2.89
Ca(0)–Ca(II)	**Ca** $\rightleftharpoons Ca^{2+} + 2e^-$	2.87
Na(0)–Na(I)	**Na** $\rightleftharpoons Na^+ + e^-$	2.714
Mg(0)–Mg(II)	**Mg** $\rightleftharpoons Mg^{2+} + 2e^-$	2.37
Al(0)–Al(III)	**Al** $\rightleftharpoons Al^{3+} + 3e^-$	1.66
Mn(0)–Mn(II)	**Mn** $\rightleftharpoons Mn^{2+} + 2e^-$	1.18
Cr(0)–Cr(II)	**Cr** $\rightleftharpoons Cr^{2+} + 2e^-$	0.91
B(0)–B(III)	**B** $+ 3H_2O \rightleftharpoons H_3BO_3 + 3H^+ + 3e^-$	0.87
Si(0)–Si(IV)	**Si** $+ 3H_2O \rightleftharpoons$ $\mathbf{H_2SiO_3}$ $+ 4H^+ + 4e^-$	0.84
Co(II)–Co(III) (as CN⁻)	$[Co(CN)_6]^{4-} \rightleftharpoons [Co(CN_6]^{3-} + e^-$	0.83
Zn(0)–Zn(II)	**Zn** $\rightleftharpoons Zn^{2+} + 2e^-$	0.763
Cr(0)–Cr(III)	**Cr** $\rightleftharpoons Cr^{3+} + 3e^-$	0.74
C(III)–C(IV)	$H_2C_2O_4 \rightleftharpoons 2CO_2 + 2H^+ + 2e^-$	0.49

Couple	*Equilibrium equation*	E^0_{298},* *volts*
	I. *Acidic Solutions*	
Fe(0)–Fe(II)	$\mathbf{Fe} \rightleftharpoons Fe^{2+} + 2e^-$	0.440
Cr(II)–Cr(III)	$Cr^{2+} \rightleftharpoons Cr^{3+} + e^-$	0.41
Cd(0)–Cd(II)	$\mathbf{Cd} \rightleftharpoons Cd^{2+} + 2e^-$	0.403
Co(0)–Co(II)	$\mathbf{Co} \rightleftharpoons Co^{2+} + 2e^-$	0.277
P(III)–P(V)	$H_3PO_3 + H_2O \rightleftharpoons H_3PO_4 + 2H^+ + 2e^-$	0.276
Ni(0)–Ni(II)	$\mathbf{Ni} \rightleftharpoons Ni^{2+} + 2e^-$	0.250
Sn(0)–Sn(II)	$\mathbf{Sn} \rightleftharpoons Sn^{2+} + 2e^-$	0.136
Pb(0)–Pb(II)	$\mathbf{Pb} \rightleftharpoons Pb^{2+} + 2e^-$	0.126
H(0)–H(I)	$H_2 \rightleftharpoons 2H^+ + 2e^-$	0.000
S(–II)–S(0)	$H_2S \rightleftharpoons 2H^+ + \mathbf{S} + 2e^-$	–0.141
Sn(II)–Sn(IV)	$Sn^{2+} \rightleftharpoons Sn^{4+} + 2e^-$	–0.15
Cu(I)–Cu(II)	$Cu^+ \rightleftharpoons Cu^{2+} + e^-$	–0.153
S(IV)–S(VI)	$H_2SO_3 + H_2O \rightleftharpoons SO_4^{2-} + 4H^+ + 2e^-$	–0.17
Sb(0)–Sb(III)	$\mathbf{Sb} + H_2O \rightleftharpoons SbO^+ + 2H^+ + 3e^-$	–0.212
As(0)–As(III)	$\mathbf{As} + 2H_2O \rightleftharpoons HAsO_2 + 3H^+ + 3e^-$	–0.248
Bi(0)–Bi(III)	$\mathbf{Bi} + H_2O \rightleftharpoons BiO^+ + 2H^+ + 3e^-$	–0.32
Cu(0)–Cu(II)	$\mathbf{Cu} \rightleftharpoons Cu^{2+} + 2e^-$	–0.337
Fe(II)–Fe(III) (as CN⁻)	$[Fe(CN)_6]^{4-} \rightleftharpoons [Fe(CN)_6]^{3-} + e^-$	–0.36
CN(–I)–CN(0)	$2HCN \rightleftharpoons (CN)_2 + 2H^+ + 2e^-$	–0.37
S(II)–S(IV)	$S_2O_3^{2-} + 3H_2O \rightleftharpoons 2H_2SO_3 + 2H^+ + 4e^-$	–0.40
S(0)–S(IV)	$\mathbf{S} + 3H_2O \rightleftharpoons H_2SO_3 + 4H^+ + 4e^-$	–0.45
Cu(0)–Cu(I)	$\mathbf{Cu} \rightleftharpoons Cu^+ + e^-$	–0.521
I(–I)–I(0)	$2I^- \rightleftharpoons I_2 + 2e^-$	–0.5355
As(III)–As(V)	$HAsO_2 + 2H_2O \rightleftharpoons H_3AsO_4 + 2H^+ + 2e^-$	–0.559
Mn(VI)–Mn(VII)	$MnO_4^{2-} \rightleftharpoons MnO_4^- + e^-$	–0.564
O(–I)–O(0)	$H_2O_2 \rightleftharpoons O_2 + 2H^+ + 2e^-$	–0.682
NCS(–I)–NCS(0)	$2NCS^- \rightleftharpoons (NCS)_2 + 2e^-$	–0.77
Fe(II)–Fe(III)	$Fe^{2+} \rightleftharpoons Fe^{3+} + e^-$	–0.771
Hg(0)–Hg(I)	$2Hg \rightleftharpoons Hg_2^{2+} + 2e^-$	–0.789
Ag(0)–Ag(I)	$\mathbf{Ag} \rightleftharpoons Ag^+ + e^-$	–0.7991
N(IV)–N(V)	$N_2O_4 + 2H_2O \rightleftharpoons 2NO_3^- + 4H^+ + 2e^-$	–0.80
Hg(0)–Hg(II)	$Hg \rightleftharpoons Hg^{2+} + 2e$	–0.854
Hg(I)–Hg(II)	$Hg_2^{2+} \rightleftharpoons Hg^{2+} + 2e^-$	–0.920
N(III)–N(V)	$HNO_2 + H_2O \rightleftharpoons NO_3^- + 3H^+ + 2e^-$	–0.94
N(II)–N(V)	$NO + 2H_2O \rightleftharpoons NO_3^- + 4H^+ + 3e^-$	–0.96
N(II)–N(III)	$NO + H_2O \rightleftharpoons HNO_2 + H^+ + e^-$	–0.996
N(II)–N(IV)	$2NO + 2H_2O \rightleftharpoons N_2O_4 + 4H^+ + 4e^-$	–1.03
Br(–I)–Br(0)	$2Br^- \rightleftharpoons Br_2 + 2e^-$	–1.087
I(–I)–I(V)	$I^- + 3H_2O \rightleftharpoons IO_3^- + 6H^+ + 6e^-$	–1.09
Cl(V)–Cl(VII)	$ClO_3^- + H_2O \rightleftharpoons ClO_4^- + 2H^+ + 2e^-$	–1.19
Cl(III)–Cl(V)	$HClO_2 + H_2O \rightleftharpoons ClO_3^- + 3H^+ + 2e^-$	–1.21
O(–II)–O(0)	$2H_2O \rightleftharpoons O_2 + 4H^+ + 4e^-$	–1.229
Mn(II)–Mn(IV)	$Mn^{2+} + 2H_2O \rightleftharpoons \mathbf{MnO_2} + 4H^+ + 2e^-$	–1.23
Cr(III)–Cr(VI)	$2Cr^{3+} + 7H_2O \rightleftharpoons Cr_2O_7^{2-} + 14H^+ + 6e^-$	–1.33
Cl(–I)–Cl(0)	$2Cl^- \rightleftharpoons Cl_2 + 2e^-$	–1.3595
Pb(II)–Pb(IV)	$Pb^{2+} + 2H_2O \rightleftharpoons \mathbf{PbO_2} + 4H^+ + 2e^-$	–1.455

Couple	*Equilibrium equation*	E^0_{298},* *volts*
	I. *Acidic Solutions*	
Mn(II)–Mn(III)	$Mn^{2+} \rightleftharpoons Mn^{3+} + e^-$	−1.51
Mn(II)–Mn(VII)	$Mn^{2+} + 4H_2O \rightleftharpoons MnO_4^- + 8H^+ + 5e^-$	−1.51
Bi(III)–Bi(V)	$2BiO^+ + 3H_2O \rightleftharpoons \mathbf{Bi_2O_5} + 6H^+ + 4e^-$	−1.6 (approx)
Cl(0)–Cl(I)	$Cl_2 + 2H_2O \rightleftharpoons 2HClO + 2H^+ + 2e^-$	−1.63
Cl(I)–Cl(III)	$HClO + H_2O \rightleftharpoons HClO_2 + 2H^+ + 2e^-$	−1.64
Ni(II)–Ni(IV)	$Ni^{2+} + 2H_2O \rightleftharpoons \mathbf{NiO_2} + 4H^+ + 2e^-$	−1.68
Mn(IV)–Mn(VII)	$\mathbf{MnO_2} + 2H_2O \rightleftharpoons MnO_4^- + 4H^+ + 3e^-$	−1.695
O(−II)–O(−I)	$2H_2O \rightleftharpoons H_2O_2 + 2H^+ + 2e^-$	−1.77
Co(II)–Co(III)	$Co^{2+} \rightleftharpoons Co^{3+} + e^-$	−1.842
Ag(I)–Ag(II)	$Ag^+ \rightleftharpoons Ag^{2+} + e^-$	−1.98
S(VI)–S(VI) (peroxo)	$2SO_4^{2-} \rightleftharpoons S_2O_8^{2-} + 2e^-$	−2.01
F(−I)–F(0)	$2F^- \rightleftharpoons F_2 + 2e^-$	−2.87
	$2HF \rightleftharpoons F_2 + 2H^+ + 2e^-$	−3.06
	II. *Alkaline Solutions*	
Ca(0)–Ca(II)	$\mathbf{Ca} + 2OH^- \rightleftharpoons \mathbf{Ca(OH)_2} + 2e^-$	+3.03
Sr(0)–Sr(II)	$\mathbf{Sr} + 2OH^- + 8H_2O \rightleftharpoons \mathbf{Sr(OH)_2 \cdot 8H_2O} + 2e^-$	2.99
Ba(0)–Ba(II)	$\mathbf{Ba} + 2OH^- + 8H_2O \rightleftharpoons \mathbf{Ba(OH)_2 \cdot 8H_2O} + 2e^-$	2.97
K(0)–K(I)	$\mathbf{K} \rightleftharpoons K^+ + e^-$	2.925
Na(0)–Na(I)	$\mathbf{Na} \rightleftharpoons Na^+ + e^-$	2.714
Mg(0)–Mg(II)	$\mathbf{Mg} + 2OH^- \rightleftharpoons \mathbf{Mg(OH)_2} + 2e^-$	2.69
Al(0)–Al(III)	$\mathbf{Al} + 3OH^- \rightleftharpoons \mathbf{Al(OH)_3} + 3e^-$	2.31
B(0)–B(III)	$\mathbf{B} + 4OH^- \rightleftharpoons H_2BO_3^- + H_2O + 3e^-$	1.79
Si(0)–Si(IV)	$\mathbf{Si} + 6OH^- \rightleftharpoons SiO_3^{2-} + 3H_2O + 4e^-$	1.70
Mn(0)–Mn(II)	$\mathbf{Mn} + 2OH^- \rightleftharpoons \mathbf{Mn(OH)_2} + 2e^-$	1.55
Cr(0)–Cr(II)	$\mathbf{Cr} + 2OH^- \rightleftharpoons \mathbf{Cr(OH)_2} + 2e^-$	1.41
Zn(0)–Zn(II)	$\mathbf{Zn} + 4OH^- \rightleftharpoons ZnO_2^{2-} + 2H_2O + 2e^-$	1.216
Cr(0)–Cr(III)	$\mathbf{Cr} + 4OH^- \rightleftharpoons CrO_2^- + 2H_2O + 3e^-$	1.2
P(III)–P(V)	$HPO_3^{2-} + 3OH^- \rightleftharpoons PO_4^{3-} + 2H_2O + 2e^-$	1.12
Cr(II)–Cr(III)	$\mathbf{Cr(OH)_2} + 2OH^- \rightleftharpoons CrO_2^- + 2H_2O + e^-$	1.1
S(IV)–S(VI)	$SO_3^{2-} + 2OH^- \rightleftharpoons SO_4^{2-} + H_2O + 2e^-$	0.93
Sn(II)–Sn(IV)	$HSnO_2^- + 3OH^- + H_2O \rightleftharpoons [Sn(OH)_6]^{2-} + 2e^-$	0.93
Sn(0)–Sn(II)	$\mathbf{Sn} + 3OH^- \rightleftharpoons HSnO_2^- + H_2O + 2e^-$	0.91
Fe(0)–Fe(II)	$\mathbf{Fe} + 2OH^- \rightleftharpoons \mathbf{Fe(OH)_2} + 2e^-$	0.877
Cd(0)–Cd(II)	$\mathbf{Cd} + 2OH^- \rightleftharpoons \mathbf{Cd(OH)_2} + 2e^-$	0.809
Co(0)–Co(II)	$\mathbf{Co} + 2OH^- \rightleftharpoons \mathbf{Co(OH)_2} + 2e^-$	0.73
Ni(0)–Ni(II)	$\mathbf{Ni} + 2OH^- \rightleftharpoons \mathbf{Ni(OH)_2} + 2e^-$	0.72
As(III)–As(V)	$AsO_2^- + 4OH^- \rightleftharpoons AsO_4^{3-} + 2H_2O + 2e^-$	0.67
Sb(0)–Sb(III)	$\mathbf{Sb} + 4OH^- \rightleftharpoons SbO_2^- + 2H_2O + 3e^-$	0.66
S(II)–S(IV)	$S_2O_3^{2-} + 6OH^- \rightleftharpoons 2SO_3^{2-} + 3H_2O + 4e^-$	0.58
Fe(II)–Fe(III)	$\mathbf{Fe(OH)_2} + OH^- \rightleftharpoons \mathbf{Fe(OH)_3} + e^-$	0.56
S(−II)–S(0)	$S^{2-} \rightleftharpoons \mathbf{S} + 2e^-$	0.48
N(II)–N(III)	$NO + 2OH^- \rightleftharpoons NO_2^- + H_2O + e^-$	0.46
Bi(0)–Bi(III)	$\mathbf{2Bi} + 6OH^- \rightleftharpoons \mathbf{Bi_2O_3} + 3H_2O + 6e^-$	0.46

Couple	*Equilibrium equation*	E^0_{298},* *volts*
	II. *Alkaline Solutions*	
Cu(0)–Cu(I)	$2\mathbf{Cu} + 2OH^- \rightleftharpoons \mathbf{Cu_2O} + H_2O + 2e^-$	0.358
Cu(0)–Cu(II)	$\mathbf{Cu} + 2OH^- \rightleftharpoons \mathbf{Cu(OH)_2} + 2e^-$	0.258
Cr(III)–Cr(VI)	$\mathbf{Cr(OH)_3} + 5OH^- \rightleftharpoons CrO_4^{2-} + 4H_2O + 3e^-$	0.13
Cu(I)–Cu(II)	$\mathbf{Cu_2O} + 2OH^- + H_2O \rightleftharpoons 2\mathbf{Cu(OH)_2} + 2e^-$	0.08
Mn(II)–Mn(IV)	$\mathbf{Mn(OH)_2} + 2OH^- \rightleftharpoons \mathbf{MnO_2} + 2H_2O + 2e^-$	0.05
N(III)–N(V)	$NO_2^- + 2OH^- \rightleftharpoons NO_3^- + H_2O + 2e^-$	−0.01
Hg(0)–Hg(II)	$Hg + 2OH^- \rightleftharpoons \mathbf{HgO} + H_2O + 2e^-$	−0.098
Co(II)–Co(III) (as ammine)	$[Co(NH_3)_6]^{2+} \rightleftharpoons [Co(NH_3)_6]^{3+} + e^-$	−0.1
Hg(0)–Hg(I)	$2Hg + 2OH^- \rightleftharpoons \mathbf{Hg_2O} + H_2O + 2e^-$	−0.123
Co(II)–Co(III)	$\mathbf{Co(OH)_2} + OH^- \rightleftharpoons \mathbf{Co(OH)_3} + e^-$	−0.17
Pb(II)–Pb(IV)	$\mathbf{PbO} + 2OH^- \rightleftharpoons \mathbf{PbO_2} + H_2O + 2e^-$	−0.248
Cl(III)–Cl(V)	$ClO_2^- + 2OH^- \rightleftharpoons ClO_3^- + H_2O + 2e^-$	−0.33
Ag(0)–Ag(I)	$2\mathbf{Ag} + 2OH^- \rightleftharpoons \mathbf{Ag_2O} + H_2O + 2e^-$	−0.344
Cl(V)–Cl(VII)	$ClO_3^- + 2OH^- \rightleftharpoons ClO_4^- + H_2O + 2e^-$	−0.36
O(−II)–O(0)	$4OH^- \rightleftharpoons O_2 + 2H_2O + 4e^-$	−0.401
Ni(II)–Ni(IV)	$\mathbf{Ni(OH)_2} + 2OH^- \rightleftharpoons \mathbf{NiO_2} + 2H_2O + 2e^-$	−0.49
Ag(I)–Ag(II)	$\mathbf{Ag_2O} + 2OH^- \rightleftharpoons 2\mathbf{AgO} + H_2O + 2e^-$	−0.57
Cl(I)–Cl(III)	$ClO^- + 2OH^- \rightleftharpoons ClO_2^- + H_2O + 2e^-$	−0.66
Br(−I)–Br(I)	$Br^- + 2OH^- \rightleftharpoons BrO^- + H_2O + 2e^-$	−0.76
O(−II)–O(−I)	$3OH^- \rightleftharpoons HO_2^- + H_2O + 2e^-$	−0.88
Cl(−I)–Cl(I)	$Cl^- + 2OH^- \rightleftharpoons ClO^- + H_2O + 2e^-$	−0.89

* Values from W. M. Latimer, "The Oxidation States of the Elements and Their Potentials in Aqueous Solutions," 2d ed., Prentice-Hall, Inc., Englewood Cliffs, N.J., 1952. The standard potentials listed follow the so-called "American Convention" in which the sign of the potential is changed when the equation for the half-reaction is written for reduction rather than oxidation (as shown in the table). There is another convention in which the sign of the potential is invariant. In the latter case the potential listed is the actual experimental value of the electrode potential as measured by comparison with the standard hydrogen electrode. These potentials have the opposite sign from those given above, whether the equation for the half-reaction is written as oxidation or reduction.

Appendix H

Appendix H_a: Reactions of Anions

Anion	*6 M* $HClO_4$	*6 M* $HClO_4$ + $AgNO_3$	*Neutral solution* + $AgNO_3$	*18 M* H_2SO_4*
Group I anions				
CO_3^{2-}	$CO_2\uparrow$ (colorless)	$CO_2\uparrow$	**Ag_2CO_3** (yellowish)	$CO_2\uparrow$ (c or h)†
NO_2^-	$NO\uparrow + NO_2\uparrow$ (brown)	$NO\uparrow + NO_2\uparrow$	**$AgNO_2$**‡ (pale yellow)	$NO\uparrow + NO_2\uparrow$ (c, h)
S^{2-}	$H_2S\uparrow$ (colorless)	**Ag_2S** (black)	**Ag_2S**	$H_2S\uparrow$ (c, h)
SO_3^{2-}	$SO_2\uparrow$ (colorless)	$SO_2\uparrow$	**Ag_2SO_3** (white) $\xrightarrow{\text{Heat}}$ **Ag** (black)	$SO_2\uparrow$ (c, h)

12 M HCl + $MnCl_2$	HCl + $K_3[Fe(CN)_6]$ + $FeCl_3$	*Specific reactions*
$CO_2\uparrow$	$CO_2\uparrow$	$CO_2 \xrightarrow{Ba(OH)_2} \mathbf{BaCO_3}$ (white)
Brown to black	$\mathbf{KFe[Fe(CN)_6]}$ (blue)	$NO_2^- \xrightarrow[H_2SO_4]{FeSO_4} [Fe(NO)]^{2+}$ (brown)
		$NO_2^- \xrightarrow[H^+,\ Fe^{3+}]{SC(NH_2)_2} Fe(NCS)^{2+}$ (red)
$H_2S\uparrow$	$\mathbf{KFe[Fe(CN)_6]}$	$H_2S \xrightarrow[(NaOH)]{NaHPbO_2} \mathbf{PbS}$ (silver to brown)
$SO_2\uparrow$	$\mathbf{KFe[Fe(CN)_6]}$	$SO_2 \xrightarrow[Ba(NO_3)_2,\ HNO_3]{KMnO_4} \mathbf{BaSO_4}$ (white)

Anion	*6 M* $HClO_4$	*6 M* $HClO_4$ + $AgNO_3$	*Neutral solution* + $AgNO_3$	*18 M* H_2SO_4*
$S_2O_3^{2-}$	$SO_2\uparrow$ + S_8 (white)	$SO_2\uparrow$ + S_8	**$Ag_2S_2O_3$** → **Ag_2S** (white) (black)	$SO_2\uparrow$ + **S_8**(c, h)
Group II anions				
Br^-	N.R.§	**AgBr** (cream)	**AgBr**	$HBr\uparrow$ (colorless) → $Br_2\uparrow$ (brown) (c, h)
Cl^-	N.R.	**AgCl** (white)	**AgCl**	$HCl\uparrow$ (colorless) (c, h)
I^-	N.R.	**AgI** (yellow)	**AgI**	HI (colorless) → I_2 (black solid, purple gas)
Group III anions				
AsO_4^{3-}	N.R.	N.R.	**Ag_3AsO_4** (red-brown)	N.R.
CrO_4^{2-}	$Cr_2O_7^{2-}$ (orange)	$Cr_2O_7^{2-}$	**Ag_2CrO_4** (brownish red)	$Cr_2O_7^{2-}$ → **CrO_3** (red) (c, h)
$C_2O_4^{2-}$	N.R.	N.R.	**$Ag_2C_2O_4$** (white)	$CO\uparrow$ + $CO_2\uparrow$ (colorless) (h)

12 M HCl + $MnCl_2$	HCl + $K_3[Fe(CN)_6]$ + $FeCl_3$	*Specific reactions*
$SO_2\uparrow + S_8$	$\mathbf{KFe[Fe(CN)_6]}$	SO_3^{2-}–free $S_2O_3^{2-} \xrightarrow{HCl} SO_2\uparrow + S_8$ $S_2O_3^{2-} \xrightarrow[(OH^-)]{[Ni(en)_3]^{3+}} \mathbf{[Ni(en)_3]S_2O_3}$ (violet)
N.R.	N.R.	$Br^- \xrightarrow[CCl_4\ (H^+)]{Cl_2} Br_2$ (yellow to brown in CCl_4)
N.R.	N.R.	$Cl^- \xrightarrow[H_2SO_4]{K_2S_2O_8} I_2\uparrow + Br_2\uparrow + Cl^- \xrightarrow{AgNO_3} \mathbf{AgCl}$ $\mathbf{MCl} \xrightarrow[H_2SO_4]{K_2Cr_2O_7} CrO_2Cl_2\uparrow$ (brown) $+ Br_2 + I_2 \xrightarrow{NaOH} Na_2CrO_4$ (yellow)
N.R.	$\mathbf{KFe[Fe(CN)_6]}$ (blue)	$I^- \xrightarrow[CCl_4\ (H^+)]{Cl_2} I_2$ (violet in CCl_4) $I^- \xrightarrow[CCl_4\ (HCl)]{NaNO_2} I_2$ (violet in CCl_4)
N.R.	N.R.	$AsO_4^{3-} \xrightarrow[HCl]{SnCl_2} \mathbf{As}$ (brown to black) $AsO_4^{3-} \xrightarrow[HCl]{CH_3C(S)NH_2} \mathbf{As_2S_3}$ (yellow) $AsO_4^{3-} \xrightarrow[NH_4NO_3,\ NH_3]{Mg(NO_3)_2} \mathbf{MgNH_4AsO_4{\cdot}6H_2O}$ (white) $\xrightarrow[AgNO_3]{CH_3COOH} \mathbf{Ag_3AsO_4}$
Brown to black	N.R.	$CrO_4^{2-} \xrightarrow[Ba(CH_3COO)_2]{CH_3COOH} \mathbf{BaCrO_4}$ (yellow) $\xrightarrow[H_2O_2]{HNO_3} CrO_5$ (blue)
N.R.	N.R.	$C_2O_4^{2-} \xrightarrow[CH_3COOH]{Ca(NO_3)_2} \mathbf{CaC_2O_4}$ (white) $\xrightarrow[KMnO_4]{HNO_3}$ colorless soln

Anion	*6 M* $HClO_4$	*6 M* $HClO_4$ + $AgNO_3$	*Neutral solution* + $AgNO_3$	*18 M* H_2SO_4*
PO_4^{3-}	N.R.	N.R.	**Ag_3PO_4** (yellow)	N.R.
Group IV anions				
CH_3COO^-	N.R.	N.R.	**$Ag(CH_3COO)$**‡ (white)	$CH_3COOH\uparrow$ (h)
F^-	N.R.	N.R.	N.R.	$HF\uparrow$ (colorless)
NO_3^-	N.R.	N.R.	N.R.	$HNO_3\uparrow$ (colorless) → $NO_2\uparrow$ (brown) (h)
MnO_4^-	N.R.	N.R.	N.R.	$Mn_2O_7\uparrow$ (explodes)
SO_4^{2-}	N.R.	N.R.	N.R.	N.R.

* Added to solid.
† c = cold, h = hot.
‡ Precipitate only if concentration of anion exceeds 5 mg ml^{-1} (approx).
§ N.R. = no observable reaction.

12 M HCl + $MnCl_2$	HCl + $K_3[Fe(CN)_6]$ + $FeCl_3$	*Specific reactions*
N.R.	N.R.	$PO_4^{3-} \xrightarrow[NH_4NO_3,\ NH_3]{Mg(NO_3)_2}$ **$MgNH_4PO_4 \cdot 6H_2O$** (white) $\xrightarrow[AgNO_3]{CH_3COOH}$ **Ag_3PO_4** (yellow) $PO_4^{3-} \xrightarrow[HNO_3,\ \Delta]{(NH_4)_2MoO_4}$ **$(NH_4)_3[P(Mo_{12}O_{40})]$** (yellow)
N.R.	N.R.	$CH_3COO^- \xrightarrow[18\ M\ H_2SO_4]{C_2H_5OH} CH_3COOC_2H_5 \uparrow$ (pleasant odor)
N.R.	N.R.	**MF** $\xrightarrow[H_2SO_4]{SiO_2} SiF_4 \uparrow$ (colorless) $\xrightarrow{H_2O}$ **SiO_2** (white)
Brown to black	N.R.	$NO_3^- \xrightarrow[18\ M\ H_2SO_4]{FeSO_4} [Fe(NO)]^+$ (brown) $NO_3^- \xrightarrow[NaOH]{Zn} NH_3 \uparrow$
Brown to black	N.R.	MnO_4^- (violet or purple) $\xrightarrow{H_2O_2}$ Soln (colorless)
N.R.	N.R.	$SO_4^{2-} \xrightarrow[BaCl_2]{HCl}$ **$BaSO_4$** (white)

Appendix H_c: Reactions of Cations

Cation	*1 M* HCl*	*0.3 M* HCl + $CH_3C(S)NH_2$	*Aq* NH_3 + NH_4^+	*Aq* NH_3 + NH_4^+ + $CH_3C(S)NH_2$
Group I cations				
Pb^{2+}	**$PbCl_2$**(c)† $\xrightarrow{x\text{'s HCl}}$ (white) $[PbCl_3]^-$ (colorless)	**PbS** (brown)	**$Pb(OH)_2$** $\xrightarrow{x\text{'s } NH_3}$ (white) N.R.‡	**PbS**
Hg_2^{2+}	**Hg_2Cl_2** $\xrightarrow{x\text{'s HCl}}$ (white) N.R.	**Hg + HgS** (black)	**Hg** + (black) **$HgNH_2Cl$**§ $\xrightarrow{x\text{'s } NH_3}$ (white) N.R.	**Hg + HgS**

Aq NH_3 + NH_4^+ + $(NH_4)_2CO_3$	NaOH	*Specific reactions*
$\mathbf{PbCO_3}$ (white)	$\mathbf{Pb(OH)_2} \xrightarrow{x\text{'s NaOH}}$ $HPbO_2$ (colorless)	$Pb^{2+} \xrightarrow{H_2SO_4} \mathbf{PbSO_4}$ (white) $Pb^{2+} \xrightarrow{K_2CrO_4} \mathbf{PbCrO_4}$ (yellow)
Hg + **HgO** (black) (yellow)	**Hg** + **HgO** $\xrightarrow{x\text{'s NaOH}}$ N.R.	$\mathbf{Hg_2Cl_2} \xrightarrow{aq\ NH_3} \mathbf{Hg} + \mathbf{HgNH_2Cl}$

Cation	*1 M* HCl*	*0.3 M* HCl + $CH_3C(S)NH_2$	*Aq* NH_3 + NH_4^+	*Aq* NH_3 + NH_4^+ + $CH_3C(S)NH_2$
Ag^+	**AgCl** $\xrightarrow{x\text{'s HCl}}$ (white) $[AgCl_2]^-$ (colorless)	**Ag_2S** (black)	**Ag_2O** $\xrightarrow{x\text{'s NH}_3}$ (black) $[Ag(NH_3)_2]^+$ (colorless)	**Ag_2S**
Group II Cations¶				
SbO^+	N.R.	**Sb_2S_3** (orange)	**SbOOH** $\xrightarrow{x\text{'s NH}_3}$ (white) N.R.	**Sb_2S_3** $\xrightarrow{NH_3 + CH_3C(S)NH_2}$ $[SbS_2]^-$ + $[SbO_2]^-$
(AsO_4^{3-})	N.R.	**As_2S_3** (yellow)	N.R.	**As_2S_3** $\xrightarrow[\ ]{(slow)\ NH_3 + CH_3C(S)NH_2}$ $[AsS_2]^-$ + $[AsO_2]^-$
BiO^+	N.R.	**Bi_2S_3** (brown)	**BiOOH** $\xrightarrow{x\text{'s NH}_3}$ N.R.	**Bi_2S_3** $\xrightarrow{NH_3\ CH_3C(S)NH_2}$ N.R.
Cd^{2+}	N.R.	**CdS** (yellow)	**$Cd(OH)_2$** $\xrightarrow{x\text{'s NH}_3}$ (white) $[Cd(NH_3)_4]^{2+}$ (colorless)	**CdS** $\xrightarrow{NH_3 + CH_3C(S)NH_2}$ N.R.
Cu^{2+}	N.R.	**CuS** (black)	**$Cu(OH)_2$** $\xrightarrow{x\text{'s NH}_3}$ (pale blue) $[Cu(NH_3)_4]^{2+}$ (deep blue)	**CuS** $\xrightarrow{NH_3 + CH_3C(S)NH_2}$ N.R.

Aq $NH_3 + NH_4^+ + (NH_4)_2CO_3$	NaOH	*Specific reactions*
$\mathbf{Ag_2CO_3}$ (yellowish) $\xrightarrow[\Delta]{H_2O}$ $\mathbf{Ag_2O}$	$\mathbf{Ag_2O} \xrightarrow{x\text{'s NaOH}}$ N.R.	$\mathbf{AgCl} \xrightarrow{aq\ NH_3} [Ag(NH_3)_2]^+ + Cl^- \xrightarrow{HNO_3} \mathbf{AgCl}$
SbOOH	$\mathbf{SbOOH} \xrightarrow{x\text{'s NaOH}} [SbO_2^-]$ (colorless)	$[SbS_2]^- + [SbO_2^-] \xrightarrow{CH_3COOH} \mathbf{Sb_2S_3}$; $\mathbf{Sb_2S_3} \xrightarrow{HCl} [SbCl_4]^-$; $[SbCl_4]^- \xrightarrow{Na_2S_2O_3} \mathbf{Sb_2OS_2}$ (orange)
N.R.	N.R.	$[AsS_2]^- + [AsO_2]^- \xrightarrow{CH_3COOH} \mathbf{As_2S_3}$; $\mathbf{As_2S_3} \xrightarrow{NaOH} [AsS_2^-] + [AsO_2]$; $[AsS_2^-] + [AsO_2] \xrightarrow{Al} AsH_3\uparrow$; $AsH_3 \xrightarrow{AgNO_3} \mathbf{Ag}$ (black); $[AsS_2^-] + [AsO_2] \xrightarrow{H_2O_2,\ CH_3COOH,\ AgNO_3} \mathbf{Ag_3AsO_4}$ (red-brown)
$\mathbf{(BiO)_2CO_3}$ (white)	$\mathbf{BiOOH} \xrightarrow{x\text{'s NaOH}}$ N.R.	$\mathbf{BiOOH} \xrightarrow{NaHSnO_2} \mathbf{Bi}$ (black)
$\mathbf{CdCO_3}$ (white)	$\mathbf{Cd(OH)_2} \xrightarrow{x\text{'s NaOH}}$ N.R.	$[Cd(NH_3)_4]^{2+} \xrightarrow{NaCN} [Cd(CN)_4]^{2-}$ (colorless) $\xrightarrow{CH_3C(S)NH_2} \mathbf{CdS}$
$\mathbf{Cu_2(OH)_2(CO_3)}$ (greenish-blue)	$\mathbf{Cu(OH)_2} \xrightarrow{x\text{'s NaOH}} [CuO_2]^{2-}$ (deep blue)	$[Cu(NH_3)_4]^{2+} \xrightarrow{NaCN} [Cu(CN)_4]^{3-} \xrightarrow{CH_3C(S)NH_2}$ N.R.; $[Cu(NH_3)_4]^{2+} \xrightarrow[K_4[Fe(CN)_6]]{CH_3COOH} \mathbf{Cu_2[Fe(CN)_6]}$ (reddish); $[Cu(NH_3)_4]^{2+} \xrightarrow[C_5H_5N,\ NH_4NCS]{CH_3COOH} \mathbf{[Cu(C_5H_5N)_2(NCS)_2]}$ (green)

Cation	*1 M* HCl*	*0.3 M* HCl + $CH_3C(S)NH_2$	*Aq* NH_3 + NH_4^+	*Aq* NH_3 + NH_4^+ + $CH_3C(S)NH_2$
Hg^{2+}	N.R.	**HgS** (black)	**$HgNH_2Cl$§** (white)	**HgS** $\xrightarrow{NH_3 + CH_3C(S)NH_2}$ N.R.
Sn^{4+}	N.R.	**SnS_2** (yellow)	**$Sn(OH)_4$** $\xrightarrow{x\text{'s } NH_3}$ (white) N.R.	**SnS_2** $\xrightarrow{NH_3 + CH_3C(S)NH_2}$ $[SnS_3]^{2-}$
Group III cations				
Al^{3+}	N.R.	N.R.	**$Al(OH)_3$** $\xrightarrow{x\text{'s } NH_3}$ (white) N.R.	**$Al(OH)_3$**
Cr^{3+}	N.R.	N.R.	**$Cr(OH)_3$** $\xrightarrow{x\text{'s } NH_3}$ (gray-green) N.R.	**$Cr(OH)_3$**
Fe^{3+}	N.R.	Fe^{2+}	**$Fe(OH)_3$** $\xrightarrow{x\text{'s } NH_3}$ (red-brown) N.R.	**Fe_2S_3** $\xrightarrow{NH_3 + CH_3C(S)NH_2}$ (black) N.R.
Fe^{2+}	N.R.	N.R.	**$Fe(OH)_2$** $\xrightarrow{x\text{'s } NH_3}$ (white→black) N.R.	**FeS** $\xrightarrow{NH_3 + CH_3C(S)NH_2}$ (black) N.R.
Mn^{2+}	N.R.	N.R.	N.R.	**MnS** $\xrightarrow{NH_3 + CH_3C(S)NH_2}$ (pink) N.R.

Aq $NH_3 + NH_4^+ + (NH_4)_2CO_3$	NaOH	*Specific reactions*
$Hg_4O_3(CO_3)$ (reddish-brown)	**HgO** (yellow) $\xrightarrow{x\text{'s NaOH}}$ N.R.	**HgS** $\xrightarrow[HNO_3]{HCl}$ $[HgCl_4]^{2-}$; $[HgCl_4]^{2-}$ $\xrightarrow{SnCl_2}$ **Hg_2Cl_2** (white) $\xrightarrow{SnCl_2}$ **Hg** (black); $[HgCl_4]^{2-}$ $\xrightarrow{Co(NO_3)_2,\ NH_4NCS}$ **$Co[Hg(NCS)_4]$** (deep blue)
$Sn(OH)_4$	**$Sn(OH)_4$** $\xrightarrow{x\text{'s NaOH}}$ $[Sn(OH)_6]^{2-}$	**SnS_2** $\xrightarrow{HCl}$ $[SnCl_6]^{2-}$ $\xrightarrow{Al}$ **Sn** (gray) $\xrightarrow{HCl}$ Sn^{2+} $\xrightarrow{HgCl_2}$ **Hg_2Cl_2** (white)
$Al(OH)_3$	**$Al(OH)_3$** $\xrightarrow{x\text{'sNaOH}}$ $[AlO_2]^-$ (colorless)	Al^{3+} $\xrightarrow[NaC_2H_3O_2]{aluminon}$ red ppt
$Cr(OH)_3$	**$Cr(OH)_3$** $\xrightarrow{x\text{'s NaOH}}$ $[CrO_2]^-$ (green)	**$Cr(OH)_3$** $\xrightarrow{NaOH}$ $[CrO_2]^-$ $\xrightarrow{H_2O_2}$ $[CrO_4]^{2-}$ (yellow) $\xrightarrow[H_2O_2]{H_2SO_4}$ CrO_5 (blue)
$Fe(OH)_3$	**$Fe(OH)_3$** $\xrightarrow{x\text{'s NaOH}}$ N.R.	Fe^{3+} $\xrightarrow{K_4[Fe(CN)_6]}$ **$KFe[Fe(CN)_6]$** (dark blue); Fe^{3+} $\xrightarrow{KNCS}$ $[Fe(NCS)]^{2+}$ (blood red)
$FeCO_3$ (white)	**$Fe(OH)_2$** $\xrightarrow{x\text{'s NaOH}}$ N.R.	Fe^{2+} $\xrightarrow{HNO_3}$ Fe^{3+}
$MnCO_3$ (white)	**$Mn(OH)_2$** (white) $\xrightarrow{x\text{'s NaOH}}$ N.R.	Mn^{2+} $\xrightarrow[HNO_3\ (heat)]{KClO_3}$ **MnO_2** (black) $\xrightarrow[HNO_3]{NaBiO_3}$ MnO_4^- (purple to violet)

Cation	*1 M* HCl*	*0.3 M* HCl + $CH_3C(S)NH_2$	*Aq* NH_3 + NH_4^+	*Aq* NH_3 + NH_4^+ + $CH_3C(S)NH_2$
Group IV cations				
Co^{2+}	N.R.	N.R.	**$Co(OH)_2$** (blue→pink) $\xrightarrow{x\text{'s } NH_3}$ $[Co(NH_3)_6]^{2+}$ (tan)	**CoS** (black) $\xrightarrow{NH_3 + CH_3C(S)NH_2}$ N.R.
Ni^{2+}	N.R.	N.R.	**$Ni(OH)_2$** (pale green) $\xrightarrow{x\text{'s } NH_3}$ $[Ni(NH_3)_6]^{2+}$ (dark blue)	**NiS** (black) $\xrightarrow{NH_3 + CH_3C(S)NH_2}$ N.R.
Zn^{2+}	N.R.	N.R.	**$Zn(OH)_2$** (white) $\xrightarrow{x\text{'s } NH_3}$ $[Zn(NH_3)_4]^{2+}$ (colorless)	**ZnS** (white) $\xrightarrow{NH_3 + CH_3C(S)NH_2}$ N.R.
Group V cations				
Ba^{2+}	N.R.	N.R.	N.R.	N.R.
Ca^{2+}	N.R.	N.R.	N.R.	N.R.

Aq $NH_3 + NH_4^+ +$ $(NH_4)_2CO_3$	NaOH	*Specific reactions*
$Co_2(OH)_2(CO_3)$ (reddish)	**$Co(OH)_2$** $\xrightarrow{x\text{'s NaOH}}$ N.R.	$Co^{2+} \xrightarrow[H_2O_2]{NaOH}$ **$Co(OH)_3$** (black) $\xrightarrow{12\ M\ HCl}$ $[CoCl_4]^{2-}$ (blue) $\xrightarrow[NH_4NCS]{NaF}$ $[Co(NCS)_4]^{2-}$ (blue in amyl alcohol-ether)
$NiCO_3$ (pale green)	**$Ni(OH)_2$** $\xrightarrow{x\text{'s NaOH}}$ N.R.	$Ni^{2+} \xrightarrow[H_2O_2]{NaOH}$ **$Ni(OH)_2$** $\downarrow$ HCl Ni^{2+} $\downarrow$ NH_3, CH_3COONa, Dimethylglyoxime red ppt
$ZnCO_3$ (white)	**$Zn(OH)_2$** $\xrightarrow{x\text{'s NaOH}}$ $[ZnO_2]^{2-}$	$[ZnO_2]^{2-} \xrightarrow{CH_3COOH} Zn^{2+}$ $[ZnO_2]^{2-}$ $\downarrow$ $CH_3C(S)NH_2$ → **ZnS** Zn^{2+} $\downarrow$ $K_4[Fe(CN)_6]$ → **$K_2Zn_3[Fe(CN)_6]_2$** (white)
$BaCO_3$ (white)	N.R.**	**$BaCO_3$** $\xrightarrow{CH_3COOH}$ Ba^{2+} $\downarrow$ K_2CrO_4, CH_3COONH_4 Ba^{2+} (yellow-green flame) $\xleftarrow{HCl}$ **$BaCrO_4$** (yellow)
$CaCO_3$ (white)	**$Ca(OH)_2$**** (white) $\xrightarrow{x\text{'s NaOH}}$ N.R.	**$CaCO_3$** $\xrightarrow{HNO_3}$ Ca^{2+} $\downarrow$ aq NH_3, $(NH_4)_2C_2O_4$ Ca^{2+} (brick-red flame) $\xleftarrow{HCl}$ **CaC_2O_4** (white)

Cation	*1 M* HCl*	*0.3 M* HCl + $CH_3C(S)NH_2$	*Aq* NH_3 + NH_4^+	*Aq* NH_3 + NH_4^+ + $CH_3C(S)NH_2$
Mg^{2+}	N.R.	N.R.	N.R.	N.R.
Group VI cations				
NH_4^+	N.R.	N.R.	N.R.	N.R.
K^+	N.R.	N.R.	N.R.	N.R.
Na^+	N.R.	N.R.	N.R.	N.R.

* Reactions shown throughout for equivalent quantities of reagents added; *x*'s indicates excess reagent.
† c = cold, h = hot.
‡ N.R. = no observable reaction.
§ Presence of Cl^- ion assumed. Products different with other anions.
¶ Simplest formulations used. Arsenic(V) present only as an anion.
** If CO_2 present or reagents contaminated with CO_3^{2-}, carbonates are precipitated. The compound $Ca(OH)_2$ is precipitated only from more concentrated solutions than those used in normal laboratory practice.

Aq $NH_3 + NH_4^+ + (NH_4)_2CO_3$	NaOH	*Specific reactions*
$\mathbf{MgCO_3 \cdot (NH_4)_2CO_3 \cdot 4H_2O}$ (white) (if C_2H_5OH present)	$\mathbf{Mg(OH)_2}$ (white) $\xrightarrow{x\text{'s NaOH}}$ N.R.	$Mg^{2+} \xrightarrow[\text{aq } NH_3 \, (NH_4^+)]{Na_2HPO_4} \mathbf{MgNH_4PO_4 \cdot 6H_2O}$ (white) $\xrightarrow{HCl,}$ Mg^{2+} $\xrightarrow[NaOH]{p\text{-nitrobenzeneazoresorcinol}}$ **blue ppt**
N.R.	$NH_3\uparrow$	$NH_4^+ \xrightarrow{NaOH} NH_3\uparrow \xrightarrow[NaOH]{[HgI_4]^{2-}}$ $\mathbf{Hg_2NI}$ (yellow to brown) $\mathbf{NH_4NO_3} \xrightarrow{\Delta} N_2O + 2H_2O$
N.R.	N.R.	$K^+ \xrightarrow{Na[B(C_6H_5)_4]} \mathbf{K[B(C_6H_5)_4]}$ (white) $K^+ \xrightarrow[C_2H_5OH]{H_2[PtCl_6]} \mathbf{K_2[PtCl_6]}$ (yellow) K^+ (reddish-violet flame)
N.R.	N.R.	$Na^+ \xrightarrow[C_2H_5OH]{HZn(UO_2)_3(CH_3COO)_9}$ $\mathbf{NaZn(UO_2)_3(CH_3COO)_9}$ (light yellow) Na^+ (bulky yellow flame)

Appendix I: Reagents; Apparatus

Solid reagents

Name	*Formula*	*Name*	*Formula*
Aluminum granules	Al	Iron wire (0.5 cm)	Fe
Aluminum powder	Al	Iron(III) chloride	$FeCl_3 \cdot 6H_2O$
Aluminum wire (0.5 cm)	Al	Iron(II) sulfate	$FeSO_4 \cdot 7H_2O$
Aluminum nitrate	$Al(NO_3)_3 \cdot 9H_2O$	Lead(II) chloride	$PbCl_2$
Ammonium acetate	$CH_3CO_2NH_4$	Lead(II) nitrate	$Pb(NO_3)_2$
Ammonium carbonate	$(NH_4)_2CO_3$	Magnesium nitrate	$Mg(NO_3)_2 \cdot 6H_2O$
Ammonium chloride	NH_4Cl	Manganese(II) nitrate	$Mn(NO_3)_2 \cdot 6H_2O$
Ammonium nitrate	NH_4NO_3	Mercury(I) nitrate	$Hg_2(NO_3)_2 \cdot 2H_2O$
Ammonium oxalate	$(NH_4)_2C_2O_4 \cdot H_2O$	Mercury(II) chloride	$HgCl_2$
Ammonium peroxodisulfate	$(NH_4)_2S_2O_8$	Mercury(II) nitrate	$Hg(NO_3)_2 \cdot 0.5H_2O$
Ammonium sulfate	$(NH_4)_2SO_4$	Nickel(II) chloride	$NiCl_2 \cdot 6H_2O$
Ammonium thiocyanate	NH_4NCS	Nickel(II) nitrate	$Ni(NO_3)_2 \cdot 6H_2O$
Barium acetate	$Ba(CH_3CO_2)_2 \cdot H_2O$	Platinum wire	Pt
Barium chloride	$BaCl_2 \cdot 2H_2O$	Potassium acetate	KCH_3CO_2
Barium hydroxide	$Ba(OH)_2 \cdot 8H_2O$	Potassium bromide	KBr
Barium nitrate	$Ba(NO_3)_2$	Potassium chlorate	$KClO_3$
Bismuth(III) chloride	$BiCl_3$	Potassium chloride	KCl
Bismuth(III) nitrate	$Bi(NO_3)_3 \cdot 5H_2O$	Potassium chromate	K_2CrO_4
Cadmium nitrate	$Cd(NO_3)_2 \cdot 4H_2O$	Potassium dichromate	$K_2Cr_2O_7$
Calcium acetate	$Ca(CH_3CO_2)_2$	Potassium fluoride	KF
Calcium nitrate	$Ca(NO_3)_2 \cdot 4H_2O$	Potassium hexacyanoferrate(III)	$K_3[Fe(CN)_6]$
Chromium(III) chloride	$CrCl_3 \cdot 6H_2O$	Potassium hexacyanoferrate(II)	$K_4[Fe(CN)_6] \cdot 3H_2O$
Cobalt(II) chloride	$CoCl_2 \cdot 6H_2O$	Potassium hydrogen sulfate	$KHSO_4$
Cobalt(II) nitrate	$Co(NO_3)_2 \cdot 6H_2O$	Potassium iodide	KI
Copper(II) nitrate	$Cu(NO_3)_2 \cdot 3H_2O$	Potassium nitrate	KNO_3
Hexamminecobalt(III) chloride	$[Co(NH_3)_6]Cl_3$	Potassium nitrite	KNO_2

Potassium peroxodisulfate	$K_2S_2O_8$	Sodium oxalate	$Na_2C_2O_4$
Potassium thiocyanate	$KNCS$	Sodium sulfate	$Na_2SO_4 \cdot 10H_2O$
Silicon dioxide (quartz, powdered)	SiO_2	Sodium sulfide	$Na_2S \cdot 9H_2O$
Silver(I) acetate	$AgCH_3CO_2$	Sodium sulfite	Na_2SO_3
Silver(I) nitrate	$AgNO_3$	Sodium tetraphenylborate(III)	$NaB(C_6H_5)_4$
Sodium acetate	$NaCH_3CO_2 \cdot 3H_2O$	Sodium thiosulfate	$Na_2S_2O_3 \cdot 5H_2O$
Disodium hydrogen orthoarsenate	$Na_2HAsO_4 \cdot 7H_2O$	Strontium nitrate	$Sr(NO_3)_2$
Sodium bismuthate	$NaBiO_3$	Sulfamic acid	HSO_3NH_2
Sodium carbonate	Na_2CO_3	Sulfur	S_8
Sodium chloride	$NaCl$	Tin(IV) chloride (liquid)	$SnCl_4$
Sodium cyanide	$NaCN$	Tin(II) chloride	$SnCl_2 \cdot 2H_2O$
Sodium fluoride	NaF	Uranyl acetate	$UO_2(CH_3CO_2)_2 \cdot 2H_2O$
Sodium hydrogen carbonate	$NaHCO_3$	Zinc dust	Zn
Disodium hydrogen orthophosphate	$Na_2HPO_4 \cdot 12H_2O$	Zinc granules	Zn
Sodium iodide	NaI	Zinc acetate	$Zn(CH_3CO_2)_2 \cdot 2H_2O$
Sodium nitrate	$NaNO_3$	Zinc nitrate	$Zn(NO_3)_2 \cdot 6H_2O$
Sodium nitrite	$NaNO_2$	Zirconyl nitrate	$ZrO(NO_3)_2 \cdot 2H_2O$

Reagent	*Concentration*	*Solute*	*Quantity of solute per liter of solution and special directions*
Acids:			
Acetic	6 *M*	CH_3CO_2H	350 ml glacial acetic acid
Hydrochloric	12 *M*	HCl	Use c.p. 37 percent acid (sp gr 1.19) without diluting
	6 *M*	HCl	500 ml 37 percent acid (sp gr 1.19)
	1 *M*	HCl	83.3 ml 37 percent acid (sp gr 1.19)
Nitric	16 *M*	HNO_3	Use c.p. 69 percent acid (sp gr 1.42) without diluting
	6 *M*	HNO_3	380 ml 37 percent acid (sp gr 1.42)
Perchloric	12 *M*	$HClO_4$	Use c.p. 72 percent acid (sp gr 1.69) without diluting
	6 *M*	$HClO_4$	500 ml 72 percent acid (sp gr 1.69)
Sulfuric	18 *M*	H_2SO_4	Use c.p. 96 percent acid (sp gr 1.84)
	3 *M*	H_2SO_4	Add 200 ml 96 percent acid (sp gr 1.84) to 800 ml water with constant stirring
Alkalies:			
Aqueous ammonia	15 *M*	NH_3	Use c.p. 28 percent ammonia (sp gr 0.90) without diluting
	6 *M*	NH_3	400 ml 28 percent ammonia (sp gr 0.90)
Barium hydroxide	Saturated	$Ba(OH)_2 \cdot 8H_2O$	80 g Stir until saturated. Filter
Sodium hydroxide	6 *M*	$NaOH$	246 g of 97 percent solid
Other reagents:			
Acetic acid–sodium acetate buffer	4 *M* CH_3CO_2H 4 *M* $NaCH_3CO_2$	CH_3CO_2H + $NaCH_3CO_2$	240 ml glacial acetic acid + 560 g $NaCH_3CO_2 \cdot 3H_2O$
Aluminon	0.1 percent	Ammonium salt of aurintricarboxylic acid	1 g
Aluminum nitrate	0.1 *M*	$Al(NO_3)_3 \cdot 9H_2O$	37.5 g
Ammonium acetate	0.1 *M*	$NH_4CH_3CO_2$	7.7 g

Ammonium carbonate	3 M	$(NH_4)_2CO_3$	250 g freshly powdered solid dissolved in enough 6 M aq ammonia to give 1 liter total volume
Ammonium chloride	1 M	NH_4Cl	53.5 g
	0.1 M	NH_4Cl	5.4 g
Ammonium molybdate–nitric acid		MoO_3 + HNO_3	Dissolve 50 g MoO_3 in 125 ml of water + 75 ml of 15 M aq ammonia. Add with stirring to mixture of 250 ml of 16 M HNO_3 and 550 ml of water. Let stand for several days, and decant from any residue
Ammonium nitrate	4 M	NH_4NO_3	320 g
	1 M	NH_4NO_3	80 g
	0.1 M	NH_4NO_3	8 g
Ammonium oxalate	0.1 M	$(NH_4)_2C_2O_4 \cdot H_2O$	14.2 g
Ammonium sulfate	1 M	$(NH_4)_2SO_4$	132 g
	0.1 M	$(NH_4)_2SO_4$	13.2 g
Ammonium tetrathiocyanatomercurate(II)	1 M (approx)	$(NH_4)_2[Hg(NCS)_4]$	300 g $HgCl_2$ + 330 g NH_4NCS (+ 0.5 percent $ZnCl_2$ for Cu^{2+} test or 0.5 percent $CuCl_2$ for Zn^{2+} test)
Ammonium thiocyanate	1 M	NH_4NCS	76.1 g
	0.1 M	NH_4NCS	7.6 g
Antimony(III) chloride	0.1 M	$SbCl_3$	Dissolve 22.8 g $SbCl_3$ in 35 ml of 12 M HCl, and dilute to 1 liter
Arsenic(III) chloride	0.1 M	As_2O_3	Dissolve 10 g As_2O_3 in 60 ml of hot 12 M HCl, and dilute to 1 liter
Barium acetate	0.1 M	$Ba(CH_3CO_2)_2 \cdot H_2O$	27.4 g
Barium chloride	0.1 M	$BaCl_2 \cdot 2H_2O$	24.4 g
Barium nitrate	0.1 M	$Ba(NO_3)_2$	26.1 g
Bismuth(III) chloride	0.1 M	$BiCl_3$	Dissolve 31.5 g $BiCl_3$ in 35 ml of 12 M HCl, and dilute to 1 liter
Bismuth(III) nitrate	0.1 M	$Bi(NO_3)_3 \cdot 5H_2O$	Dissolve 48.5 g $Bi(NO_3)_3 \cdot 5H_2O$ in 50 ml 15 M HNO_3, and dilute to 1 liter

Reagent solutions *(Continued)*

Reagent	*Concentration*	*Solute*	*Quantity of solute per liter of solution and special directions*
Cadmium nitrate	0.1 *M*	$Cd(NO_3)_2 \cdot 4H_2O$	30.8 g
Cadmium sulfate	0.1 *M*	$3CdSO_4 \cdot 8H_2O$	80 g
Calcium acetate	1 *M*	$Ca(CH_3CO_2)_2$	158.2 g
Calcium nitrate	0.1 *M*	$Ca(NO_3)_2 \cdot 4H_2O$	23.6 g
Chlorine water	Saturated	Cl_2	Bubble Cl_2 through water until saturated. Store in brown bottle
Chromium(III) chloride	0.1 *M*	$CrCl_3 \cdot 6H_2O$	Dissolve 26.6 g $CrCl_3 \cdot 6H_2O$ in water, add 20 ml of 12 *M* HCl, and dilute to 1 liter
Cobalt(II) chloride	0.1 *M*	$CoCl_2 \cdot 6H_2O$	23.8 g
Cobalt(II) nitrate	0.1 *M*	$Co(NO_3)_2 \cdot 6H_2O$	29.1 g
Copper(II) nitrate	0.1 *M*	$Cu(NO_3)_2 \cdot 3H_2O$	24.2 g
Dimethylglyoxime	1 percent	Dimethylglyoxime	Dissolve 10 g of solid in 1 liter of 95 percent ethyl alcohol
Formaldehyde		HCHO	Use formalin (40 percent HCHO)
Hexamminecobalt(III) chloride	0.1 *M*	$[Co(NH_3)_6]Cl_3$	26.7 g; add 6 *M* HCl until acidic
Hexachloroplatinic(IV) acid	0.5 percent	$H_2[PtCl_6] \cdot 6H_2O$	Dissolve 0.5 g $H_2[PtCl_6]$ in 50 ml water and 50 ml ethyl alcohol
Hydrogen peroxide	3 percent	H_2O_2	Use c.p. 3 percent H_2O_2 solution
Iron(III) chloride	0.1 *M*	$FeCl_3 \cdot 6H_2O$	Dissolve 27 g $FeCl_3 \cdot 6H_2O$ in water, add 10 ml of 12 *M* HCl, and dilute to 1 liter
Iron(II) sulfate	0.1 *M*	$FeSO_4 \cdot 7H_2O$	27.8 g. Add 5 ml 18 *M* H_2SO_4. Prepare fresh each week
Lead(II) nitrate	0.1 *M*	$Pb(NO_3)_2$	33.1 g

Magnesia mixture		$Mg(NO_3)_2 \cdot 6H_2O$ + NH_4NO_3 + NH_3	Dissolve 100 g $Mg(NO_3)_2 \cdot 6H_2O$ and 300 g NH_4NO_3 in 500 ml of water. Add 180 ml of 15 *M* aq ammonia and dilute to 1 liter
Magnesium nitrate	0.1 *M*	$Mg(NO_3)_2 \cdot 6H_2O$	25.6 g
Manganese(II) chloride–hydrochloric acid	0.1 *M*	$MnCl_2 \cdot 4H_2O$ + HCl	Dissolve 2 g $MnCl_2 \cdot 4H_2O$ in 100 ml of 12 *M* HCl
Manganese(II) nitrate	0.1 *M*	$Mn(NO_3)_2 \cdot 6H_2O$	28.7 g
Mercury(I) nitrate	0.1 *M*	$Hg_2(NO_3)_2 \cdot 2H_2O$	Dissolve 56.1 g $Hg_2(NO_3)_2 \cdot 2H_2O$ in a mixture of 60 ml of 16 *M* HNO_3 and 60 ml of water. Dilute to 1 liter and add excess mercury
Mercury(II) chloride	Saturated	$HgCl_2$	80g. Stir until no more dissolves. Filter
Mercury(II) nitrate	0.1 *M*	$Hg(NO_3)_2 \cdot \frac{1}{2}H_2O$	33.4 g
Methyl orange	0.1 percent	Methyl orange	Dissolve 0.1 g in 100 ml of water
Methyl violet	0.1 percent	Methyl violet	Dissolve 0.1 g in 25 ml 95 percent ethyl alcohol and 75 ml water
Nickel(II) chloride	0.1 *M*	$NiCl_2 \cdot 6H_2O$	23.8
Nickel(II) nitrate	0.1 *M*	$Ni(NO_3)_2 \cdot 6H_2O$	29.1 g
p-Nitrobenzeneazoresorcinol	0.1 percent	*p*-Nitrobenzeneazo-resorcinol	Dissolve 0.1 g of reagent in 100 ml of 2 *M* NaOH, and dilute to 1 liter
Phenolphthalein	0.1 percent	Phenolphthalein	Dissolve 0.25 g in 125 ml of ethyl alcohol and 125 ml of water
Potassium bromide	0.1 *M*	KBr	11.9 g
Potassium chloride	0.1 *M*	KCl	7.5 g
Potassium chromate	0.1 *M*	K_2CrO_4	19.4 g
Potassium cyanide	0.1 *M*	KCN	6.5 g
Potassium dichromate	0.1 *M*	$K_2Cr_2O_7$	29.4 g
Potassium fluoride	0.1 *M*	KF	5.8 g. Store in polyethylene bottle
Potassium hexacyanoferrate(III)	0.1 *M*	$K_3[Fe(CN)_6]$	32.9 g
Potassium hexacyanoferrate(II)	0.1 *M*	$K_4[Fe(CN)_6] \cdot 3H_2O$	42.2 g
Potassium iodide	0.1 *M*	KI	16.6 g

Reagent solutions *(Continued)*

Reagent	*Concentration*	*Solute*	*Quantity of solute per liter of solution and special directions*
Potassium nitrite	0.2 *M*	KNO_2	17.0 g
Potassium permanganate	0.1 *M*	$KMnO_4$	15.8 g
Potassium thiocyanate	0.1 *M*	KNCS	9.7 g
Silver(I) nitrate	Saturated	$AgNO_3$	Stir 30 g $AgNO_3$ with 100 ml of water. Decant
	0.1 *M*	$AgNO_3$	17.0 g
Sodium acetate	Saturated	$NaCH_3CO_2$	Heat 130 g. $NaCH_3CO_2$ with 100 ml of water. Cool. Filter
	6 *M*	$NaCH_3CO_2$	492 g
	0.1 *M*	$NaCH_3CO_2$	8.2 g
Sodium bromide	0.1 *M*	NaBr	10.3 g
Sodium carbonate	Saturated	Na_2CO_3	Warm excess Na_2CO_3 with 1 liter of water. Cool. Filter
	0.1 *M*	Na_2CO_3	10.6 g
Sodium chloride	Saturated	NaCl	Heat 40 g NaCl with 100 ml of water. Cool. Filter
Sodium cyanide	1 *M*	NaCN	49 g
Sodium hydrogen orthoarsenate	0.1 *M*	$Na_2HAsO_4 \cdot 7H_2O$	31.2 g
Sodium hydrogen carbonate	Saturated	$NaHCO_3$	Warm excess $NaHCO_3$ with water. Cool. Filter
Sodium hydrogen orthophosphate	0.1 *M*	$Na_2HPO_4 \cdot 12H_2O$	35.8 g
Sodium hypochlorite	0.5 *M*	$NaOH + Cl_2$	Dilute 135 ml of 6 *M* NaOH to 1 liter, and pass in chlorine until the mixture is nearly neutral to litmus. Store in a brown bottle
Sodium iodide	0.1 *M*	NaI	15 g

Sodium nitrate	Saturated	$NaNO_3$	Warm 100 g $NaNO_3$ with 100 ml of water. Cool. Filter
	0.1 *M*	$NaNO_3$	8.5 g
Sodium nitrite	1 *M*	$NaNO_2$	69 g
	0.1 *M*	$NaNO_2$	6.9 g
Sodium oxalate	0.1 *M*	$Na_2C_2O_4$	13.4 g
Sodium polysulfide		$Na_2S \cdot 9H_2O$ + NaOH + S_8	Dissolve 480 g $Na_2S \cdot 9H_2O$ and 40 g NaOH in minimum amount of water. Dissolve 16 g powdered S in this solution, and dilute to 1 liter
Sodium sulfate	0.1 *M*	Na_2SO_4	14.2 g
Sodium sulfide	0.1 *M*	$Na_2S \cdot 9H_2O$	24.0 g
Sodium sulfite	0.1 *M*	Na_2SO_3	12.6 g
Sodium tetraphenylborate(III)	0.1 *M* (approx)	$NaB(C_6H_5)_4$ + $AlCl_3 \cdot 6H_2O$	Dissolve 4 g $NaB(C_6H_5)_4$ and 1 g $AlCl_3 \cdot 6H_2O$ in 100 ml of water. Add 6 *M* NaOH until the solution is pink to phenolphthalein. Allow to stand for several hours, filter, dilute to 200 ml
Sodium thiosulfate	0.1 *M*	$Na_2S_2O_3 \cdot 5H_2O$	24.8 g
Strontium nitrate	Saturated	$Sr(NO_3)_2$	Warm excess $Sr(NO_3)_2$ with 100 ml water. Cool. Filter
Thioacetamide	1 *M*	$CH_3C(S)NH_2$	75 g
Tin(IV) chloride	0.1 *M*	$SnCl_4$	Dissolve 26 g $SnCl_4$ in 50 ml of 12 *M* HCl, and dilute to 1 liter
Tin(II) chloride	Saturated	$SnCl_2 \cdot 2H_2O$	Stir excess $SnCl_2 \cdot 2H_2O$ with 100 ml of 6 *M* HCl. Decant. Add granulated tin

Reagent solutions *(Continued)*

Reagent	*Concentration*	*Solute*	*Quantity of solute per liter of solution and special directions*
	0.1 *M*	$SnCl_2 \cdot 2H_2O$	Dissolve 22.6 g $SnCl_2 \cdot 2H_2O$ in 35 ml of 12 *M* HCl, and dilute to 1 liter. Add granulated tin
Zinc nitrate	0.1 *M*	$Zn(NO_3)_2 \cdot 6H_2O$	29.7 g
Zinc uranyl acetate	0.25 *M*	$UO_2(CH_3CO_2)_2 \cdot 2H_2O$ + $Zn(CH_3CO_2)_2 \cdot 2H_2O$ + CH_3CO_2H	Dissolve 10 g $UO_2(CH_3CO_2)_2 \cdot 2H_2O$, 10 g Zn $(CH_3CO_2)_2 \cdot 2H_2O$, and 16 ml of glacial acetic acid in 80 ml of water. Allow to stand 24 hr. Filter
Zirconyl nitrate	0.4 *M* (approx)	$ZrO(NO_3)_2 \cdot 2H_2O$ + HNO_3	Heat 10 g $ZrO(NO_3)_2 \cdot 2H_2O$ and 100 ml of 1 *M* HNO_3 to boiling with stirring. Allow to stand for 24 hr. Decant

ORGANIC SUBSTANCES

Acetic acid (glacial)
Aluminon
Amyl alcohol
Amyl alcohol–ether (1:2 by volume)
Benzyl alcohol
Carbon tetrachloride
Chloroform
Dimethylglyoxime
Dowex 1-X8 resin
Ether
Ethyl alcohol (95 percent)
Ethylenediamine (10 percent)
Formaldehyde (40 percent)
Litmus paper (red and blue)
Methyl orange
Methyl violet
p-Nitrobenzeneazoresorcinol
Phenolphthalein
pHydrion paper
Pyridine
Thioacetamide
Thiourea
Urea

LIST OF APPARATUS

1 bath rack, aluminum
2 beakers, 50ml
2 beakers, 100 ml
1 beaker, 250 ml
13 bottles, reagent, dropper, 15 ml
1 brush, funnel
1 brush, test tube, small
6 bulbs for dropper pipettes
1 burner, small (with tubing)
1 casserole, porcelain, 30 ml
1 clamp, test tube, small
1 clamp, utility
2 crucibles, porcelain, 15 ml (with covers)
1 cylinder, graduated, 10 ml
1 cylinder, graduated, 100 ml
1 dish, porcelain, 80 mm
1 file, triangular
2 flasks, Erlenmeyer, 50 ml
1 flask, Erlenmeyer, 125 ml
1 flask, Erlenmeyer, 300 ml
1 flask, Florence, 500 ml
1 forceps, steel
2 funnels, 40 mm
1 funnel, 75 mm
1 gauze, wire, asbestos center
2 glasses, cobalt or didymium
5 glasses, watch, 50 mm
1 box matches, safety
1 mortar, porcelain, 55 mm (and pestle)
1 rack, test tube, semimicro
4 rods, glass, 4 mm
1 roll pHydrion paper
1 sponge
1 spot plate, black
1 spot plate, white
1 stand, filter
1 stand, test tube, wooden
12 test tubes, Pyrex, 13 × 100 mm
6 test tubes, Pyrex, 18 × 150 mm
1 tongs, crucible
1 tray, for dropper bottles
1 triangle
1 tube, rubber, 8 mm
1 tube, thistle
1 vial litmus (blue)
1 vial litmus (red)
1 wing top

Also

1 centrifuge (for 13- × 100-mm test tubes) for every six students. Supplies of 6-mm glass tubing, corks, labels

1 15-cm length of 10-mm OD glass tubing per student

Answers to Numerical Problems

CHAPTER 1

Exercise 1-11. 46 percent ethanol by weight
Exercise 1-12. 2.3 *M*; 4.6 *N*; 2.6 *m*
Exercise 1-13. 91 g
Exercise 1-14. (*a*) 8.333 ml
Exercise 1-15. approx 130 kg
Exercise 1-16. 1.0 ml of 5.00 *F* $KMnO_4$ per 2.5 liters solution
Exercise 1-17. (*a*) 147.10 (*b*) 49.03
Exercise 1-18. 3.1 gal
Exercise 1-20. 431

CHAPTER 2

Exercise 2-6. 30 g
Exercise 2-7. 720 ml
Exercise 2-8. $3\frac{1}{2}$ liters
Exercise 2-9. 286 g
Exercise 2-10. 6.02 g

Exercise 2-11. 89 ml
Exercise 2-12. 18 mg
Exercise 2-13. 1.4 mg ml^{-1}
Exercise 2-14. (*a*) 0.07858 *N* (*b*) 0.00982 *N*

CHAPTER 3

Exercise 3-4. (*a*) 0.75 *n* (*b*) 100 *n*
Exercise 3-7. 80 moles $liter^{-1}$

CHAPTER 4

Exercise 4-6. (*a*) 7 (*b*) 4.52 (*c*) −0.84 (*d*) 10.89 (*e*) 7.26
Exercise 4-7. (*a*) 1.6×10^{-7} (*b*) 4.0×10^{-12}
(*c*) 1.0×10^{-7} (*d*) 8.0×10^{-3}
(*e*) 1.8×10^{-11}
Exercise 4-8. 4.2×10^{-4}
Exercise 4-9. (*a*) 0.52 (*b*) 13.48 (*c*) 2.63 (*d*) 2.94 (*e*) 11.37
Exercise 4-10. (*a*) 1.4×10^{-3} *M* (*b*) 1.5×10^{-4} *M*
(*c*) 5.6×10^{-7} *M* (*d*) 3.2×10^{-7} *M*
Exercise 4-11. (*a*) 7.9 (*b*) 8.8 (*c*) 5.6
Exercise 4-13. 3.29
Exercise 4-14. (*a*) 3.21 (*b*) 3.34
Exercise 4-15. 4.52
Exercise 4-16. 99.5 ml (approx)
Exercise 4-19. (*d*) 7.7 (*e*) 4.2

CHAPTER 5

Exercise 5-15. 1.8×10^{-14}; 1.6×10^{-18}; 2.5×10^{-13}; 5.7×10^{-5}
Exercise 5-16. $C_{Pb^{2+}} = 1.6 \times 10^{-2}$ *M*; $C_{Ag^+} = 4.6 \times 10^{-5}$ *M*
Exercise 5-17. 10^{-10}; 2.5×10^{-7}; 1.3×10^{-5}; 1.3×10^{-4}
Exercise 5-18. (*a*) 1.4×10^{-3} *M* (*b*) 5.4×10^{-4} *M*
(*c*) 3×10^{-5} *M*
Exercise 5-19. 2.1 *M*
Exercise 5-23. 10.23
Exercise 5-24. (*a*) $BaSO_4$ when $C_{SO_4^{2-}} = 7.9 \times 10^{-9}$ *M*;
$PbSO_4$ when $C_{SO_4^{2-}} = 1.3 \times 10^{-6}$ *M*
(*b*) 8.4×10^{-5} *M*
Exercise 5-25. 550 mg
Exercise 5-28. 1.2×10^{-8} *M*
Exercise 5-29. essentially 100 percent (6×10^{-20} percent remains)
Exercise 5-31. 6.2×10^{-22} *M* calculated (experimental > calculated)
Exercise 5-32. $C_{Fe^{2+}} = 10^{-2}$ *M*; $C_{Zn^{2+}} = 1.6 \times 10^{-6}$ *M*;
$C_{Cd^{2+}} = 1.8 \times 10^{-11}$ *M*; $C_{Cu^{2+}} = 8.0 \times 10^{-20}$ *M*;
$C_{S^{2-}} = 10^{-17}$ *M* (approx)

CHAPTER 6

Exercise 6-16. 3.6×10^{-8} *M*
Exercise 6-17. 11 (approx)

CHAPTER 7

Exercise 7-6. 500 liters (approx)
Exercise 7-7. (*a*) +1.62 (*b*) +0.34 (*c*) +0.27 (*d*) +0.27
Exercise 7-11. +0.88
Exercise 7-12. 1.3 volts
Exercise 7-14. 9.2
Exercise 7-16. 4×10^{231}

CHAPTER 10

Exercise 10-3. Ca^{2+} salt, 4.2×10^{-1} mole liter^{-1}; Pb^{2+} salt, 156 mole liter^{-1}
Exercise 10-16. None
Exercise 10-17. I^-, 5.0×10^{-15}; Br^-, 1×10^{-3} (each liter^3 mole^{-3})
Exercise 10-29. No; C_{Ag^+} at precipitation $= 1.0 \times 10^{-6}$ (AgNCS), 6.3×10^{-7} (AgBr) mole liter^{-1}
Exercise 10-39. 43.0 ml
Exercise 10-40. 1.2×10^{-4} mole liter^{-1}
Exercise 10-51. 3 mole liter^{-1}
Exercise 10-52. 18.8 ml
Exercise 10-53. 30 ml
Exercise 10-54. 1,600 ml
Exercise 10-55. $NaNO_2$, 0.002 mole; HOAc, 0.006 mole; NaOAc, 0.003 mole; KOAc, 0.003 mole
Exercise 10-56. 1.27 g
Exercise 10-57. $2MgCO_3 \cdot MgO \cdot 2H_2O$
Exercise 10-58. H_3PO_4, 1.7×10^{-16}; $H_2PO_4^-$, 6.4×10^{-6}; HPO_4^{2-}, 2.07; PO_4^{3-}, 2.21 (all mole liter^{-1})
Exercise 10-59. H_3O^+, 2.23×10^{-4}; A^-, 4.48×10^{-5}; HA_2^-, 1.79×10^{-4} (all mole liter^{-1})
Exercise 10-60. SO_4^{2-}, 5.7×10^{-4}; Sr^{2+}, 5.7×10^{-4}; Ba^{2+}, 1.4×10^{-7} (all mole liter^{-1})

CHAPTER 11

Exercise 11-1. $C_{Pb^{2+}} = 1.8 \times 10^{-4}$; $C_{Ag^+} = 6.7 \times 10^{-10}$; $C_{Hg_2^{2+}} = 1.2 \times 10^{-17}$ (all mole liter^{-1})
Exercise 11-2. $C_{Pb^{2+}} = 7.0 \times 10^{-8}$; $C_{Ag^+} = 2.8 \times 10^{-16}$; $C_{Hg_2^{2+}} = 5.0 \times 10^{-28}$ (all mole liter^{-1})
Exercise 11-16. $C_{Cu^{2+}} = 6.2 \times 10^{-16}$; $C_{Cd^{2+}} = 1.0 \times 10^{-7}$; $C_{Pb^{2+}} = 5.4 \times 10^{-8}$ (all mole liter^{-1})
Exercise 11-39. $C_{Fe^{2+}} = 5.6 \times 10^{-4}$; $C_{Fe^{3+}} = 1.0 \times 10^{-20}$ (both mole liter^{-1})
Exercise 11-42. $KClO_3$, 0.0205 g; $NaNO_2$, 0.0345 g

Exercise 11-55. Co(II), 3.7×10^{-6}; Ni(II), 4.3×10^{-6}; Zn(II), 2.0×10^{-6} (all mole liter^{-1})
Exercise 11-59. 2.6 ml
Exercise 11-116. $Ag_2O > Ag_2CO_3 > Ag_2CrO_4 > Ag_3PO_4 > AgBr > AgI > Ag_2S$
Exercise 11-117. 3.5 ml
Exercise 11-118. 317 ml
Exercise 11-119. $[SnCl_6]^{2-}$, 0.03 mole; Al^{3+}, 0.02 mole
Exercise 11-120. Ag^+, 9×10^{-3}; NO_3^-, 6×10^{-2}; H_3O^+, 5×10^{-2}; OH^-, 2×10^{-13}; Cl^-, 9×10^{-6} (all mole liter^{-1})
Exercise 11-121. 0.43 mole, or 122 g
Exercise 11-122. 1.5×10^{-2} mole liter^{-1}
Exercise 11-123. $Hg(ClO_4)_2 > Na_2[HgI_4] > Hg(OH)_2 > HgS$
Exercise 11-124. $K_{sp} = 5.0 \times 10^{-13}$ mole liter^{-1}
Exercise 11-125. Pb^{2+}, 6×10^{-3} mole liter^{-1}

CHAPTER 12

Exercise 12-26. 51.9 ml
Exercise 12-27. 11.8 percent
Exercise 12-28. 268 ml
Exercise 12-29. 120 ml
Exercise 12-30. 150 ml

Index